Date Due

City

of

Light

THE SLOAN TECHNOLOGY SERIES

Dark Sun: The Making of the Hydrogen Bomb
Richard Rhodes

Dream Reaper:
The Story of an Old-Fashioned Inventor in
the High-Stakes World of Modern Agriculture
Craig Canine

Turbulent Skies: The History of Commercial Aviation
Thomas A. Heppenheimer

Tube: The Invention of Television
David E. Fisher and Marshall Jon Fisher

The Invention that Changed the World:
How a Small Group of Radar Pioneers Won the Second
World War and Launched a Technological Revolution
Robert Buderi

Computer: A History of the Information Machine
Martin Campbell-Kelly and William Aspray

Naked to the Bone: Medical Imaging in the Twentieth Century
Bettyann Kevles

A Commotion in the Blood:
A Century of Using the Immune System to
Battle Cancer and Other Diseases
Stephen S. Hall

Beyond Engineering: How Society Shapes Technology
Robert Pool

The One Best Way:
Frederick Winslow Taylor and
the Enigma of Efficiency
Robert Kanigel

Crystal Fire: The Birth of the Information Age
Michael Riordan and Lillian Hoddesen

Insisting on the Impossible:
The Life of Edwin Land, Inventor of Instant Photography
Victor McElheny

City of Light: The Story of Fiber Optics
Jeff Hecht

Visions of Technology:
A Century of Provocative Readings edited by Richard Rhodes

Last Big Cookie
Gary Dorsey (forthcoming)

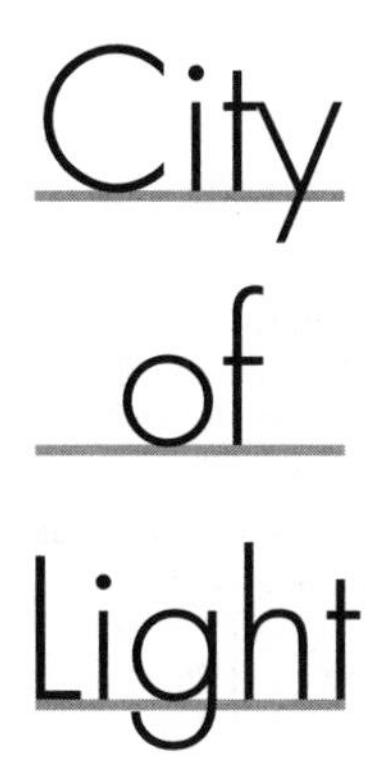

City of Light

The Story of Fiber Optics

JEFF HECHT

New York Oxford

Oxford University Press

1999

Oxford University Press

Oxford New York
Athens Auckland Bangkok Bogotá Buenos Aires Calcutta
Cape Town Chennai Dar es Salaam Delhi Florence Hong Kong Istanbul
Karachi Kuala Lumpur Madrid Melbourne Mexico City Mumbai
Nairobi Paris São Paulo Singapore Taipei Tokyo Toronto Warsaw

and associated companies in
Berlin Ibadan

Published by Oxford University Press, Inc.
198 Madison Avenue, New York, New York 10016

Oxford is a registered trademark of Oxford University Press

Library of Congress Cataloging-in-Publication Data
Hecht, Jeff.
City of light : the story of fiber optics / by Jeff Hecht.
p. cm.
Includes bibliographical references and index.
ISBN 0-19-510818-3
1. Fiber optics. I. Title.
TA1800.H42 1999
621.36'92—dc21 98-6135

1 3 5 7 9 8 6 4 2

Printed in the United States of America
on acid-free paper

For Lois,

Leah,

and Jolyn

Preface to the Sloan Technology Series

Technology is the application of science, engineering, and industrial organization to create a human-built world. It has led, in developed nations, to a standard of living inconceivable a hundred years ago. The process, however, is not free of stress; by its very nature, technology brings change in society and undermines convention. It affects virtually every aspect of human endeavor: private and public institutions, economic systems, communications networks, political structures, international affiliations, the organization of societies, and the condition of human lives. The effects are not one-way; just as technology changes society, so too do societal structures, attitudes, and mores affect technology. But perhaps because technology is so rapidly and completely assimilated, the profound interplay of technology and other social endeavors in modern history has not been sufficiently recognized.

The Sloan Foundation has had a long-standing interest in deepening public understanding about modern technology, its origins, and its impact on our lives. The Sloan Technology Series, of which the present volume is a part, seeks to present to the general reader the stories of the development of critical twentieth-century technologies. The aim of the series is to convey both the technical and human dimensions of the subject: the invention and effort entailed in devising the technologies and the comforts and stresses they have introduced into contemporary life. As the century draws to an end, it is hoped that the series will disclose a past that might provide perspective on the present and inform the future.

The Foundation has been guided in its development of the Sloan Technology Series by a distinguished advisory committee. We express deep gratitude to John Armstrong, Simon Michael Bessie, Samuel Y. Gibbon, Thomas P. Hughes, Victor McElheny, Robert K. Merton, Elting E. Morison (deceased), and Richard Rhodes. The Foundation has been represented on the committee by Ralph E. Gomory, Arthur L. Singer, Jr., Hirsh G. Cohen, and Doron Weber.

Alfred P. Sloan Foundation

Author's Preface

When I started to explore the history of fiber optics, I had no idea the origins of what seemed such a young technology dated back more than a century and a half. Yet the more I dug, the farther the roots stretched, back in time, around the world, and across disciplines. I found forgotten heroes, discovered mistakes in the sketchy standard histories, and tried to untangle a few lingering mysteries. I learned how a powerful new technology evolved to fill the needs of our society.

The basic concept behind fiber optics began as a thing of beauty, but Victorian scientists saw it as little more than a parlor trick to play with light. Over the decades, others borrowed the idea, inventing and re-inventing ways to guide light. The trickle of innovation reached a critical mass in the 1950s, and the young technology slowly emerged into the world. More advances followed, including a series of breakthroughs that in twenty years transformed a crazy idea into the backbone of the global telecommunications network.

My job here is to tell the story of fiber optics. If I were writing a novel, my hero might make an elegant invention in her basement, struggle for years to perfect and market the idea, and ultimately become the multibillionaire head of an industrial empire. Modern technology doesn't work like that, making the tale both more complex and more fascinating.

No one genius did it all. It took a cast of thousands to develop the essential pieces and assemble them into working systems. Think of it as a city of light, a still-growing community building a structure elegant in concept and useful in function. I've had the good fortune to spend many years watching and writing about that richly textured place. It's full of struggles and successes, bright and beautiful ideas, and fireside tales told relaxing with old friends. I have tried to fill this book with that spirit.

I owe many people thanks for making this book possible. A generous grant from the Alfred P. Sloan Foundation gave me the time and resources for the job. Art Singer of the Sloan Foundation supplied help and encouragement well beyond the financial support. Victor McElheny generously helped me with my proposal; he and John Armstrong gave thoughtful feedback on my manuscript. My Oxford editor, Kirk Jensen, patiently guided me to tell the story and explain the technology clearly. Thanks also to my agent, Jeanne Hanson, and to Helen Gavaghan for telling me about the Sloan program.

The story of the origins of light guiding draws heavily on careful research by the late Kaye Weedon, who found the earliest accounts of demonstrations by Daniel Colladon and Jacques Babinet. Sadly, I never met Weedon, and he never published his findings. However, he did give several lectures and gave copies of his notes to Jeofry Courtney-Pratt, who kindly sent me copies. I wish I could have thanked Weedon in person for pointing me in the right direction.

Many others have given generously of their time, lent me documents and pictures, and commented on earlier drafts of parts of the book. For insight into the early days of imaging fiber optics, I thank especially Lem Hyde, Brian O'Brien, Jr., Walt Siegmund, Martin Carey, Willem Brouwer, Larry Curtiss, Bob Potter, Will Hicks, and Holger Møller Hansen. Many others helped me understand the development of fiber-optic communications, including Charles Kao, George Hockham, Don Keck, Murray Ramsay, Dick Dyott, John Midwinter, Martin Chown, Roger Heckingbottom, Jack Cook, David Hanna, Charlie Sandbank, Richard Epworth, Laszlo Solymar, Rich Cerny, Al Kasiewicz, Paul Lazay, Ray Jaeger, Rob Cassetti, and Jack Kessler. Max Riedl, Jean-Louis Trudel, Jonathan Beard, and Julian Carey translated articles. To The Point Graphics in Newton, Massachusetts, helped me with drawings. Phyllis Smith, Donna Cunningham, Joan Finamore, Mary Wright, Connie Coburn, Patricia Thiel, Andy Goldstein, Tim Proctor, René Sigrist, and Lesley Hepden helped me track down details, people, and pictures. Fred Abbott, Rick Martin, and Steve Salt trusted me with rare references that have sat too long in my office. Countless others talked with me in person or on the phone, sharing stories and answering questions about friends, relatives, and former co-workers. I've tried to keep everything straight, but if I haven't it's my fault, not theirs.

I've had a good time writing this book. I hope you enjoy reading it.

Newton, Massachusetts *J. H.*
December 1997

Contents

City

of

Light

1

Introduction

Building a City of Light

> I . . . managed to illuminate the interior of a stream [of water] in a dark space. I have discovered that this arrangement . . . offers in its results one of the most beautiful, and most curious, experiments that one can perform in a course on optics.
>
> —J. Daniel Colladon, 1842[1]

The Sunday evening phone call sounded completely ordinary. A friend who lives a few miles away wanted me to call a few other Caltech graduates for the alumni association. I had helped him before, and I agreed to help him again. We chatted a bit, and as we finished, he added, "You know how dedicated I am? I'm calling from my hotel room in Cairo at two in the morning."

I was stunned. When I was growing up in the 1950s and 1960s, it was rare to get long-distance calls from relatives a few hundred miles away. They didn't call to chat but to report news in static-toned voices that sounded far, far away. My friend's voice sounded as clear and sharp as a telephone can be, as if he was calling from his home in the next town, not 5400 miles away in Egypt. Such miracles are the daily work of fiber optics.

A Global Fiber-Optic Network

I have grown accustomed to such miracles. As a correspondent for the London-based weekly magazine *New Scientist*, I use the telephone network to reach across the world. International dialing is easy. My fingers have memorized the codes for the London office. I start with 011, the code for international direct calling. Then comes 44, the country code for the United Kingdom, and 171, the city code for central London. Those eight digits route the call to the right region; four digits—a "1" and the area code—suffice for calls in North America. Then I push seven more buttons and after a breath or two a phone rings on an editor's desk.

You cannot mistake a telephone voice for a live one; the telephone is not a high-fidelity instrument. Some distort voices much more than others. With a little practice, you can recognize speaker phones, cellular phones, or $9.95 discount-store specials. Yet without the effects of the telephone itself, it is hard to tell a call from London from one from down the block. I can recognize the editors' voices; they can tell when sinus trouble clogs my nose, or laryngitis roughens my throat. The telephone lines are as transparent as they sound. A few miles from my home, my telephone calls shift from copper wires to glass optical fibers.

The part of the telephone network we see is electrical. A microphone converts the vibrations of air molecules shaken by my voice into electronic signals. The phone sends the electronic signals through copper wires in my house, which connect to wires that run to a telephone pole across the street. From there, more wires carry the signals down the pole and through underground ducts to a building a few miles away, where electronics convert them to digital code—a string of ones and zeroes. The circuits decode the numbers I dialed, figure out the call's destination, switch the signals to a cable headed in that direction, and shuffle the bits of my voice together with the digitized signals from other phone calls being sent along the same digital highway on their routes to their separate destinations.

The electronic bit stream switches off and on a tiny semiconductor laser no larger than a grain of salt, turning my voice into pulses of invisible infrared light. A hair-thin optical fiber collects the millions of pulses a second that carry my words—and other voices and facsimile messages and computer data—and sends them on their way south to New Jersey, where domestic phone lines connect to fiber-optic cables that cross the Atlantic. At the international switching center, other circuits amplify and reroute the bits of my voice along with thousands of other digitized conversations. Then they travel thousands of miles through optical fibers protected from the abyssal depths by the layers of white plastic and metal shielding that make up a 0.827-inch (21-millimeter) submarine cable. On the other side of the Atlantic, a British international switching center reroutes them to other fibers that carry them to London. In the British capital, more fiber-optic cables carry the light pulses through a maze of underground ducts to the building that houses *New Scientist*. An electronic receiver in a box somewhere in the building turns the light pulses back into electrical signals that go through wires to the phones on the editors' desks.

We don't see the optical fibers in the telephone system, any more than we see the electronic chips that control a videotape recorder. If we were to see them, they would not look spectacular. A single optical fiber is a tiny clear filament, coated with plastic that makes it a little thicker and stiffer than a human hair, like monofilament fishing line. Your eye can easily miss a single bare fiber, and your fingers fumble at trying to pick one up. The semiconductor lasers that generate the signals that pass through the fibers resemble tiny flakes of metallic confetti, the size of grains of salt.

Yet these technological wonders are far more sophisticated than they look. The fibers that span the globe are the clearest, purest glass. Ordinary plate glass is very different stuff, old-fashioned pea-soup London smog compared to the sharp, clear mountaintop air of fibers. A pane of plate glass transmits only about 90 percent of the light that reaches it, no matter how clean you scrub it. The surfaces reflect most of the remaining 10 percent, but the glass itself absorbs some. We are so used to the reflections and dirt that we use them as cues to tell if the glass is there; clean a glass door too well, and someone may try to walk through it. We often don't notice that any light is lost inside the glass. Yet look into the edge of a piece of plate glass and it seems dark and green, because little light can pass through the several inches of glass between the edges of the pane.

Plate glass is good enough for windows. Binocular and camera lenses are made of more costly optical glasses that contain fewer light-absorbing impurities. The glass for optical fibers must be even purer—with no more than ten impurities per billion atoms. The fibers are as clear as glass can be, so clear that even after passing through 100 kilometers (60 miles) of fiber, one percent of the light emerges from the other end. It's as if you could duck your head into the water below the cliffs of Dover and peer through the English Channel to dimly see the feet of bathers wading along the French beach at Calais.

The tiny lasers that send the signals are not simple crystals like grains of salt. They are made of thin layers of semiconductor, with compositions designed to control how light and electric current flow through them. Their structures are delicate and elaborate; the layers are thinner than a thousandth of a millimeter, or 1/25,000th of an inch. The tiny chips are cheap, but it is expensive to package them to direct their light unerringly into the tiny cores that carry light in optical fibers.

Watching the Breakthrough

Optical fibers didn't seem likely to go so far when I first encountered them a quarter century ago. They seemed mere optical toys, which guided light along their lengths. Bend a fiber, and the light turns the corner. Hold one end of a bundle of fibers close to a light bulb and let the other end splay out, and the tips of the fibers sparkle with light as a decorative lamp. Back then, I bought my sister one as a Christmas present. She used it for a couple of years, but it ended up as a techno-artifact in my attic.

I started hearing about fiber optic communications a couple of years later in 1974, when I started working at a little laser-industry magazine called

Laser Focus. It was one of dozens of new ideas the industry was investigating, and it didn't even seem the most promising at the time. Lots of people tested lasers for various applications, but few of the ideas proved practical.

The editor wanted me to chase news, so I called around America, asking California companies and Bell Labs in New Jersey about fiber-optic communications. However, I didn't call overseas when I heard that a British police department had installed the first practical fiber-optic communication system in 1975. I didn't have a phone number, and I knew international calls were expensive. A few years earlier a cousin stationed in South Korea by the Army had made a collect call to a girl friend in New York. They argued, and spent about a half hour on the line—and ran up a bill of $750. That was about three weeks of my salary, and I knew better than to run up bills like that. I based my story on a report in another magazine; it wasn't good journalism, but I had to fill the magazine.

Progress over the next few years was rapid. In 1976, Bell Labs ran a field trial in Atlanta. The next year, Bell Labs, GTE, and British Telecom used fiber optics to carry live telephone traffic in their networks. They poked and prodded and cautiously examined the new technology, as if they expected a horde of gremlins to emerge and shout "gotcha," and all the potential attractions of fiber optics to evaporate. Again and again they reported the same welcome but monotonous litany—the system worked as promised.

There are few more exhilarating experiences in journalism than riding the crest of a technology breakthrough. Everywhere I turned there were advances. Fibers improved, lasers improved, detectors improved. Each new system sent signals faster and farther than old ones. New people and companies came into the field. The little village of fiber-optic pioneers grew. I was caught up in the excitement of a fast-growing field. I still remember the thrill of hearing Peter Runge of Bell Labs outline plans for laying the first transatlantic fiber-optic cable at a 1980 meeting. I felt the same thrill I remembered as a boy reading Werner Von Braun's plans for space travel in *Across the Space Frontier*. Fiber-optic communications was on a roll.

Laser Focus had grown, too, and after seven years I was managing editor with four people working for me, but it wasn't where I wanted to be. I left to freelance and naturally started writing about fiber optics. I wrote an article on the fiber-optics market for a short-lived magazine called *Technology*, and one about futuristic fiber communications to the home for *Omni*. Meanwhile, the technology grew apace. I watched three generations of fiber-optic systems appear in five years. Laboratory systems kept pushing to higher and higher performance. Optical fibers became the medium of choice for long-distance telephone transmission, the backbone of national networks. The first transatlantic fiber cable, called TAT-8, came into service at the end of 1988. Naturally, I wrote about it for *New Scientist*.

I also noticed the difference it made in international calls. Before TAT-8, satellite links carried most calls. The signals took a fraction of a second to make the round trip to and from the satellite, parked 22,000 miles above the equator, just long enough to disturb the timing of a conversation. Satellite channels sometimes brought other irritations. I recall many one-way conversations, when London could hear me but I couldn't hear them, or vice versa. Sometimes silence, or a hum or whine from presumably unhappy electronics, suddenly replaced the British voices. Other times echoes of my words would return a second after I spoke them, often louder than the voice on the other end. "Sorry," I would say, "let's try again," and one of us would call back to try to get a better line.

The fiber cable gave a new alternative. The bad lines and the satellites were still there, but TAT-8 added 40,000 good circuits through echo-free cable—a thousand times more than the first transatlantic telephone cable carried in 1956. Since then, more fiber cables have crossed the Atlantic, and poor connections to Britain are as rare as within North America. Even Egypt, Israel, Hong Kong, and Australia come in loud and clear. When something goes wrong with the phone line, it's almost always in the few miles of aged copper wire that connect my house to the nearest fibers in the telephone network.

Wires, cables, fibers, and other transmission media are not the whole story of the revolution in communications. Telephone calls must be routed from point to point, by switches that make temporary connections in the global lattice of fibers, wires, and radio signals. At the start of the century, telephone switching was by hand, with ranks of operators plugging wires into sockets in the sort of big black switchboards that survive only in old movies. Later clattering banks of electromechanical switches replaced them, which flipped, flopped, and stepped from point to point, making and breaking connections in response to control pulses that originated in telephone dials. Sometimes you could hear the mechanical switches clicking on their way to making connections on the line, just as you could hear the pulses as a telephone dial rotated. Later came transistorized electronic switches, special-purpose computers designed to send signals along complex routes. Without them, phone calls could never weave their ways through the maze of big-city networks. But they're another story, to be told another day.

How Fiber Optics Changed Communications

What is so good about fiber optics? Looking at the telephone network will help you understand. The old wires running to my house are both its strength and weakness—new equipment has to work with the same wires used for decades. With a screwdriver and very little ingenuity, I can hook up a massive 45-year-old dial phone in Western Electric basic black. Someone dropped it a few years back and broke a corner off the case, but it still works. So do my modem, answering machine, and fax machine. It's a simple, versatile system of information pipes that is standard throughout North America.

Phone companies laid the wires to provide what they call "POTS," plain old telephone service. They designed the network to carry electrical signals that replicate sound waves at the frequencies we must hear to understand speech—300 to 3000 hertz (or cycles per second), if you like specifications. That's only part of the human ear's nominal range of 20 to 20,000 hertz, so sound quality isn't good. But it does the job because you don't need high fidelity to understand speech.

Phone wires can carry other audio signals as well as speech. The push buttons on a Touch-Tone phone whistle musical notes at audio frequencies. Modems and fax machines warble their codes as tones in the audible speech range. Our ears don't understand their electronic speech, but the telephone network carries their signals nonetheless, and other modems and fax machines decode the sounds.

The phone wires in your walls are strands of copper, coated with plastic insulator and grouped in pairs (or sometimes fours). When they have to go a long way, they're often wound around each other, making what the phone industry calls twisted pairs. The problem with standard twisted-pair phone wires is that they were designed to carry a single conversation, which doesn't amount to much information. Adjust and condition them properly, and if all goes well they can carry the equivalent of a few dozen conversations a few miles, or of many more over a thousand feet. But that's it. So far, no more information can fit through the pipe.

Modern telephone systems contain electronics that convert the whistles, warbles, and words into digital signals within a few miles of your home. Each "voice channel" becomes a series of 56,000 bits per second. More electronics interleave those bits with the bits that encode dozens, hundreds, or even thousands of other telephone calls. Phone companies have a hierarchy of levels, each one combining more digitized voice channels to generate a higher-speed signal. Sometimes the slower signals go through special metal cables or are relayed by microwave towers, but typically they go through optical fibers.

Most modern fibers concentrate light signals in tiny cores about nine micrometers—0.009 millimeter—across. That might seem a tiny pipe, but in fact it has a huge information capacity. Its small size keeps short pulses of light from spreading out and interfering with each other as they travel down the fiber. With the best state-of-the-art transmitters phone companies can buy, a single fiber can carry up to 400 billion bits per second, equivalent to five million telephone conversations.[2] Pairs of fibers carry two-way traffic, one carrying signals in each direction.

The clarity of glass fibers is another advantage. Depending on the design, they can carry signals tens of miles without any internal amplification. That's vital if you're building a cable crossing a continent or an ocean, or a cable-television network serving a small city. Cable television network use coaxial cable ("coax"), with a central wire surrounded by a plastic layer that is encased in a metal sheath, and usually covered with a protective jacket. Coax thinner than a pencil hooks home video systems together; thicker coax—costlier but able to carry signals farther—links cable companies to homes. Yet even those costly coaxial cables can't carry signals very far. Cable companies have to install signal-amplifying repeaters about every half kilometer

(500 yards) because the signal weakens with distance. Those amplifiers keep cable technicians busy. You don't need them with fiber, so cable-television companies are replacing their old coax with fiber-optic cables.

The huge information capacity of optical fibers changes the ground rules of telecommunications, which the telephone defined early in the twentieth century. What seemed adequate pipeline then today delivers only an electronic trickle in a world parched for information. The telephone network was engineered to manage information as carefully as nomads husband the scarce waters of the Sahara. The huge capacity of fibers promises to irrigate our dry lands, but first we must find uses for the flow that will justify the costs of extending the data pipeline all the way to our homes. We value information almost instinctively, like desert nomads think water is a "good thing," but faced with an overwhelming abundance, we are as lost as a Bedouin by the sea. We have long dreamt of making the desert bloom, but we don't know how.

The Roots of Fiber Optics

I discovered fiber-optic communications when the technology was still young and the City of Light was but a small town. Some early settlers are still around, active developers of new technology or elder statesmen. Others have died or moved on. As the field has grown, many new people have joined it. Awareness of fiber optics has spread around the world like the glass threads themselves. The field and the technology have not stopped growing. Even as I put the finishing touches on this book, I heard of new advances that promise even better fiber-optic systems in coming years.

Where did this technological revolution come from? How did it grow so fast, out-competing older and better-funded projects? This book tells the story, one more fascinating than I imagined when I started.

In studying fiber optics, I have learned how technology evolves, like life itself. The idea began as little more than a parlor trick, guiding light in jets of water or bent glass rods to make physics lectures more entertaining. Brighter electric lights powered illuminated fountains that awed visitors to the great Victorian exhibitions, who were accustomed to faint gas lights. Later, bent rods of glass or plastic guided light to illuminate teeth for dental exams.

Another generation of engineers adapted light guiding, assembling arrays of thin, flexible glass fibers to look into inaccessible places. The market niche was not filled until developers found how to keep light from leaking between fibers. Once that innovation was perfected, the technology of fiber bundles evolved rapidly, like animals that had just arrived in a new, unpopulated land. New applications proliferated in the late 1950s and 1960s, from looking down the throat to military imaging systems.

Flexible bundles of optical fibers let physicians reach into otherwise inaccessible parts of the body. Surgeons threaded fiber-optic endoscopes down patients' throats to examine the interiors of their stomachs without surgery. Similar instruments probed the other way through the intestines, seeking cancers or precancerous lesions so physicians could remove them before they spread. "Fiber optics saved my life," a geologist told me, by allowing his surgeon to direct radiation therapy precisely at a tumor inside his skull.

Optical communications required different types of optical fibers, and the evolutionary leap was not easy. Some people who have worked in both areas consider medical imaging bundles a different technology than communication fibers. It took breakthroughs in transmission concepts and glass production, and like the first dinosaurs to take wing as birds, the flight of fiber-optic communications was at first uncertain. But the evolutionary potential was there, and an outpouring of innovations launched fiber communications so it soared high.

Why a "City of Light"?

The title "City of Light" first came to mind as a catchy phrase. The more I worked on the book, the more I liked it.

One reason is that the roots of fiber optics lie in the nineteenth century, a time when scientists entertained the public by devising elegant lecture demonstrations. First they guided light along jets of water in lecture halls. Later engineers picked the same principle to show glittering displays of light in the night-time "fairy fountains" of the great Victorian exhibitions. The most spectacular was in Paris, the City of Light, in 1889.

Another is that a veritable city of people helped create fiber-optic technology, with their work spanning decades. Too often we gloss over the many people behind some great innovation, to focus on a single "inventor." Look at the wall of inventions at the subway stop serving the Massachusetts Institute of Technology,[3] and you will be told that Narinder Kapany invented fiber optics in 1955. In reality, Kapany was only one of many contributors, and 1955 marks only the awarding of his doctoral degree for a project conceived by his thesis adviser, Harold H. Hopkins. The technology grew from the interplay of many people and many ideas, often competing with each other. I have tried to focus on the most important, but there is no room in a book like this to credit everyone.

In addition, the best-known role of fiber optics has become as a technology of communication, and communication is an essential part of any city. Both communication and the failure to communicate play important roles in the story. And this is a story played out on a grand scale, in laboratories around the world. Most of the work was done in Europe, North America, and Japan, but some developers came from other places, including China, India, and South Africa. The fruits of their labor now link all the continents except Antarctica. We can explore the world through optical fibers, and talk with people half a world away. That still amazes me when I stop to think about it.

Finally, the idea of a City of Light looks toward the future that we hope can in some way be brighter. The fiber-optic revolution is not over. New wonders continue emerging from the labs, and skilled engineers keep converting them into practical hardware. A quarter century has passed since the first visionary suggested stringing fiber optics all the way to homes, and they haven't arrived at my door yet. Perhaps I have become too much a fan of fiber optics, but I still think that day should come before another quarter century passes.

2

Guiding Light and Luminous Fountains

(1841–1890)

> Among the most wonderful displays, electric and visual, at the recent French exposition were those pertaining to the luminous fountains, which were arranged on a grand scale and occupied a large portion of the plateau in front of the main entrance. The chameleon-like changes of color in the fountain water were something astonishing to behold. It was not accomplished by the mere throwing of colored lights upon the exterior of a spouting jet, but was due to an interior electric illumination of the molecules of the water; the beams of light being, so to speak, thrown into and imprisoned within the crystal walls of the water and then carried along with it, becoming visible by interior reflection during the discharge of the water.
>
> —*Scientific American*, December 14, 1889[1]

It is 400 kilometers (250 miles) as the crow flies from Geneva to Paris, a leisurely day's drive on modern highways. The train ride took a much longer day in 1889, but to the 86-year-old Daniel Colladon that marked tremendous progress. In 1825, he had spent four bumpy days and nights on the same route in a horse-drawn coach on his way to study physics in the center of European science. Paris had been called the City of Light since it became the intellectual center of the Age of Enlightenment in the late eighteenth century.[2]

Colladon made his 1889 trip to see the Universal Exhibition, which celebrated the centennial of the French Republic and the nineteenth-century progress that eased the old man's journey. Honored, eminent, and proud, Colladon was still an active scientist. He was famous for inventing compressed air, which powered machines deep underground and delivered fresh air to miners, making their job faster and safer. Yet he felt cheated because Italian engineers won the contract to build the first long railroad tunnel through the Alps using his idea.

The fountains that lit up the City of Light in 1889 used another of Colladon's inventions. Forty-eight years earlier, he had shone a beam of light along a jet of water, to entertain lecture halls in Geneva and London with an optical trick called total internal reflection (see box, pp. 24–26). Now a new generation of engineers had made his invention a thing of beauty on a much grander scale, building a tableau of illuminated fountains before the main entrance to the Universal Exhibition.[3] The public marveled at how electric lights—still a novelty—illuminated the fountains at night. Some lights played on the water from above; others shone up from the base of the fountain or along jets emerging from the sides of sculptures. Trapped within the water, the light passed along the jets until it emerged sparkling, as it had in Colladon's lectures, and as it would in optical fibers.

The Man Who Guided Light

Daniel Colladon was a 38-year-old professor at the University of Geneva when he first demonstrated light guiding in 1841.[4] He wanted to show fluid flow through various holes and the breaking up of water jets observed earlier by French physicist Felix Savart.[5] Colladon's experiments worked, but at first the audience couldn't see the flowing water in his lecture hall, particularly with the poor lighting available in the mid-nineteenth century.

He solved the problem by collecting sunlight and piping it through a tube to the lecture table. A lens focused the light through the water tank and along a jet squirting out a hole in the other side. When the light rays in the water hit the edge of the jet at a glancing angle, total internal reflection trapped them in the liquid. They bounced along the curving arc of water until the jet broke up, as shown in figure 2-1. Instead of traveling in a straight line, the light followed the curve of the water.

In a dark room, the effect was impressive, "one of the most beautiful, and most curious experiments that one can perform in a course on optics," Colladon wrote. "If the water is perfectly clear, and the opening of the diaphragm very [smooth], the stream is scarcely visible, even though a very intense light circulates inside it. But whenever the stream encounters a solid body that obstructs it, the light that it contains escapes, and the point of contact becomes luminous. . . . If the stream falls from a great height, or if its diameter is only of some millimeters, it breaks apart into drops in the lower region. Then only does the liquid give light, and each point of rupture in the stream casts a bright light."[6]

Proud of his new trick, Colladon used it in the public talks to the urban intelligentsia that were an important sideline for mid-nineteenth-century sci-

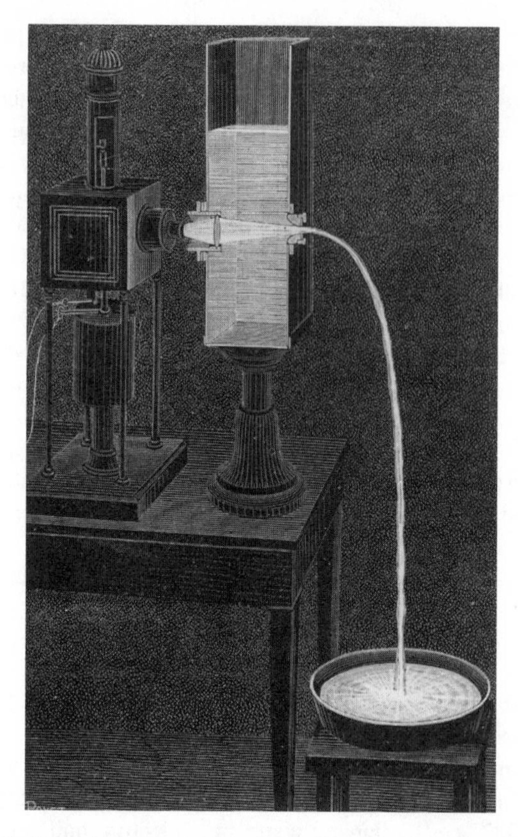

Figure 2-1: Colladon's fountain sparkles with light from an arc lamp, illustrated in his 1884 article. (from Daniel Colladon, "La Fontaine Colladon," *La Nature* 2nd half year 1884, p. 325)

entists. Auguste de la Rive, another Geneva physicist, duplicated Colladon's experiments, replacing the sun with an electric arc light. Colladon had a craftsman make an electric-arc version for the Conservatory of Arts and Sciences of Paris in October 1841, and Colladon or an associate demonstrated light guiding in London.[7] Only the following summer did Colladon get around to writing up his work for publication. The likely impetus was a rumor that a Belgian scientist was doing something similar, and Colladon pointedly insisted his work was "previous by several months." In fact, the Belgian's experiments were quite different.[8]

The Curious Coincidence of Prof. Babinet

Colladon sent his report to François Arago, an old friend who headed the French Academy of Sciences and edited its journal, *Comptes Rendus*. Arago recalled that Jacques Babinet, a French specialist in optics elected to the academy in 1840, had made similar demonstations in Paris. Babinet focused candlelight onto the bottom of a glass bottle as he poured a thin stream of water from the top. Total internal reflection guided the light along the jet, illuminating a china plate or sheet of paper at the end. Arago asked Babinet to write up his work.

Well established as a lecturer and instrument maker,[9] Babinet complied, but he wasn't as eager for recognition as Colladon. His brief account suggests he didn't think the whole thing was very important, and Colladon probably put on a better show. Yet, in passing Babinet mentioned something else. The idea also "works very well with a glass shaft curved in whatever manner, and I had indicated [it could be used] to illuminate the inside of the mouth."[10]

Colladon mentioned only water jets; Babinet, who specialized in optics, extended the principle to guiding light along bent glass rods. Since glass fibers are merely very thin glass rods, that meant he anticipated the idea of fiber optics. He even suggested a practical application that would resurface a half century later: dental illuminators. Yet Babinet also knew the limits of glass technology. Examining lighthouse lenses, he had found that the best glasses of his time were not very clear, so light couldn't go far through them. Look into the edge of a sheet of plate glass today, and you can see the same green color Babinet saw a century and a half ago. It's no wonder he thought light guiding little more than a parlor trick.

The Magic of Glass

Babinet probably also doubted light guiding in glass was new. It almost certainly was not. Glass dates back at least 4500 years to ancient Egypt and Mesopotamia.[11] By 3500 years ago, Egyptians were sculpting miniature heads of transparent glass, which the years have since turned milky white.[12]

Ancient and medieval glass workers must have seen glass rods guide light. For 2000 years, glass blowers have thrust glass tubes into glowing furnaces to soften them. Workers pausing to relax from the hot, hard job must have seen the fiery glow of the furnace emerging from the glass. Master glass makers noted total internal reflection as they made sparkling ornaments for the bright glass chandeliers that illuminated the great rooms of the rich. But they kept their secrets to themselves. Scientists began to study total internal reflection in medieval times as they sought to understand the mystery of the rainbow,[13] but they did not understand it until the laws of refraction were formulated in the seventeenth century.

By the mid-nineteenth century the cutting edge in physics had moved elsewhere. The industrial revolution made glass commonplace in windows and bottles. Babinet may have felt his brief report in *Comptes Rendus* only restated the obvious. After sending his letter to Arago, Babinet apparently never returned to the guiding of light before he died in 1872.

Special Effects for *Faust*

Daniel Colladon's interests turned toward the practical side of science we now call engineering. He designed a plant to extract gas from coal for the spreading network of gas lamps in Geneva. Later he helped build other plants in Basel, Bern, Lausanne, and Naples. But he didn't forget light guiding.

Lighting was a problem in nineteenth-century theaters; candles, gas lights, and lanterns cast only feeble illumination and raised the specter of fire. In 1849, the Paris Opera began testing electric arcs, which passed a strong current between a pair of carbon electrodes to create a blindingly bright light. Producing the current was not easy, but the opera found it worth the effort because audiences liked the bright lights. Four years later, Colladon helped the opera duplicate his light-guiding trick as a special effect to catch audience attention in a ballet called "Élias et Mysis." Gounod's opera *Faust* followed, with light from an arc lamp focused along a red-glass tube filled with water in a scene where the devil (Mephistopheles) makes a stream of fire flash from a wine barrel. It wouldn't hold a candle to modern special effects, but it was impressive to 1853 audiences who saw bright lights as a novelty.[14]

The opera recognized a good crowd pleaser and in 1855 put L. J. Duboscq, who had worked with Colladon, on the full-time payroll.[15] He brought mock suns, rainbows, and lightning to the opera stage. He also designed illuminated playthings for the rich, which he eventually offered in an 1877 illustrated catalog. People could spend up to 1000 francs for luminous fountains, where light played on dancing water, its color changing as a wheel rotated color filters in front of a lamp.[16]

Illuminated Fountains and Great Exhibitions

Luminous fountains took on a much grander scale in the series of great exhibitions the Victorians held to celebrate technological progress. Light was a symbol of that progress to people who grew up with only the feeble light of gas lamps and lanterns to fend off the night. Electric arcs were rare before electric power generators began to spread, but the fairs had their own generators. The Victorian public came to marvel at bright outdoor lights, and illuminated fountains were among the most spectacular evening displays.

The first great fountains were at the International Health Exhibition, held in the South Kensington district of London in 1884. The Prince of Wales had urged the eight water companies serving London to build a pavilion showing how they gave the city clean, pure water. That inspired Sir Francis Bolton, an engineer who was the city's water examiner, to design giant illuminated fountains to highlight the display.[17] The new incandescent light bulb was far too faint to light the fountains. Bolton lit them with the fiery streaks of controlled lightning in electric arcs. Arc lights at the base of the fountain shone through glass plates onto the rising water jets. Other arc lights illuminated water jets falling from columns. Changeable filters colored the beams, tinting the sparkling reflections from the water.

Bolton's design probably did not guide light as efficiently as Colladon had 40 years earlier. Each jet spouted from a pipe above the center of a focusing lens, so the light initially surrounded the jet instead of being confined inside by total internal reflection.[18] Bolton may not have grasped this nuance, or he may have found it easier to aim light along jets spouting over a thousand gallons a minute.

Five operators ran the lights and pumping machinery from inside a miniature island. To produce the splendid spectacle, they worked in a cramped and miserable space, wearing dark blue goggles to protect their eyes. Heat and fumes from the arc discharges made several of them ill.[19] Yet outside, the audience happily watched the sparkling fountains like we watch modern fireworks. Bolton personally controlled the changing colored lights during the shows, limited to half an hour by the time the arcs burned. Initially operated only two nights a week, the fountains were so popular that the shows were made nightly.[20]

Word of the London exhibition soon reached Geneva. Still angry that others had gained credit for his tunneling innovations, Colladon did not want his light jet to be forgotten. Nominally at the request of the editor, he slightly updated his 1842 *Comptes Rendus* paper for the French journal *La Nature*, which published it in late 1884 as "The Colladon Fountain."[21] The Swiss physicist must have thought his place secure.

The South Kensington fountains operated for two more years, as the fairgrounds hosted other exhibits. Bolton remained in charge but died suddenly at 56 on January 5, 1887.

By then others were planning bigger and better fountains for the 1887 Royal Jubilee Exhibition, held in Manchester to celebrate the 50th year of Queen Victoria's reign. The organizers paid W. and J. Galloway and Sons, a Manchester engineering firm, £3,943[22] to build a 120-foot "Fairy Fountain"[23] with triple the illumination and water flow used in South Kensington. Two lamps illuminated a massive central jet; single separate lamps illuminated each of sixteen others, placed in two rings. Fairgoers loved the two daily half-hour shows, and the organizers claimed the fountain was "the largest and most magnificent that has ever been erected."[24] The next year, Glasgow had the Galloway firm build a similar sized fountain.[25] The public was thrilled, but the electrical industry wanted fountains bigger than Manchester.[26] Colladon watched from afar as he labored on his 643-page autobiography.[27]

The Universal Exposition

The Eiffel Tower is the most visible legacy of the 1889 Universal Exposition, but the ambitious planners of the Paris fair also wanted spectacular illuminated fountains. They sent G. Bechmann, chief water engineer for Paris, to the British exhibitions, and he returned to design fountains that stood outside the main entrance, shown in figure 2-2.

As in Britain, changing filters colored the light electric arcs cast on the waters. Flood lamps lit the fountains from above. Mirrors and lenses aimed light up jets that spouted vertically, much like earlier fountains. However, Bechmann designed nozzles for horizontal jets with lenses in their middle, so the light emerged in the middle of a hollow cylinder of flowing water, shown in figure 2-3. Viewers saw luminous water spout from the mouths of sculpted dolphins, then descend in graceful parabolas to the pools beneath.[28] As in *Faust*, the water seemed alive with light.

Colladon must have been delighted on his visit to Paris. Strictly speaking, total internal reflection did not guide the light; it was trapped within a flowing tube of water. The simpler scheme of shining light inside a water jet was impractical with fast-flowing water and bright lights. However, the effect looked the same to the audience, and that was good enough for anyone but an optical purist. The fountains at the Universal Exposition were the most spectacular yet. Paris recognized Colladon as the father of the illuminated fountains, and Bechmann immortalized him in an article in *Le Grande Encyclopedie*.[29]

Light Pipes in America

Luminous fountains eventually brightened nights at the 1894 World's Fair in Chicago,[30] but before then a resolutely practical American had more mundane ideas for light guiding. A Yankee engineer named William Wheeler wanted to illuminate homes by piping light from an electric arc in the basement. It wasn't a crazy idea in 1880 when he filed for a patent. People were piping gas, heat, and water through buildings—why not light?

Wheeler had just finished an eventful few years. He was trained as an engineer in the first class to attend what is now the University of Massachusetts, graduating at 19 in 1871. He settled in his native Concord, Massachusetts, and ran an engineering practice in Boston until his college mentor, William Smith Clark, lured him to Japan to help organize an American-style agricultural college on the northern island of Hokkaido. Clark became a Japanese legend, and Wheeler served as college president while still in his twenties.[31] Wheeler came home full of ideas in 1880, and in the next few years filed over a dozen patents, most on reflectors to concentrate light. In 1881, after resuming his engineering practice, he founded the Wheeler Reflector Company to manufacture the things.

The ambitious young engineer carefully plotted the logic of light-pipe illumination. His patent[32] explains:

> It is well understood that electric light may be produced through the use of dynamo-electric machines cheaper than light from gas or any known method by combustion. This requires, however, that the light be produced

Figure 2-2: *Top.* Luminous fountains brighten the night at the close of the Universal Exposition in Paris November 6, 1889. (*Scientific American*, Dec. 14, 1889, p. 376) *Bottom.* Daytime view of illuminated fountains (from *La Nature*, first half year 1889, Fig. 3 on p. 593)

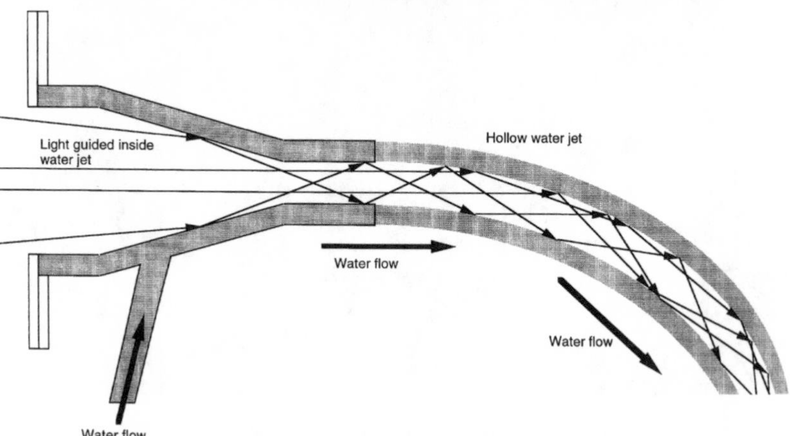

Figure 2-3: A mirror aimed light from an arc lamp into the center of the parabolic jets at the Universal Exposition. The hollow nozzle generated a tube of flowing water, lit from inside by the reflected light. Large-scale picture (top) from *La Nature,* first half year 1889, p. 408; cutaway (below) by author.

> in voltaic arcs of great power and intensity, for if more than one arc is maintained with the same current of electricity, the light generated becomes rapidly less, and therefore dearer, as the number of arcs or lights is increased. It is in the nature of the electric current, therefore, that for the economical generation of light it should be used in great intensity in one point instead of small intensity in many points.[33]

Wheeler examined every detail, down to the placement of screws holding pipe segments together. At the heart of his idea were hollow glass pipes, clear on the inside, coated with silver on the outside, then covered with asphalt to prevent scratches and tarnish. They did not guide light by total internal reflection, but instead relied on the strong reflection of light striking clear glass surfaces at a glancing angle in air. Tilt a piece of glass as you look along its surface, and you can see the effect. The glass doesn't reflect all the light, but neither do the shiniest metal mirrors. His drawing, shown in figure 2-4, is an elegant masterpiece of Victorian patent art.

Solid glass rods didn't enter the picture; Wheeler knew the clearest glass on the market couldn't carry light through a house. What he didn't know was that the incandescent bulb, invented the year before he filed his patent, would make small lamps practical. His light pipes never got off the ground, but the Wheeler Reflector Company did and made street lamp reflectors until the late 1950s.[34]

Illuminating Rods

A handful of other inventors tried guiding light short distances through glass in the late 1800s. Most wanted to deliver light to hard-to-reach places. That was difficult with gas or oil lamps, and even with early incandescent bulbs. All generated far more heat than light, making them dangerous to put close to most objects.

That was a serious problem for surgeons and dentists. They needed light to see but didn't want a hot lamp to burn themselves or their patients. In late 1888, two men in Vienna adapted "the well-known experiment for showing total reflection of light in a jet of water or in a glass rod" to illuminate the inside of the nose and throat. They attached an electric lamp to one end of a glass rod, which carried the light but not the heat, and put the other end against the side of a patient's throat. Enough light passed through the skin for them to examine the larynx, and they thought the same approach could illuminate body cavities during surgery.[35]

A decade later, an Indianapolis man patented a dental illuminator that used a curved glass rod to deliver light from a lamp into the mouth.[36] Interestingly, he designed versions for both incandescent and acetylene gas lamps, a reminder that electric lighting had not become standard even by the turn of the century. Scientists likewise illuminated microscope slides with bent glass rods to avoid drying out their specimens. In the 1930s, DuPont devel-

(No model.) 4 Sheets—Sheet 1.

W. WHEELER.

APPARATUS FOR LIGHTING DWELLINGS OR OTHER STRUCTURES.

No. 247,229. Patented Sept. 20, 1881.

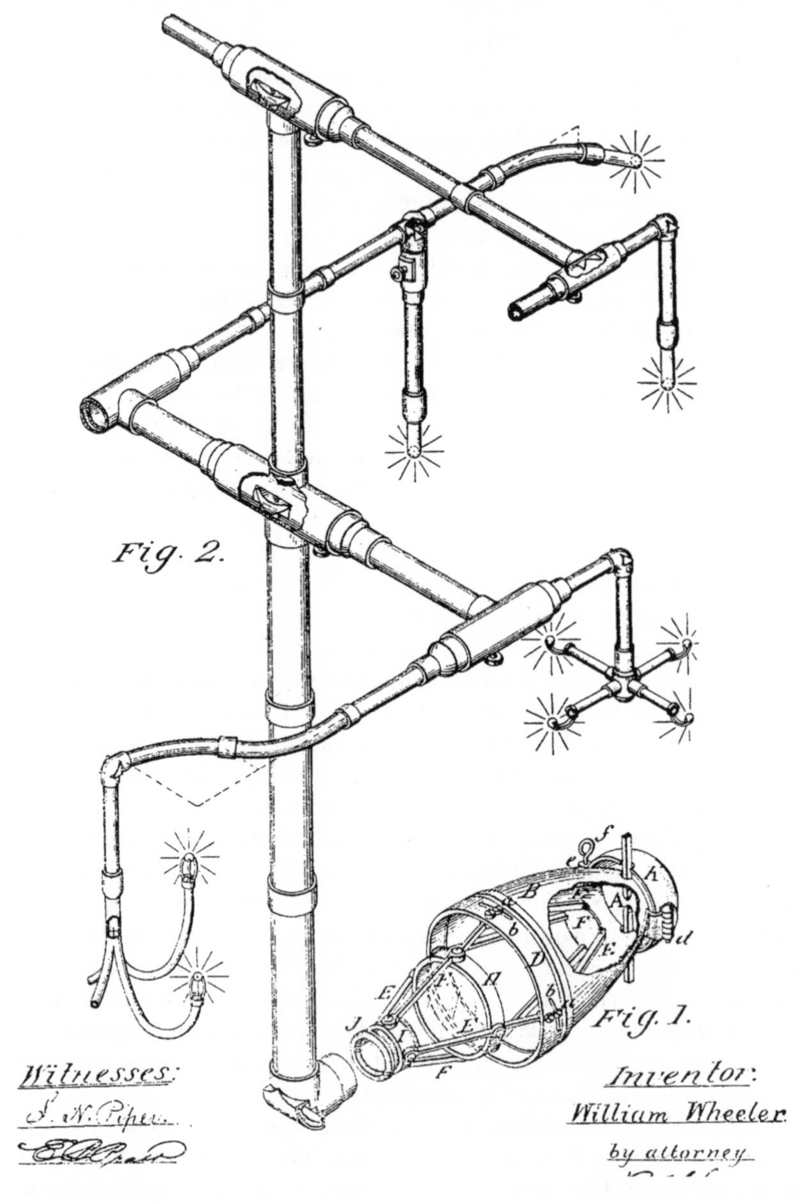

Figure 2-4: William Wheeler's light pipes, from his patent drawing. (US Patent 247,229, Figure 2, on sheet 1)

oped a clear plastic that quickly replaced glass and quartz illuminators because it is lighter, cheaper, more durable, and easier to bend.[37]

The light-guiding effect that Colladon had conceived as a thing of beauty was evolving into a useful technology with mundane applications. That would have pleased Colladon, who took pride in his practical inventions such as compressed air. He probably intended his paper in *La Nature* and his autobiography to ensure that future generations recognized his pioneering role. Yet he would have been furious to see the pioneers of fiber optics assigned the credit for light guiding in a water jet not to him but to John Tyndall, who first demonstrated it thirteen years after Colladon.

Professor Tyndall's Performance

A small man with bright gray eyes, Tyndall joined the Royal Institution in London in 1853. He quickly made a name for himself by giving informative and entertaining talks in a long-running series of Friday evening lectures.[38] However, he found himself in trouble as time came for his talk on May 19, 1854. The demonstration he had planned wasn't ready for the talk, and as the deadline approached he talked with his mentor, Michael Faraday, who had given similar talks for years. Faraday suggested he demonstrate the flow of water jets and and how total internal reflection could guide the light along the flowing liquid.

In his handwritten notes, Tyndall apologized for not showing "something entirely new,"[39] but published accounts of the lecture omit that apology. He used the water jet to conclude his lecture by demonstrating "the total reflexion of light at the common surface of two media of different refractive indices"—water and air. Initially, the light emerged from a glass tube in the side of a tank. Then he turned a valve so water could flow through the tube. Thanks to total internal reflection, he wrote, the light "seemed to be washed downward by the descending liquid, the latter being thereby caused to present a beautiful illuminated appearance."[40]

It must have been a good show, and Tyndall made it part of his repertoire. But he saw light guiding more as a parlor trick than a new scientific concept and made no effort to properly attribute it to Faraday or anyone else in his published account. Indeed, total internal reflection was a well-known phenomenon, and Tyndall may have considered light guiding too obvious an application to be new. However, Tyndall also probably didn't know whose idea it was.

Faraday was one of the greatest scientists of his time, but about 1840 his memory started to fail him, a problem that slowly worsened over the years.[41] That fact that Tyndall didn't credit him implies that Faraday remembered a demonstration by someone else but that Faraday's failing memory couldn't come up with the name and Tyndall didn't want to embarrass him. It was a common problem for Faraday by that time, especially on bad days. The name Faraday forgot probably was Daniel Colladon.

Total Internal Reflection: What makes a diamond sparkle

The same effect that makes a diamond sparkle guides light along a jet of water or an optical fiber. It also helps to create the rainbow. It's called total internal reflection, and it's something we rarely recognize. Total internal reflection has been known for hundreds of years; light guiding is a more recent discovery.

Total internal reflection is a side effect of refraction, the bending of light that passes from one transparent material into another. Lenses depend on refraction, for example, to focus the words from this page onto the back of your eye or to focus a movie film image onto a screen at the front of the theater.

Refraction occurs because the speed of light in transparent materials is less than the universal speed limit of 299,792 kilometers per second in a vacuum. In air, light is just a little bit slower, but in glass it slows to about 200,000 kilometers per second. In diamond, light is even slower, about 125,000 kilometers per second.

This slowing down causes light to bend as it passes from air into glass (figure A) or vice versa. The light waves keep oscillating at the same frequency, but they are closer together in slower materials. The degree of bending depends on the refractive index, which is the speed of light in vacuum divided by the speed in the material. The refractive index of air is 1.0003, of pure water 1.33, of ordinary glass about 1.5, and of diamond about 2.4.

The amount of refraction at the surface depends on the difference in refractive index; the bigger the difference, the more refraction. Light rays passing from air

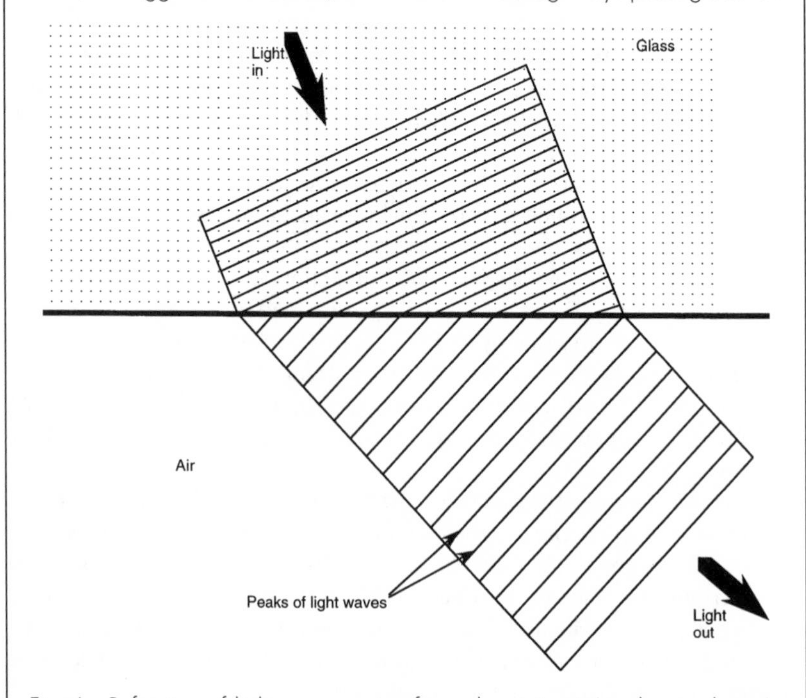

Fig. A: Refraction of light waves going from glass into air (peaks are shown).

→

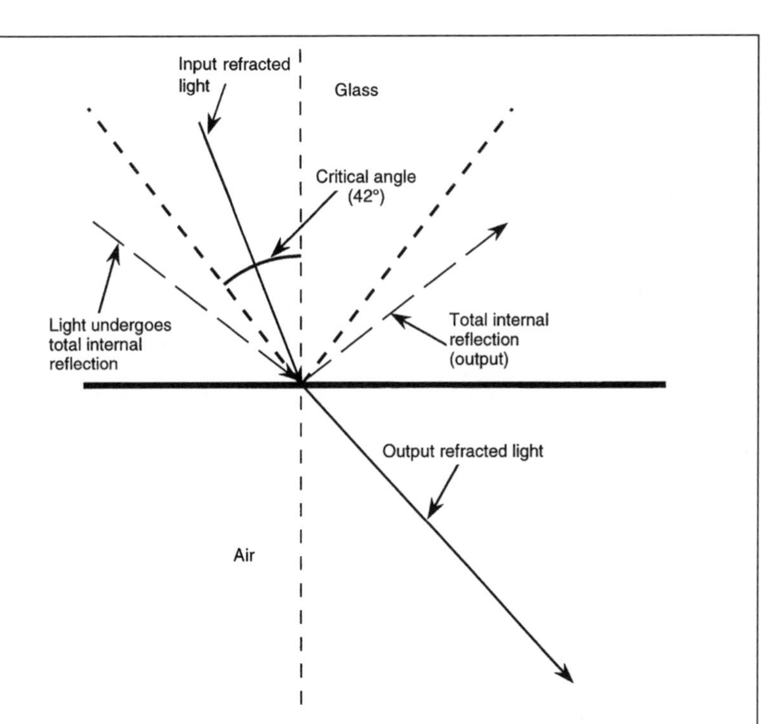

Fig. B: Total internal reflection occurs where light rays hit glass at a glancing angle outside the critical angle. Light that hits the surface nearly straight on is refracted into the air.

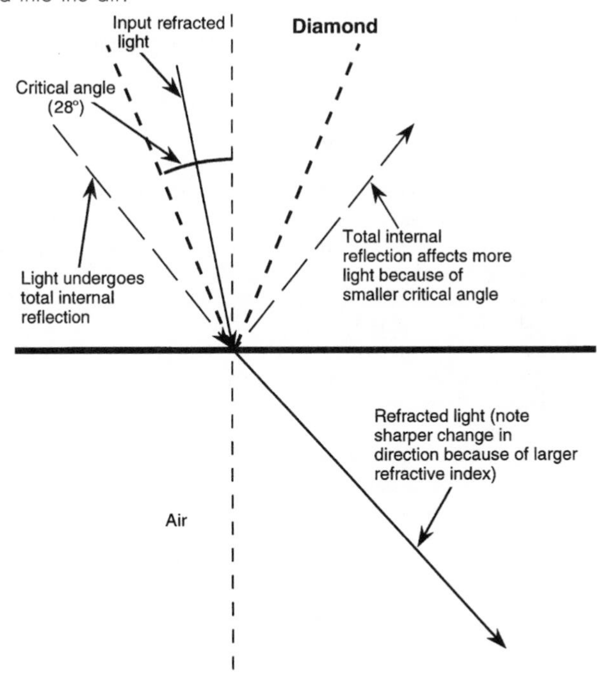

Fig. C: Diamond has an index of refraction much higher than glass, so total internal reflection occurs over a much wider range of angles—making the gem glitter. (The bright colors come from a related effect: the refractive index varies strongly with wavelength, spreading out miniature rainbows in the diamond.)

into glass bend into the glass, while light going from glass into air bends toward the surface.

As light in the glass hits the surface at a steeper angle, it emerges into air almost parallel to the surface. Eventually, it reaches a "critical angle" where it cannot emerge into the air, as shown in figure B. (This critical angle is measured from a line called the "normal" which is perpendicular to the surface.) All the light outside the critical angle is reflected back into the glass.[A] For glass with a refractive index of 1.5 in air, this angle is about 42 degrees. The higher the refractive index, the larger the angle and the easier it is to see reflections sparkling from within the glass.

Diamond has a refractive index of 2.4, so the critical angle is 25 degrees, and much more light entering it undergoes total internal reflection as shown in figure C. Diamond cutters take advantage of this effect and cleave the gems so they collect light entering the top facet and reflect it back to the eye, making them appear to shine. Because the refractive index changes with wavelength, colors return at slightly different angles, making the stone glitter with color.[B] (Zircon crystals have a refractive index of 2.1, so they glitter more than glass, but not as brightly as diamond.) The variation of refractive index with wavelength also breaks light passing through a prism into a spectrum and forms a rainbow when sunlight strikes tiny water droplets.[C]

You can't see total internal reflection in a window pane; you can only see it in a large block of glass when you look at a surface at less than the critical angle. Turn it in your hands and you can see total internal reflection start and stop as you pass through the right angle. Fine crystal glass has a higher refractive index than normal glass, so it shows total internal reflection more easily.

If you shine light along a glass rod or a jet of water, total internal reflection will keep it within the glass or liquid. This light-guiding effect is the basis of fiber optics.

A. This is a consequence of the law of refraction, which says the sine of the angle (I) from the normal times the refractive index in one material (n_i) must equal the sine of the angle (R) from the normal in the other material times its refractive index (n_r): ($n_i \sin I = n_r \sin R$). Plug in numbers, and you find that light trying to leave the high-index material at large angles from the normal (I.e., small angles from the surface) would have to emerge at an angle with sine greater than one. That's impossible, so the light doesn't get out. For a more thorough explanation, see Chapter 2 in Jeff Hecht, *Understanding Fiber Optics 3rd ed.* (Prentice Hall, Upper Saddle River, N.J., 1999).

B. For a more complete explanation of how the high refractive index of diamond affects its optical properties, see David Falk, Dieter Brill, and David Stork, *Seeing the Light: Optics in Nature, Photography, Color Vision, and Holography*, Harper & Row, New York, 1985, pp. 63–64.

C. For more on the rainbow and other atmospheric phenomena, see Robert Greenler, *Rainbows, Halos, and Glories* (Cambridge University Press, Cambridge, UK, 1980).

The circumstantial evidence is strong. Faraday spent the summer of 1841 in Switzerland and was a friend of de la Rive, who duplicated Colladon's demonstration at exactly that time. Faraday probably saw either de la Rive or Colladon perform the water-jet demonstration, and the physics would have lingered in his memory longer than the name. According to Colladon, someone demonstrated the water jet in London about that time but the person is unknown. It's very unlikely it was Faraday, but it could have been de la Rive.[42] In addition, Tyndall cites the fluid-flow research of Felix Savart, which Colladon mentioned. Faraday was not trying to steal the idea; he had been involved in a nasty controversy earlier and was afterward very careful to credit others. But poor Faraday couldn't remember, and Tyndall diplomatically ignored the issue.

An outspoken man of wide-ranging interests, Tyndall developed a strong public persona and became a Victorian version of Carl Sagan, scientist, popularizer, and public figure. He described the water-jet demonstration in one of his popular books, which circulated widely and stayed in print for many years. When later physicists went back to look for the origins of light guiding, they came upon Tyndall's account. By then, Colladon's papers were buried on the back shelves of scholarly libraries, unreferenced and forgotten, so Tyndall was credited with the invention over half a century after he and Colladon had died.[43]

3

Fibers of Glass

> I do not believe, if any experimentalist had been promised by a good fairy that he might have anything he desired, that he would have ventured to ask for any one thing with so many valuable properties as these fibers possess.
>
> —Charles Vernon Boys, 1889[1]

The difference between glass rods and glass fibers is merely a matter of diameter. Anyone who can be trusted with a chemistry set can draw a fiber easily from a glass rod. Hold the two ends of a rod several inches long and put the middle in a hot flame. The heat softens the glass, melting the rigid solid into a thick, viscous liquid. After the rod becomes flexible, pull the two ends apart while removing the rod from the flame. The molten glass stretches into a long, tapered thread, which solidifies almost instantly as air cools it. Although the material remains glass, the thin filament is flexible and seems much less brittle than the rod.

Glass is an unusual material; few others form thick liquids easy to stretch into fibers. Most liquids are like water, far too thin to make fibers from. Only if you load water with sugar and heat the mixture do you get a hot, thick syrup that can be spun into threads of cotton candy.

It is the malleability of hot glass, as well as the transparency of the solid, that have made it a material for the ages. The ancients probably made the first glass by accident, when they mixed sand, soda, lime, and ash in a fire and discovered their creation when the embers cooled. Ancient glass makers learned how to mold and blow hot glass, shaping it into works of art and things of utility. Generations of artisans, inventors, engineers, and scientists have refined glass technology, making it cheaper, clearer, and more durable. The recipes for glass have changed, but it still surrounds us, taking many forms including insulating fiberglass, fine crystal, the transparent shells of light bulbs, drinking glasses, and the transparent windows through which we look upon the world.

A Long History of Short Fibers

Glass fibers go back a long way. The Egyptians made coarse fibers by 1600 BC, and fibers survive as decorations on Egyptian pottery dating back to 1375 BC.[2] In the Renaissance, Venetian glass makers used glass fibers to decorate the surfaces of plain glass vessels. However, glass makers guarded their secrets so carefully that no one wrote of glass fibers until the early seventeenth century.[3]

The eighteenth century brought the invention of "spun glass" fibers. Réné de Réaumur, a founder of the French iron and steel industry, was trying to make artificial heron feathers from glass.[4] He made fibers by rotating a wheel through a pool of molten glass, pulling threads of glass where the hot thick liquid stuck to the wheel. His fibers were short and brittle, but he predicted that spun glass fibers as thin as spider silk would be flexible and could be woven into fabric.[5]

By the start of the nineteenth century, glass makers learned how to make longer, stronger fibers by pulling them from molten glass with a hot glass tube. Inventors wound the cooling end of the thread around a yarn reel, then turned the reel rapidly to pull more fiber from the hot glass. Wandering tradesmen spun glass fibers at fairs, and it became a novelty for decorations and ornaments. Powdered wigs were fashionable at the time, and wigs made of spun glass became novelties for the "naturalia cabinets" of royalty.[6] However, the stuff was of little practical use; the fibers were brittle, ragged, and no longer than ten feet (three meters), the circumference of the largest drums or reels. Others tried different approaches, such as forcing molten glass through a tiny nozzle.[7]

By the mid-1870s, the best glass fibers were finer than silk and could be woven into fabrics or assembled into imitation ostrich and egret feathers to decorate hats.[8] Cloth of white spun glass resembled silver; fibers drawn from yellow-orange glass looked golden when woven into fabric. Although the glass fabrics looked alluring, even enthusiasts noted that "the spinning requires extraordinary dexterity and constant attention . . . [and] is said to be very trying to the sight."[9]

Hanging by a Thread of Glass

The properties of spun glass caught the eye of Charles Vernon Boys, a young demonstrator of physics at the Royal College of Science in the South Kensington district of London. His passion was building sensitive scientific instruments; the classes he had to teach were unwelcome interruptions. Writer H. G. Wells, who suffered through them, called Boys "one of the worst teachers who has ever turned his back on a restive audience, messed about with the blackboard, galloped through an hour of talk, and bolted back to the apparatus in his private room."[10]

In 1887, Boys was on the verge of a measurement breakthrough, the start of a few brilliant years that would make him a giant of British physics. He wanted to measure the effects of delicate forces on objects. He knew that one way to sense weak forces was to hang an object from a thread. The problem was that the thread had to be thin, strong, and elastic to measure the forces. Silk and spider line were the best fibers of the time, but they were not good enough. Nor were metal wires, because they stayed bent if twisted too much. Boys tested spun glass fibers a thousandth of an inch (0.025 millimeter) thick but found them wanting because they stayed bent after being twisted, making accurate measurements impossible.[11]

After he had disposed of his classes, Boys retreated to his laboratory to try drawing glass into strong, elastic fibers. He wanted them long, thin, and uniform, so he wanted to pull the molten glass very quickly along a straight path. He built a miniature crossbow, and made light arrows made by fastening a needle to a piece of straw a few inches long. He stuck the arrow to one end of a glass rod with sealing wax, and heated the glass until it softened. Then he fired the arrow through two long rooms with a foot trigger. The bow propelled the little arrow so forcefully that it could pull a fiber tail from a blob of molten glass that hung briefly behind in mid-air before falling to the ground. When the arrow landed, Boys found attached to it "a glass thread 90 feet long and 1/10,000 inch in diameter, so uniform that the diameter at one end was only one sixth more than that at the other."[12]

The delighted physicist then tried his new toy on other materials. He had started with ordinary glass, probably whatever lay readily at hand in his laboratory, but like bread, glass can be made from many recipes. Strictly speaking, glass is a solid inorganic material that never crystallized as it cooled from molten form. Some materials form glasses when cooled quickly; many others do not. The most important ingredient in common glasses is silicon dioxide, a durable mineral known as silica. Nature can mold silica into clear crystalline quartz, but we usually find it as sand.

You can make glass from pure silica, but not easily; it melts at about 1600°C (2900°F). Since antiquity, glass makers have added other compounds including soda, lime, and ash to lower the melting point and give glass other desirable properties. Glass specialists have developed particular recipes for many purposes: fine glasses used in crystal ware, clear and uniform glasses for precision optics, heat-resistant glasses for cookware, plate glass for windows, and colored glasses for bottles. Silica is the key ingredient in all, but glasses can be made in countless ways from innumerable mixtures of ingredients.

Armed with a high-temperature oxy-hydrogen flame, Boys tried a variety of minerals that resembled glass in some ways. Some, like sapphire and ruby, did not stretch into fibers. Yet others did work. The best of these was natural quartz, crystals of almost pure silica. Much thicker than molten glass, molten quartz dragged heavily on the miniature arrows, so they usually fell far short of their target. It took many tries before Boys could stretch quartz into fibers, but success proved exciting. Some fibers were "so fine that I believe them to be beyond the power of any possible microscope," he told the Physical Society in London on March 26, 1887.[13]

The best of his threads were as strong as steel wires the same size, a marvel for a material always considered fragile. Their thinness and strength made the quartz fibers ideal for suspending objects on torsion balances to measure delicate forces. Boys used them in an instrument he called a "radio-micrometer," with which he could detect the heat from a candle nearly two miles (three kilometers) away. He was utterly delighted with the fibers and with the measurements they let him perform.

Boys stretched quartz fibers in front of him and found that in the right light they glittered with colors. He stretched many fibers parallel to each other and they acted like a diffraction grating, an array of parallel lines that spreads out a spectrum like a prism. He must have seen Sir Francis Bolton's illuminated fountains; the exhibit grounds were close to his college. Yet Boys did not record for posterity any experiments with light guiding, even the results of pointing one end of a fiber at a lamp and looking into the other end. Perhaps he tried and saw nothing, because his fibers were not clear enough. Perhaps it was impractical with the fine fibers that fascinated him the most. Or perhaps he thought the idea too trivial to mention at a time when anyone could marvel at illuminated fountains displaying the same principle far more spectacularly.

Glass Fabrics

A German immigrant named Herman Hammesfahr patented glass fibers in America, but it took him years to interest others in his vision of glass fabric. He succeeded when the Libbey Glass Company wanted something spectacular to show at the 1892 World's Fair in Chicago. At first, two young women wove glass fibers into fabric for lamp shades. Then the fabric caught the eye of actress Georgia Cayven, who wanted a dress made of it. Hammesfahr succeeded by combining satin thread with the glass fibers. Libbey showed the dress at the fair, where it caught the eyes of many fairgoers, including Princess Eulalie of Spain, who paid $30,000 for a copy.[14]

Glass fabrics glittered brightly at the turn of the century, but they were hardly practical. Many years later, Hammesfahr's granddaughter recalled modeling at the 1904 World's Fair in St. Louis a glass dress her mother had made: "All that I remember is that it scratched, and I couldn't sit down for fear of breaking the glass threads."[15]

The real role of the glass dresses, neckties, and other fabrics was to publicize new glass technology. Nineteenth-century industry had made glass a better and cheaper product, commonplace from windows to whiskey bottles. Glass fabrics promised new marvels for the twentieth century. Part of the attraction was the way light glittered from the glass, but the very idea of flexible glass threads was itself a wonder.

Industry saw a host of uses for glass fibers besides clothing. Glass fibers could withstand corrosive chemicals, so chemists and druggists used them to filter solid particles out of liquids. Woven glass fibers were used as bandages. Industry realized that tangled glass fibers, called glass wool, made a good insulator, and packed them around steam pipes. Industry would pay more per square inch for glass fabric than people would pay for clothing—and didn't worry if it was scratchy.

As interest grew in glass fibers, engineers and scientists devised new ways to make them. One was to suspend a glass rod vertically, heat the lower end in a furnace, and draw a fine filament downward from the molten zone. In one early scheme, the experimenters hung a weight from the bottom of the rod, which dangled below the hot zone of a cylindrical furnace. Rapid heating softened the glass quickly, and the weight stretched it into a thin fiber. The scheme also stretched glass tubes into fine capillaries, allowing British scientists to make 0.001 millimeter tubes with walls thinner than 0.0001 mm.[16]

Boys's instruments interested other experimenters in quartz fibers. A few companies started manufacturing them, but they were expensive, so *Scientific American* published a do-it-yourself recipe. Readers wanting long, fine fibers were told how to make their own miniature catapults—a design that owed much to Boys's crossbow.[17]

Nonetheless, glass fibers remained a specialty item for the first three decades of the twentieth century. Only in Germany did an industry develop, and that only because Allied embargoes cut the country's supply of asbestos during World War I. Natural fibers were better—and cheaper—for insulation and filtering, and in that innocent age the health hazards of asbestos were unknown.

In December 1931 engineers at the Owens-Illinois Glass Company in Newark, Ohio, demonstrated the first commercially viable technique for mass-producing inexpensive glass fibers. They abandoned the idea of drawing one fiber at a time. Instead, they blew hot air into molten glass, splattering short, coarse threads of liquid into the air, where they quickly solidified. The process yielded a soft, fleecy mass of flexible fibers—glass wool, the fiberglass used in modern insulation.

The Corning Glass Works was also working on glass fibers, and the two companies formed a joint venture, the Owens-Corning Fiberglas Corporation.[18] By 1935, they were producing fibers so strong, fine, and flexible that they could be woven into cloth, which could be bent and folded without breaking the fibers—as had worried the little girl in the glass dress 30 years earlier.[19] The fiberglass industry has been growing ever since.

Nobody paid much attention to the optical properties of glass fibers, other than when they glittered in fabrics on display. However, engineers did learn about the mechanics of glass fibers and the properties of molten glass. In time, that would prove valuable knowledge for developers of optical fibers.

4

The Quest for Remote Viewing

Television and the Legacy of Sword Swallowers

(1895–1940)

No great pioneer invention has been made in the laboratory of a great corporation; and by pioneer inventions I mean new mechanisms which started new industries.

—C. Francis Jenkins, American pioneer in motion pictures and television, 1927[1]

Europe and America were fertile ground for new ideas at the end of the nineteenth century. The great Victorian exhibitions illustrated the wonders of the age. Their illuminated fountains displayed light guiding for all to see; in the halls, visitors saw glass fibers and fabrics. Scientists marveled at the fine quartz fibers Charles Vernon Boys made for his instruments. Innovations in lighting tickled the imaginations of inventors. They sought ways to make moving pictures, and to view events remotely, by sending pictures as well as the sounds they could hear on the telephone. A few realized that arrays of many glass rods or fibers could carry the patterns of light we call images.

With hindsight, the idea of image transmission seems simple. Hold a handful of drinking straws together such that they all point in the same direction and look through the bundle at a picture a little beyond the tips. Each straw shows one piece of the picture. In your mind's eye, shrink the straws so that each one shows a single spot on the page. Then replace them with thin glass rods, stacked together the same way, but guiding light by total internal reflection instead of through central holes. The bundle transmits an image point by point. Replace the rods with flexible fibers, and the bundle can guide light around bends, reaching into otherwise inaccessible places.

In practice, you need some refinements. If a bundle of fibers is to reproduce an image, the fibers must be arranged in the same pattern on both ends. If the fibers are not aligned, light collected from one spot on the picture ends up somewhere else on the display end, scrambling the image.

The fibers also must be small, because the light passing through them is homogenized. Look through a single long thin, bent rod or fiber that covers a printed letter and you see not the black-and-white pattern of the letter, but a gray spot that mixes the black and white. If you want to read the letter, you have to cover it with many fibers, with each showing a dot representing part of the letter. The thinner the fibers, the smaller the dots and the clearer the image. Specialists in printing and imaging call this resolution and measure it in dots per inch. Pack small dots tightly and you don't notice that the letters or pictures are made of dots.

The Problem of Remote Viewing

From the Victorian era into the early twentieth century, inventors took many approaches to achieving what they called remote viewing. They began with a naive optimism. Telegraphs had been sending messages for decades; Alexander Graham Bell had launched the telephone era in 1876. Sending pictures from one point to another seemed just another logical step.

Still pictures are much simpler to send than moving ones, particularly if you write on treated paper or photographic film. The first facsimile machine was tested in the 1840s,[2] although the image quality was awful. True remote viewing of a constantly changing scene is a much bigger challenge. You need a sensitive camera that responds quickly, a transmission medium to carry the picture, and a screen that can display a changing image.

Ingenious inventors devised a host of schemes, quickly adapting the latest scientific discoveries. They waxed enthusiastic over the 1870s discovery that light made selenium conduct electricity better; the more light, the more readily the material carried current. A single selenium cell could monitor changes in a beam of light. A Boston inventor proposed building an array of selenium cells to sense light intensity at different points on its surface.[3] He thought electrical signals from the selenium cells could reproduce an image by controlling brightness of an array of incandescent bulbs.

Such a system would be simple to build today, but it was beyond 1880 technology. Selenium cells and incandescent bulbs were new; assembling them in arrays would have been difficult. Connecting them would have required a complex tangle of wires—at least 10,000 to feed a display 100 bulbs wide and 100 high. Modern solid-state electronics can handle the problem by combining many signals into one, but even the vacuum tube was decades away in 1880. William Wheeler's light pipes were practical in comparison.

An Array of Light Pipes

A host of other inventors also dabbled with remote viewing, including Henry C. Saint-René, who taught physics and chemistry at the agriculture school in Crezancy, a town of about 500 people in the French district of Aisne. Most of his ideas came from other inventors, but he proposed a novel display based on moving tinted glass slides in front of lighted openings. Grading the tint from top to bottom would control brightness, but Saint-René worried that gaps between the moving slides would ruin the effect.

As he searched for a way to eliminate the gaps, he recalled the principle of light guiding in fountains. He thought curved glass rods could collect light from each opening and deliver it to a screen, where their other ends could be packed together without gaps to form an image. He wrote:

> A bent glass rod—its two ends cut perpendicularly to its axis—will receive on its lower, vertical cross-section the light which it will conduct (acting like the liquid conduit in an illuminated fountain) to the upper, horizontal cross-section, where it is perceived by the eye.[4]

Saint-René realized that each rod homogenized the light passing through it, so he needed rods with small ends to show details. He wrote: "The whole array gives a complete illusion of the object if the diameter of each point does not exceed 1/3 millimeter when the viewer is at a distance of one meter from the image."[5]

It is unclear whether Saint-René built anything, but he liked the idea. However, he evidently lacked the money to file for a patent, or doubted he could earn back his investment. Instead, in the summer of 1895 he wrote a description of his invention and sent it in a sealed packet to the French Academy of Sciences in Paris, which has a novel system for documenting claims of priority. Inventors can wait years, then ask the Academy to open the packet and publish a summary in *Comptes Rendus* to prove their priority. Saint-René bided his time for over 14 years before asking the Academy to publish his summary.

No one else shared Saint-René's enthusiasm, even in 1910 when his idea finally saw print. His display was the first use of a bundle of glass rods or fibers to transmit images. Yet it remained obscure because his remote-viewing system was impractical, one of countless schemes that came and went before the new technology of television evolved at the hands of a new generation of engineers and inventors. Saint-René himself vanished from the scene, and his records probably were destroyed with those of the agriculture school when World War I ground through the area.[6]

The Birth of Television

Invention and technology changed in subtle but crucial ways in the decades between the Paris exposition and the 1920s. Corporate research labs emerged, competing with inventor-entrepreneurs like Wheeler. General Electric and AT&T recruited cadres of engineers and scientists to deal with increasingly complex technologies. With large teams and healthy corporate bank accounts, the new research labs had resources that old-fashioned inventors couldn't match. Even as lone inventors complained that the corporate labs lacked the spark of genius, the new teams cranked out new developments. The Radio Corporation of America, founded as a trust to hold radio patents, developed increasingly powerful transmitters that spanned greater distances.

Television would prove a crucial battleground. By the early 1920s, electronic technology had brought radio broadcasting to the threshold of practicality. When conditions were right, radio waves could carry voices across oceans, and the technology was improving steadily. To a handful of men, television seemed a logical extension of radio.

The early leaders were John Logie Baird in Britain and C. Francis Jenkins in America. Their approaches shared many crucial features. They agreed that images had to be built up by scanning the scene in a way that sampled light intensity at one point at a time, then reproducing the image point by point and line by line on a display. Baird and Jenkins gained early leads by building mechanical scanners that looked through holes in spinning disks. Both men were old-school inventor-entrepreneurs with enough charisma to raise money to pursue their visions.

Born in Scotland in 1888, Baird was a blend of visionary engineer, entrepreneur, and self-promoter, with perhaps a dash of crackpot to add spice. He had made a series of inventions—none gloriously successful—before he turned to television in the winter of 1922–23. Success took a while. His landlord evicted him for failing to pay the rent the following year,[7] but in April 1925 he conducted the first public demonstration of television, at Selfridge Department Store in London. The image showed only eight lines per frame—but it was witnessed by the public, and Baird made the most of the publicity to raise money for further development. Later he built the first television system used by the British Broadcasting Corporation, using a spinning disk to scan a 30-line picture.[8]

Like many other inventors, Baird filed patent applications prolifically to protect his ideas. He accumulated an impressive 178 patents from the start of his television work until his death in 1946. Inevitably, most never proved practical, but they covered a wide range of ideas. One was sending an image through an array of parallel tubes, transparent rods, or clear fibers.

Baird was groping for ways to scan images. His patent proposes a screen that uses a honeycomb device assembled from a large number of short tubes "to produce an image without the use of a lens." Collect the tubes like a bundle of drinking straws and they can transmit images point by point. Baird had learned the patent game well, so he wrote his application to cover many variations on the idea. Among them were "thin rods or tubes of glass quartz or other transparent material. . . . The rods . . . need not necessarily be straight, but could be bent or curved, or in the case of very fine quartz fibers, could be flexible."[9]

He never got that far. In 1927, he collected 340 metal tubes two inches long and a tenth of an inch in diameter, and stacked them in a 17 × 20 array that he used to scan an image. He abandoned that approach without trying fibers or glass rods.[10] The mechanical television system he pioneered in Britain used spinning disks.[11]

C. Francis Jenkins: An American Inventor

Born on an Ohio farm in 1867, C. Francis Jenkins began thinking about television in 1894 and dabbled with it most of the rest of his life. A founder and the first president of the Society of Motion Picture Engineers, Jenkins helped develop early motion picture projectors,[12] but his most profitable invention was a spiral-wound waxed paper container.[13] He was neither rich nor beyond embellishing his achievements, but nobody was about to throw him into the streets. Compared to the impoverished Baird, he had made a successful business of invention, and turned to television with a solid track record behind him.

In late 1923, he brought two leading radio-magazine editors[14] to his Washington, D.C. laboratory, where they saw the shadow of Jenkins's hand waving on the screen. However, Jenkins fell behind Baird. His first public demonstration came two months after Baird's, and Jenkins transmitted only black-and-white silhouettes while Baird showed shades of gray.

Thanks to his years of movie experience, Jenkins knew more about optics.[15] He tested several types of television cameras and receivers, and one of the receivers used quartz rods to carry light.[16] He mounted 48 three-inch rods in a screw-thread pattern around the inside of a cylinder that spun 3600 times per minute. As they spun, they collected light from a central stationary bulb with four filaments turned on and off by radio signals. The other ends swept by the viewer's eyes, creating 15 new images every second. A magnifying lens and mirror showed a flickering silhouette about six inches square made up of one line per quartz rod. The images weren't very good, but at the time any image was impressive.[17]

The rotating drum receiver proved impractical; the quartz rods transmitted only 40% of the lamp light, and the lamps lasted just a few hours. Full of energy in his sixtieth year, Jenkins quickly devised a different receiver, which he used to transmit "Radio-Movies" across a room on May 5, 1928. He boasted to the *New York Times* that he would soon be able to project pictures on the wall.[18] Jenkins coaxed vacuum-tube pioneer Lee De Forest and New York bankers to invest $10 million in the Jenkins Television Company. The company made some receivers and built television stations in Washington, D.C., and Jersey City, but soon ran out of money. In September 1929, a month before the stock market crash, De Forest Radio took over Jenkins Television. Ill with heart trouble, Jenkins sold his laboratories at the end of 1930. De Forest Radio and Jenkins Television both failed during the Depression, and Jenkins died in 1934.[19]

By then, Baird had become a reluctant convert to a newer technology, electronic television, in which electron beams instead of spinning disks scanned images. The impetus came from America, where a young inventor-businessman named Philo Farnsworth was struggling to stay ahead of a well-funded team at RCA.

C. W. Hansell Sees the Problem Clearly

The magical allure of radio attracted a host of bright young engineers after World War I. Many of the best of them joined the fast-growing RCA. It was a golden opportunity to ride the breaking wave of a new technology at a company with ample resources to support innovative engineers. Among them was Clarence Weston Hansell, who went by his initials C. W. He was in the first generation of engineers trained in electronics, a man with a gift for invention and a wide-ranging mind. Pondering the problem of reading an instrument dial that was annoyingly out of sight in late 1926, Hansell realized that a flexible bundle of transparent glass fibers could do the job.

That was a vital step. Colladon, Babinet, Saint-René, Baird, and Jenkins all saw the potential of light guiding by total internal reflection. But none of them grasped the elegant simplicity of image transmission through a flexible bundle of thin glass fibers. Align the fibers in the same way on both ends of the bundle, and they should carry a pattern formed on one end point by point, fiber by fiber, to the other end. Bend the bundle, and you could see into otherwise inaccessible places—from automobile gas tanks to the human stomach.

Born January 20, 1898, in Medaryville, Indiana, Hansell was the oldest of eight children in a poor farm family.[20] After earning a degree in electrical engineering at Purdue in an Army training program, he worked briefly for General Electric before it divested its radio business to the newly formed RCA. In 1922, he developed and installed the first vacuum-tube transmitter for commercial wireless telegraphy across the Atlantic.[21] In 1925, the young engineer founded the RCA radio transmission lab at Rocky Point, Long Island, which he headed for over 30 years. He bought a house on Long Island Sound and raised a family there.

Short and stocky, Hansell had a mind that bubbled with new ideas.[22] In his 44-year engineering career, he collected over 300 US patents, plus many more issued overseas. His specialty was radio transmission, and he was good at it, helping perfect short-wave radio antennas and building the world's largest radio transmitter at Rocky Point. Like other forward-looking radio engineers, Hansell hoped to adapt radio technology to sending pictures, including the "wirephoto" services that evolved into what we now call facsimile, as well as television.

Sometimes his inventive mind wandered. On December 30, 1926, he filled two pages in his engineer's notebook tight to their edges with a concise proposal, neatly written with a fountain pen: "Method for transferring a dial reading to a distance." He started with his initial goal: "reading the dial of an instrument from points at some distance away and from points obstructed for vision. It would be a great help if the vision of such an instrument could be piped from one place to another around corners, curves, etc."

He did not say where he got the idea of piping light. It could have been a decorative lighting fixture or fountain, or perhaps an illuminator he saw at his dentist's office. He must have seen curved glass rods guide light; his patent application says "experimentally, the remote end of such a rod may be used to ignite paper even though the conductor is quite cool along its surface, and may be held without any sensation of heat."[23] It's easy to imagine him focusing sunlight into a bent glass rod to burn holes in pieces of paper.

The inspiration added up to a simple solution to his problem:

> It is only necessary to make up a cable of parallel laid quartz hairs or strands of similar material, the ends of which can be cut off plane. One end of this cable can face the instrument and another can face toward the observer. The light from the instrument will fall on the ends of the quartz fibers and be transmitted through them coming out as an image in the other end visible to the observer. No light will pass from one quartz hair to the other due to the total reflection from the walls of the quartz. Of course, the area corresponding to the cross section of each hair will have its particular part of the image mixed up so that its detail is lost, but the image as a whole can be made quite good by using a great many very small hairs.[24]

Except for one crucial detail, Hansell had nailed dead-on the idea of bundling glass fibers together to transmit an image. His agile mind saw many other ways to use imaging bundles. He jotted a few in the notebook. One was "a flexible periscope, which could be quite useful in the army and navy." He proposed relaying "the reading of a gasoline meter from the tank on the back of an automobile to the dashboard." The same idea could be extended to airplanes and ships, though in some cases a lamp might have to illuminate the meter to get enough light. Surgeons could thread similar cables down the throat to examine the stomach.

By the time RCA's patent department filed an application the following August, Hansell had added another potential use that was a closer match to RCA's electronics business—a "picture transfer cable" for facsimile transmission. Newspapers were the main customers; they wanted to transmit news photos around the globe, just as they could wire articles by telegraph. RCA, AT&T, and telegraph giant Western Union had demonstrated wirephoto systems in 1924.[25] Hansell thought he could speed image scanning by using a several fibers instead of a single light collector. Scanning with 20 fibers spaced along a straight line delivering light to 20 separate detectors could collect information 20 times faster than a single detector. Likewise, 20 fibers in the receiver could deliver light from separately modulated bulbs and write 20 times faster. In addition to speeding the system, Hansell hoped to simplify alignment.

(Hansell's patent also contains another prescient concept. He realized that the signal transmitted depended on how the fibers were arranged in the bundle. If they were scrambled at the transmitter, they would have to be scrambled in the same way at the receiver or different lines would appear in the wrong places, jumbling the entire image. He suggested that intentional scrambling could prevent nonpaying customers from "stealing" pictures by picking up RCA's radio-facsimile transmissions.)

Yet despite Hansell's uncanny vision, there is no evidence he went much further than writing down his ideas for the RCA patent department. Colleagues recall him as more likely to sketch new ideas on paper than to build anything.[26] He was an energetic man but a busy one; not yet 30, he directed research at the RCA Rocky Point Lab. Imaging was peripheral to his job of developing new radio transmission systems, and the often-modest Hansell never worked up enough enthusiasm to push it. But the time his patent was issued in 1930, he was on to new inventions.

Fiber optics wasn't the only bright idea that Hansell never found the time to pursue. When Edwin Land showed him the young Polaroid Corporation's plastic polarizing material, Hansell suggested it might make good sunglasses. His family recalls that Land rewarded him with the first pair off the production line.[27] And Hansell was the first person to write on paper by controlling the flow of an ink jet. His prototype printer could record 7500 words per minute from a radio telegraph—an astounding speed in the 1930s. It was far too fast for RCA's radio-transmission manager, who took one look and decided the miles of paper tape would overwhelm his operation.[28] Hansell reluctantly put the idea on the shelf. Today, millions of low-cost inkjet printers are used with personal computers.

Heinrich Lamm and the Legacy of Sword Swallowers

As Hansell and RCA moved on to other things, a medical student in Germany came upon the same idea from a different direction. Heinrich Lamm wanted to build a flexible gastroscope that a physician could thread down a patient's throat to peer into the stomach.

Physicians had tried to build instruments to look into the human body since the early nineteenth century, with no real success. The feats of sword swallowers inspired one physician to try looking down the throat through a rigid tube, but it wasn't a viable approach.[29] The esophagus is reasonably straight, but the entrance and exit are not, so thrusting an inflexible tube through delicate body passages could be catastrophic. One physician later called the rigid gastroscope "one of the most lethal instruments in the surgeon's armamentarium."[30]

Some progress came after World War I. Rudolf Schindler, a specialist in gastrointestinal disease, spent a decade developing a semiflexible gastroscope that could be bent up to 30 degrees.[31] It was an improvement, but not much from the patient's viewpoint.[32] Lamm took a course from Schindler at the University of Munich and decided the instrument wasn't flexible enough after watching Schindler use it.

Lamm decided a bundle of glass fibers would be much more flexible. That would be a boon both to physicians and to patients, few of whom took well to having a slightly bendable pipe thrust down their throats. Lamm probably had seen bent glass rods guide light; it was a common physics demonstration.[33] He knew about total internal reflection in rods and knew it also occurred in "a very thin rod, e.g., a glass or quartz thread."

Like Hansell, Lamm realized a bundle of fibers could carry an image point by point if the fibers were arranged so they sat at corresponding points at each end. He wrote:

> If we then project upon one terminal surface of this bundle a real image, this image will be transmitted point by point (or more exactly, minute area by minute area) to the other terminal surface of the bundle. There one can view it directly or through an ocular lens. It will be a halftone image. If the threads get all jumbled in the middle part of the bundle (as they invariably do if the bundle is bent), the transmission of the image remains unchanged.[34]

The medical student then went a step further than the engineer had. Lamm liked to tinker with instruments, so he talked Schindler out of enough money to test his idea. It was a brash and ambitious project for a third-year medical student, but Lamm was bright, enthusiastic, and independent minded. The busy Schindler didn't work much on the project, but Lamm got help and laboratory space from Walther Gerlach, a prominent physicist later involved in the abortive German atomic bomb program. He bought glass fibers from the G. Rodenstock Optical Works (curiously, the name means "clear stick" in German).

Lamm painstakingly combed the glass threads into a fat, awkward bundle a few inches long, aligned them and clamped them together at each end. He bent the bundle to show the light could follow a curved path. He probably tried looking through the bundle, but it wasn't very transparent, so little light emerged. To show the idea worked, he focused the bright light from the V-shaped filament of an unfrosted bulb onto one end of the bundle. It's an easy optical trick, like focusing an image of the label on top of a light bulb onto the ceiling with a lens held above the bulb. Looking through the other end of the bundle, Lamm saw a faint V, an image of the filament. He recorded the image on photographic film for posterity and his professor, with dark regions showing the areas exposed by the light (see figure 4-1).

The 22-year-old medical student had high hopes and was somewhat disappointed by the results. "I have not succeeded in my attempts to transmit adequately bright and sharp images," wrote Lamm. "Nonetheless the pictures . . . show that one can transmit pictures through a flexed bundle, even with variable degree of bending."[35] Although the glass was not very clear, the fibers were not perfectly aligned, and the ends probably were not polished smooth, the image is recognizable. Heinrich Lamm had transmitted the first fiber-optic image. The year was 1930.

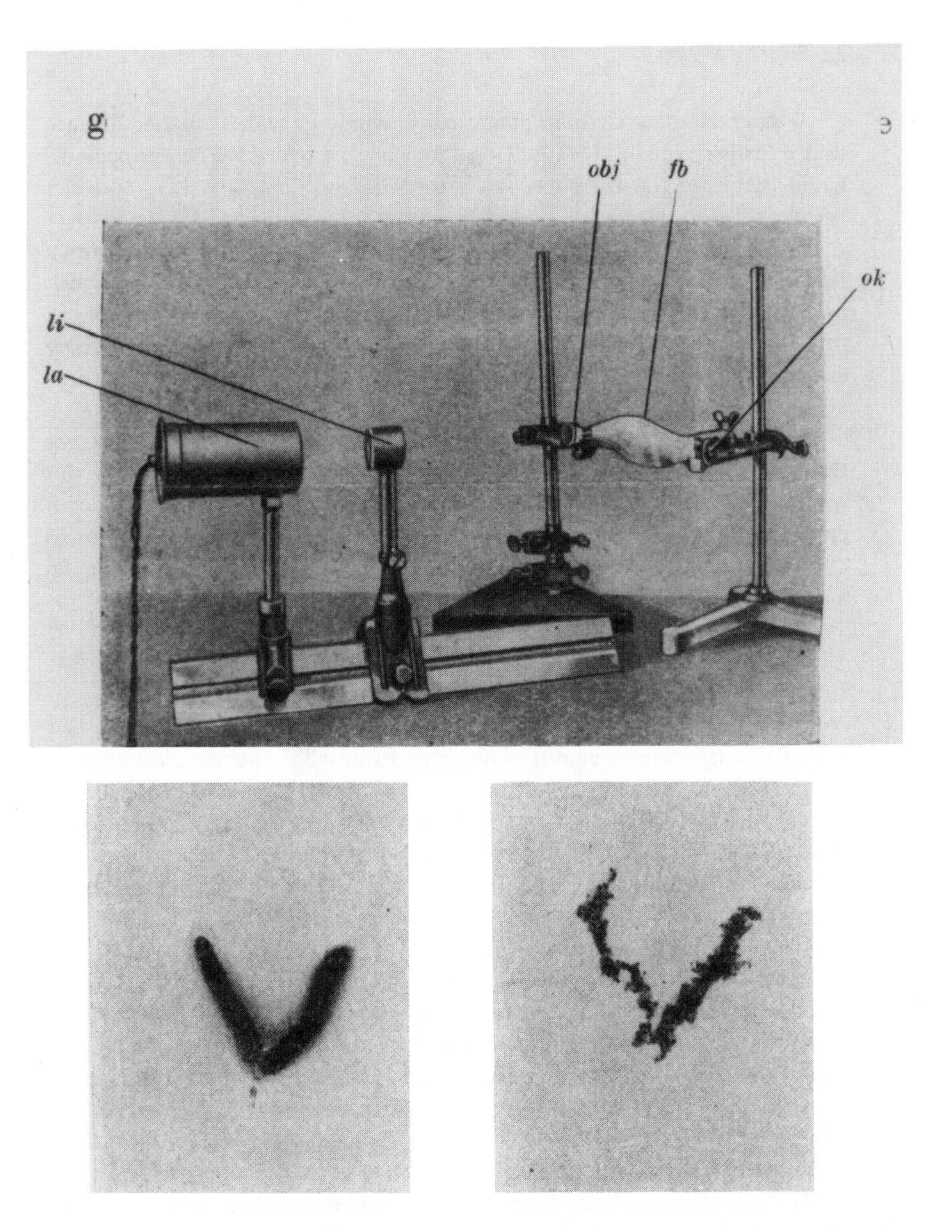

Figure 4-1: Heinrich Lamm assembled an awkward bundle of glass fibers (top) and showed it could carry the V-shaped image of a light-bulb filament around a bend. The image at left is the bare filament; at right is the filament viewed through the fiber bundle. If you look closely you can see the spots where light emerged from individual fibers in the bundle. The filament is dark because the image is a photographic negative. (Copyright Michael Lamm; reproduction courtesy of Corning Glass Center)

He hurried to the German patent office, where he was astounded to learn that a British version of Hansell's patent had just issued.[36] The disappointed Lamm wrote to the British licensee, the Marconi Company, who told him neither they nor Hansell "have tried to utilize this principle."[37]

Most students in his position would have given up, but not Heinrich Lamm. He wrote a paper, which the journal *Zeitschrift fur Instrumentenkunde* published in its October 1930 issue. It appears under Lamm's name alone, unusual at a time when European professors typically insisted they share credit for student work. Indeed, Schindler thought Lamm's experiment had failed.[38]

Lamm insisted "the experimental proof of the possibility to transmit images through a flexible multi-fiber conductor of radiant energy . . . justifies this communication." However, he knew he could go no further by himself, concluding: "I also hope that some optical firm possessed of more means, sources of supply, and experience than I have, could be induced by this report to build a serviceable flexible gastroscope."[39]

Two Decades of Stillness

Lamm's hopes were in vain. His paper and Hansell's patent sank without a trace. Crucial pieces of the technological puzzle were still missing. Only a feeble, fractured image had emerged from Lamm's bundle. His glass fibers, like Jenkins's quartz rods, absorbed too much light, and although no one realized it at the time, they didn't confine light well either. Other technologies seemed far more practical for looking inside the body, for television, and even for displaying meter readings in more convenient locations.

The rise of Adolf Hitler brought trouble for Lamm, a Jew.[40] Survival became more important than following up an old experiment. He and his physician wife fled Germany after their residencies at the Jewish Hospital in Breslau (now Wroclaw, Poland). They reached America in 1937, where an uncle got Heinrich a job at a psychiatric hospital in Kansas City.[41] The next year, Lamm traveled through Texas, New Mexico, and Oklahoma with a salesman friend, seeking an exotic place to settle down as a surgeon. He found it in a town of 1500 in the southern tip of Texas, where the old town physician was retiring. The town lacked a Jewish community, but the native prejudices didn't include antisemitism.[42] It was a welcome relief from Germany.

Lamm tinkered a bit, and his restless mind explored many areas, but mostly he practiced medicine. He was an early and enthusiastic crusader for automobile seat belts.[43] He put on weight and his wavy hair thinned, but he retained his intelligence and passions, and kept a reprint of his paper from *Zeitschrift fur Instrumentenkunde.*

The Radio Corporation of America had the resources Lamm lacked, but the company filed Hansell's patent and forgot it. No company can do everything. Broadcast radio was a tremendous success; television was promising but costly to develop. Fiber optics was but a footnote to the work at Rocky

Point, and the importance of Rocky Point faded as other RCA ventures grew much faster.

The use of light-guiding rods continued to spread slowly. In the 1930s, DuPont invented the transparent plastic called Lucite in America and Perspex in England, and the durable, lightweight material replaced quartz. Dentists used lamps with curved Lucite rods to illuminate mouths; doctors used Lucite tongue depressors. Yet the idea of imaging through bundles of thin fibers was stillborn. While Hansell conceived of the idea at RCA, the company did nothing with it. In that sense, Jenkins was right in saying no "great pioneer invention" could come from a giant corporation. RCA—and perhaps Hansell himself—never saw the potential of fiber optics, and wandered away, leaving its development to others.

A Critical Insight

The Birth of the Clad Optical Fiber (1950–1955)

Genius is one per cent inspiration and ninety-nine per cent perspiration.

—Thomas Alva Edison

Heinrich Lamm and C. W. Hansell lacked not merely resources but one crucial bit of inspiration. They didn't realize that light was leaking between bare glass fibers where they touched each other. The solution was disarmingly simple: cladding the light-carrying fiber with a transparent material that had a lower refractive index. However, like most other critical insights, it's obvious only in hindsight. Two decades after Hansell and Lamm, the idea sprouted separately in the fertile minds of two very different men. One was an eminent professor, a maker and shaker in mid-century American science. The other was a Danish inventor laboring in a small home workshop.

A Fateful Cocktail Discussion

Optics was a quiet backwater of physics in October 1951,[1] when two of the biggest fish in that small pond shared cocktails in the large living room of a stucco-tile house at the end of Harwood Lane in East Rochester, New York. The host was Brian O'Brien, president of the Optical Society of America and director of America's leading school of optics at the University of Rochester. His guest was Abraham van Heel, president of the International Commission for Optics and professor of physics at the Technical University of Delft in the Netherlands.

Born at the close of the nineteenth century, both men were in their early fifties, near the peaks of eminent careers. Although not close friends, they shared the camaraderie of long-time workers in the same small field. When van Heel toured American optics labs, it was natural for O'Brien to invite him for dinner at his large two-story house, build just before World War I and tastefully fitted with mahogany furniture and heirloom oriental rugs.[2] They expected to relax and chat about optics in postwar Europe and America.

The Cold War military research that deeply involved both men also was on the agenda. The West was in a technological arms race with the Soviet Union, still ruled by Josef Stalin. Holland was trying to rebuild a submarine fleet that had been among the world's largest before World War II, and they wanted better periscopes. The German companies that had supplied Holland before the war lay in ruins; America and Britain wanted to keep their latest technology for their own subs. Knowing how curved glass or plastic rods could carry light, van Heel thought a bundle of thin fibers might relay the image in a periscope.[3]

Van Heel's assistant Willem Brouwer had suggested that fiber bundles could scramble images as well as transmit them—an idea that excited the Dutch security agency. As for image transmission, the fibers first would be aligned to form the same pattern at both ends. The trick was to scramble the fibers in the middle, fix them in place with glue, then cut the bundle in half at the scrambled point. Each half would convert an image into an unintelligible pattern, but mating the severed halves properly would reassemble the image. You could encode a message by taking a photograph through one half of a scrambler, because only the person with the matching half could decode it.

The ideas sounded good, but van Heel was making little progress. The Dutch government asked its American allies to recommend a well-qualified and highly cleared US scientist who could help.[4] American officials picked O'Brien, whose work on military optics during the war had earned him the Medal for Merit, the nation's highest civilian award, from President Harry Truman.[5] It was a singularly prescient choice.

Van Heel explained that bare glass fibers didn't work well. If two fibers touched, light leaked between them because total internal reflection didn't work. The fiber surface scratched easily, causing even more losses. Plastic fibers suffered the same problems. Seeking an alternative to total internal reflection, van Heel coated fibers with a reflective silver film. Yet almost no light emerged from the bundle of silver-coated fibers he showed to O'Brien.[6]

O'Brien was surprised van Heel was working on light guides, but was not surprised his approach didn't work. The American was studying single fibers himself and knew metal coatings would not work. A shiny metal surface always absorbs a little incident light, and losses build up with each reflection. Bounce light 100 times off a 99 percent reflective surface, and only 36.6 percent of the light remains; after 1000 reflections, only 0.0043 percent is left. You might think so many reflections unlikely, but the thinner the fiber, the more times light bounces off its walls—and you need thin fibers to carry

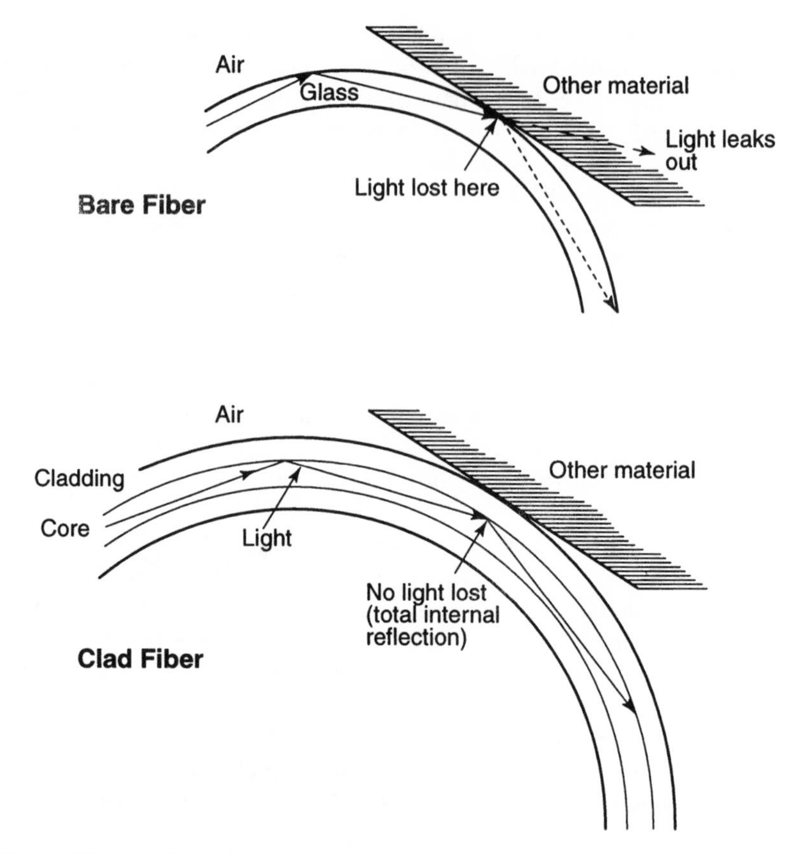

Figure 5-1: Total internal reflection at the boundary between core and cladding guides light along a clad optical fiber (bottom); light can leak from unclad fibers where they touch other objects (top).

recognizable images. In fact, over 1000 reflections are likely in a single meter (39 inches) of fiber.[7]

"That's not the way to do it, Bram," O'Brien said, surprised van Heel had not recognized the limits of metal coatings. The trick was to use total internal reflection in a way subtly different than everyone else had since Daniel Colladon demonstrated light guiding a century earlier. Total internal reflection works as long as the surrounding material has a lower refractive index than the light guide. Those who followed Colladon thought automatically of the outside material as air, but other materials also have refractive indexes lower than glass or plastic. Coating a bare fiber with such a low-index material (see figure 5-1) would maintain total internal reflection while protecting the optical surface. The secret to making fibers transmit light well, O'Brien said, was to clad them.[8]

Van Heel immediately realized O'Brien had a good idea, although neither knew quite how good it was or how hard it would be to implement. Total internal reflection requires a virtually perfect surface—clean, smooth, and touching nothing. The surfaces of bare glass or plastic fibers were far from perfect after they were assembled into a bundle. The fibers touched, and everywhere they touched light could pass from fiber to fiber. The fibers also rubbed each other, forming surface scratches that scattered light out of the fibers. Dirt and fingerprints on the surface also let light leak out.

The cladding offered a simple and elegant way around those problems because it protected the surface where total internal reflection occurred. At a stroke, it kept light from leaking between the light-guiding cores and protected the vital reflecting surface from scratches and fingerprints. The trade-offs were added complexity and a slight reduction in how much light the fiber could collect.[9] Van Heel probably was annoyed that he hadn't thought of the idea, but that didn't stop him from embracing the crucial conceptual breakthrough that launched modern fiber optics.

An Insight from Vision

The idea of cladding did not come instantly to Brian O'Brien. It grew from his study of an esoteric visual effect discovered in the 1930s. The visual center of the eye responds more strongly to light if it comes through the center of the pupil than if it enters at an angle near its edge.[10] Specialists were puzzled until O'Brien found the reason lay in the structure of the light-sensing cells in the retina, a layer of cells at the back of the eyeball.

The retina contains two types of light-sensing cells, cones concentrated near the center that give us color vision, and rods spread over a larger area that provide night vision. Both are long and thin, with one end exposed to light and the other attached to nerve fibers. Light must travel through the whole cell to reach the sensing elements at the back. As far back as 1844, a German physiologist had suggested that total internal reflection might guide light along rods and cones, perhaps inspired by Colladon's demonstration.[11] A century later, O'Brien recognized that was exactly what was happening. Light arriving from the center of the pupil was guided to the back of the cones more efficiently than light coming from the sides of the pupil. This made light coming from the center of the pupil look brighter.

To test his theory, O'Brien made a model cone cell from plastic foam. His model was 20,000 times larger than a cell, so he tested it with microwaves 20,000 times longer than visible light. That's how physicists think. Both visible light and microwaves are electromagnetic waves; they should behave similarly if you multiply all the dimensions by the same factor. The log-sized blocks of lightweight plastic drew some surprised stares at Rochester, but the experiments validated O'Brien's model.[12] It was an elegant example of scientific deduction, and earned O'Brien the Optical Society's highest award.[13]

O'Brien had not stopped with the microwave experiments. He had seen transparent plastic rods guide light for illumination, and wondered what would happen in thin transparent fibers. He tested clear glass fibers from the

Corning Glass Works. To see what would happen in a two-layer structure, like the retina, he had coated them with a transparent layer of instrument lacquer, which has a lower refractive index than glass. That was about all he had done when van Heel arrived, but it was enough to get the idea.

Before they sat down to dinner, the two men agreed that each would contact the other before publishing anything.[14] Each thought he had made an important contribution. Van Heel had brought the ideas of image transmission and scrambling in fiber bundles. O'Brien contributed the idea of cladding the fibers.

Their discussion might have become the basis of a fruitful collaboration, but transatlantic communications were limited at mid-century. In 1951, only radio-telephones could carry voices across the Atlantic, and people had to book channels in advance. Most people had to rely on the mails and an occasional telegram. Busy men like O'Brien and van Heel could easily fall out of touch, and they soon did.

An Independent Inventor

Brian O'Brien was not the only person thinking of applying a clear cladding to guide light along transparent fibers in 1951. On April 11, Holger Møller Hansen applied for a Danish patent on a "flexible picture transport cable." Like Heinrich Lamm, Møller Hansen thought that a bundle of parallel glass or plastic fibers would make a much more flexible gastroscope than rigid tubes and lenses. Unlike Lamm, he was trained in engineering. That helped Møller Hansen realize that bare fibers would not work well. He thought that circular fibers would minimize points of contact where the light could leak out. His application also showed he realized the importance of a cladding: "Eventually the threads will be coated with a substance whose index of refraction approaches one."[15] In essence, Møller Hansen had come close to the same idea that O'Brien suggested to van Heel six months later.

The Danish inventor had a background very different from the two eminent professors of physics. The son of a machinist, Møller Hansen was born in 1915 in Assens on the Danish island of Funen. He spent only five years in school before starting work as a farmhand at 12. He apprenticed as an electrician, then as a gardener, before joining the Danish army. There the young man showed his bent for invention, improving a cryptographic machine used to code secret messages. Code machines were important for a small country bordering Nazi Germany, and the invention earned newspaper headlines for the 23-year-old soldier. A rich mill owner saw the articles and paid Møller Hansen's way to engineering school in Copenhagen. It was a struggle for a man with little formal education, but Møller Hansen takes pride in being one of four from a class of 23 who finished the program.[16]

The Danish subsidiary of Philips N. V. hired Møller Hansen to develop telephone systems, and he continued tinkering in a small home workshop. In late 1949,[17] the segmented eye of the fly[18] inspired him to consider using a flexible array of transparent fibers to look into inaccessible places. He struggled to draw his own glass fibers but found it too hard to keep their size constant. He tested other glass and plastic fibers and found the best were fibers made by a company called Extrusion to insulate undersea cables. However, he had trouble bundling them together for image transmission.

He also had trouble finding the right coating material. Like van Heel, Møller Hansen tried metal coatings, but the coated fibers didn't transmit light. He then tried coating fibers with materials having a lower refractive index. There aren't many solids, so he tried oils and got much better transmission. Canada balsam oil worked, but he got the best results with margarine, which he knew was impractical.[19]

By 1951, he had made enough progress to seek a patent. His application[20] is short, but it shows he had thought carefully about making an instrument to look inside the body. He realized illumination was needed, and suggested delivering light through some fibers in the bundle. He also realized the need to seal the ends against contamination.

Enthusiastic about his new idea, Møller Hansen talked to reporters. The press loved him; bright, colorful, and outspoken, he still makes a good interview. In May, the story hit some European newspapers; the following month, the *Los Angeles Times* carried a Reuters report.[21] He had experimented mostly with glass, but he envisioned making future fibers of clear plastics. A magazine photographed him looking into a bundle inserted into his ear,[22] but Møller Hansen says the photo was faked because that bundle could not transmit an image.[23] He accumulated a fat file of Danish and German clippings.

That was the high point of his fiber-optic career. The Danish patent office rejected Møller Hansen's application after discovering the Hansell patent, which anticipated most—but not all—of his claims. Møller Hansen should have had a clear priority on the crucial idea of applying a low-index cladding to the fiber. His wording was vague by modern standards, but at the time O'Brien had made no move to publish or patent his idea. Unfortunately, neither Møller Hansen nor the Danish patent office realized the importance of the cladding, and his patent claims died.

With limited resources, Møller Hansen could not go much further. He lacked the equipment to make good fibers and to investigate better cladding materials. He couldn't buy good fibers. He couldn't convince British or Danish companies that anyone wanted fiber-optic bundles to look inside the body. Frustrated, in 1952 he turned to another idea he could better pursue in his cluttered workshop—plastic-bubble "shock absorbers" for mailing envelopes. He did patent that idea, but it didn't make him rich.[24]

An Urbane European Professor of Optics

Abraham Cornelis Sebastiaan van Heel was a cultured member of the European professoriat, in marked contrast to the ingenious Danish farm boy. Born in Java in 1899, when it was a Dutch colony, van Heel earned a doctorate in 1925. He was a lively, cheerful man whose interests reached well beyond physics. He fell in love with the French language and literature while working in Paris,[25] knew Latin and Greek, and was fluent in English. He also became fascinated with optics.

The study of optics was hardly fashionable in the 1920s, but van Heel was keenly aware of its deep roots in Holland. A Dutch spectacle-maker built the first telescope in the early seventeenth century, and a Dutch scientist formulated the laws of refraction.[26] Van Heel collected early books on optics; on one American trip, he made a point of visiting a Library of Congress exhibit that showed four early volumes on optics missing from his personal library.[27]

Bram van Heel was not a trailblazing superstar of science; you must dig deep to find his name today, although former students remember him fondly. His genius lay in the demanding fields of precision optical measurement and lens design.[28] He helped build the Dutch optical industry in the 1930s. After the war, he salvaged old mechanical telephone relays to build one of the world's first computers, which he used for the complex calculations needed to design lenses.[29]

Dutch officials sought his help to rebuild the country's war-shattered optical industry and to develop new military optical systems. Willem Brouwer, a Resistance fighter who had returned to school after the war, joined van Heel on the periscope project. In late 1949, they intially considered an array of thin reflective tubes, but soon decided transparent fibers with mirror coatings would be more efficient.[30] In January 1950, Brouwer fired arrows attached to drops of molten glass down a long hall, drawing their first glass fibers the same way Boys had done 60 years earlier. They also made plastic fibers. Initially, none of them transmitted much light.

Van Heel returned from America enthusiastic about transparent claddings.[31] He and Brouwer first pulled plastic fibers through liquid beeswax, a material common in mid-century laboratories. It improved fiber transmission more than they expected.[32] Next they pulled bare fibers through liquid plastic, stringing them around the lab so the plastic could cure and painting them black to keep light from leaking between them.

They made imaging bundles by winding a single long fiber around and around a reel, and cementing the fibers together at one spot. Cutting the cemented area gave them a flexible bundle with fibers in matching positions at both ends. An image focused on one end appeared on the other end. By April 1952, they had sent images through bundles of about 400 fibers as long as 20 inches (half a meter).[33]

Progress was interrupted in early 1953 when Brouwer left for America. Young, impulsive, and without strong ties to Holland, Brouwer had applied to immigrate during a moment of frustration in 1950. When approval finally came in early 1953, it was good for only three months. Van Heel asked him to stay, but with Europe still recovering from the war, Brouwer saw more opportunity in America. On the other side of the Atlantic, he abandoned fiber optics to work on military reconnaissance cameras, and the aerial sextant for the U2 spy plane.

With Brouwer gone, van Heel turned fiber work over to two other students.[34] Meanwhile, a Dutch colleague who later that year would win the Nobel Prize in physics, Fritz Zernicke,[35] returned from England with disturbing news. Harold H. Hopkins, an optics specialist at Imperial College in London, claimed he had invented a way to send images through bundles of glass fibers.

There's nothing like competition to get scientists moving. Van Heel decided to rush a paper into print to establish priority over Hopkins.[36] He must have gotten approval from the Dutch National Defense Research Council. He also air mailed a letter to O'Brien.[37] With transatlantic phone calls out of the question, van Heel could only wait for a reply by air mail or telegram. He didn't know O'Brien had been distracted.

Major Distractions in America

If European culture marked Bram van Heel, American drive and restless energy were the hallmarks of Brian O'Brien. Tall, thin, and bespectacled, O'Brien was a master grantsman before the word was coined, who moved in the highest circles of American government science. Born January 2, 1898, in Denver, he grew up in Milwaukee, the son of a prominent mining engineer. He attended Yale, receiving an undergraduate degree in electrical engineering in 1918 and a doctorate in physics in 1922. He arrived at the University of Rochester in 1930, initially specializing in vision. In 1938 the university named him the first permanent director of what was then America's only institute of optics.

O'Brien made pioneering measurements of atmospheric ozone from high-altitude manned balloons, sending one to a then-record height of 72,000 feet (21.8 kilometers) in 1935.[38] He applied his engineering talent to an ingenious system that enriched the vitamin D content of milk by flowing the liquid past an ultraviolet lamp. World War II brought new challenges. O'Brien designed an instrument that produced visible images from invisible infrared light, so pilots and soldiers could see at night. He built another instrument that allowed soldiers to spot bombers trying to hide their attack by flying out of the sun. After the war, he built a camera that could record an image in one ten-millionth of a second and used it to photograph the Able atom bomb test at Bikini Atoll.[39]

Intense, energetic, and adventuresome, O'Brien kept many irons in the fire. His students recall his technological eloquence. He "had this wonderful way with his hands. He would show you an instrument and caress it" as he described its workings, says one.[40] His energy was remarkable; another recalls, "He never walked but he ran."[41] *The Saturday Evening Post*, then one of America's biggest magazines, profiled him as a colorful wizard of optics who delighted in flying military planes while testing his equipment.[42]

Optical fibers were only one of many projects O'Brien juggled at the university and in his home laboratory. Inspired by plastic illuminating rods[43] and by his discovery of light guiding in the eye,[44] he concentrated on transmission in single fibers. He drew glass fibers about the same diameter as cells in the eye and had a student dip them in liquids to study light guiding. He coated fibers with instrument lacquers, which dried to a clear plastic with lower refractive index than glass.[45] After talking with van Heel, O'Brien told his military and security contacts about the possibilities of fiber-optic imaging and encoding.

Impatience came with O'Brien's driving energy. He liked to explore new ideas and develop new instruments, but he rarely got around to writing papers about them and couldn't be bothered with details. A thick layer of papers covered his desk.[46] The head of Rochester's physics department praised O'Brien's intelligence, but told the *Post*: "He won't fill out expense accounts. His reports get in late or not at all. He never gets around to answering his mail."[47]

Growing restless at Rochester, O'Brien was open to an offer from American Optical Company, then one of the country's three largest optics companies. He agreed to join the century-old maker of spectacles and microscopes as a vice president and head of a new research division at its headquarters in Southbridge, Massachusetts, in early 1953, after Rochester completed a fund-raising drive. Well versed in patent law, O'Brien retained rights to inventions in progress, including optical fibers. However, before he got to Southbridge, O'Brien met an unexpected distraction—the flamboyant promoter Mike Todd, best remembered today as Elizabeth Taylor's third husband.[48] The impresario needed an expert on the "optical dodge."[49]

Born Avrom Hirsch Goldbogen in 1907, Todd had "the soul of a carnival pitchman and the ambition of a Napoleon" in the words of a contemporary writer.[50] His formal education stopped when he was caught running a schoolyard crap game and expelled from the sixth grade, but he was bright and energetic. A born salesman, he made and lost fortunes, produced Broadway musicals, and had ties to Hollywood.

In 1950, Todd decided the future of motion pictures lay in wide-screen pictures that put the audience into the action. The idea dated back to the 1920s, but in the 1950s the movies needed a new gimmick to compete with television. The industry also needed to solve an old optical problem—the distortion inevitable when projecting a small image onto a big screen. Todd found a little company called Cinerama that used three cameras and three projectors to fill a huge screen that wrapped partway around the audience to cover the whole field of vision. Cinerama was broke, but Todd became a partner and almost single-handedly revived it. He raised money, sent crews around the world to film a documentary, and booked theaters around the country. Audiences flocked to *This Is Cinerama* after its September 1952 opening.[51]

However, by then Todd had been eased out after clashing with the other owners. He also saw Cinerama's limitations. Its three projectors left two seams in the picture, and occupied valuable theater space. A complete installation cost $75,000,[52] big money then, so only 17 theaters around the world showed the film. Todd wanted a better, cheaper system, "Cinerama out of one hole." The night after *This Is Cinerama* opened, he asked his son Michael Todd, Jr., to find the "Einstein of optics." The trail led to Brian O'Brien.[53]

O'Brien didn't know what to make of the promoter, who called insistently, typically in late evening. He finally agreed to meet Todd, who chartered a plane to Rochester and explained what he wanted. When O'Brien said such a system was probably possible, Todd tried to hire him. The physicist demurred; he thought the job belonged at a big optical company. After three weeks of calls failed to persuade O'Brien, Todd chartered a plane to meet with O'Brien and Walter Stewart, the president of American Optical, in Southbridge. After shaking hands, Todd laid a certified check for $60,000 on Stewart's desk and said, "Let's talk business."[54]

The result was a joint venture called Todd-AO. American Optical supplied optical designers, engineers, and manufacturing expertise. Todd applied his considerable persuasive powers to secure film rights to the Broadway musical *Oklahoma!*, the first time Rogers and Hammerstein allowed anyone to film one of their shows. It seemed a golden opportunity to make money and rub shoulders with the rich and famous. Todd-AO preempted O'Brien's working hours, leaving little time to establish a research division. The company put its best engineers on the project, where they worked at a frantic pace.

The little project on guiding light in glass fibers couldn't compete with the movie business. Except for asking company lawyers to work on a patent, O'Brien almost forgot about fiber optics in the months after he arrived in Southbridge. Nothing seemed urgent; he expected van Heel to warn him in plenty of time of any developments that might affect the patents.

Few people realized something else was draining O'Brien's tremendous personal energy. His wife of 30 years, Ethel, was seriously ill and dying.[55]

A Controversy over Publication

In the midst of this chaos, van Heel's letter arrived in early May 1953, announcing his plans to publish on image transmission through clad fibers.[56] The fate of that crucial document is unclear. Although O'Brien was notorious for not replying to letters, van Heel's should have been an exception. Perhaps the letter never reached O'Brien's attention. Perhaps O'Brien saw it, asked military clearance to respond, and never got a reply. Perhaps he set it aside for a moment, only to be distracted by some other emergency. It may have slipped through the cracks at a time made difficult by joining American Optical, launching Todd-AO, and dealing with his wife's illness. Or perhaps O'Brien did reply but his letter was lost in the international mail. In any case, something went wrong.

After a brief wait for a response that never came, van Heel sent a short letter to the prestigious British weekly *Nature*, which received it May 21, 1953. He also sent a longer article to *De Ingenieur*, a Dutch-language weekly little known outside Holland. He heard nothing from *Nature* for months, but *De Ingenieur* rushed his article into the June 12, 1953 issue, documenting van Heel's priority.

Van Heel didn't call his invention fiber optics. His article was titled "Optical representation of images without use of lenses or mirrors," and his English summary called the fibers "wires coated with a substance of lower refractive index."[57] He described cladding fibers and sending images through bundles up to 8 inches (20 centimeters) long. He wrote that plastic fibers[58] 0.1 to 0.13 millimeter thick worked best, and cited potential uses in medicine and image encoding.

With the resources of a university and some military funding, the little group at Delft[59] went further than Møller Hansen, Lamm, or Hansell. Indeed, van Heel had not heard of the others. Lamm's paper and Hansell's patent were hidden in the oblivion of dusty archives, and Møller Hansen never published a scientific paper. Van Heel's report in *De Ingenieur* could have suffered the same oblivion because almost no one saw it outside Holland.

With its global circulation and prestigious reputation, *Nature* was a bid for much more attention. However, the journal dawdled an exceptionally long seven months before publishing van Heel's letter on January 2, 1954.[60] Barely a column long, it covered the same ground in much less detail, but it was visible around the world. So was a longer paper by Hopkins. Brian O'Brien saw both and was furious that van Heel failed to mention O'Brien had suggested the cladding.

Why didn't van Heel credit O'Brien? Remembered by all as a man of integrity, van Heel held to high standards and was not one to steal ideas. His Dutch paper carefully lists the three men who worked for him and credits Brouwer with suggesting image scramblers. His omission of O'Brien is odd. Did he expect the American to write his own paper? Did he think O'Brien's contribution was not crucial? Did he omit O'Brien's name to avoid upsetting American security officials? Had he taken O'Brien's failure to respond as a sign he didn't want credit? Or did he omit O'Brien's name at the American's request, as Brouwer believes?[61]

The most likely answer is that van Heel asked O'Brien to tell him if he wanted credit. Eager to beat Hopkins, and probably aware of O'Brien's poor correspondence habits, van Heel may have written that he would publish unless he heard otherwise from O'Brien. When no reply came, van Heel may have decided O'Brien didn't care, or didn't want his name mentioned because he lacked clearance to discuss the idea in public. Perhaps O'Brien's carelessness about correspondence annoyed the more meticulous European. Van Heel may have decided that if O'Brien didn't get around to writing, the American would have only himself to blame.

In retrospect, O'Brien deserves credit for the cladding, which helped launch modern fiber optics. Yet that was only a first step. As Thomas Edison testified, genius is only one percent inspiration. Many others would supply the perspiration.

A Surprise from Britain

Another unsolved puzzle is *Nature*'s juxtaposition of van Heel's paper with one that Hopkins submitted six months later.

Harold Horace Hopkins was a young rising star in the little world of European optics. Born December 6, 1918, in Leicester, England, the son of a bakery worker, his quick intelligence earned him a university scholarship. When World War II started shortly after his graduation in 1939, the government sent Hopkins to work for an optics company, diverting his interests from theoretical physics to optical design. During the war he earned a doctorate from the University of London, and soon afterward he was named to the faculty at the Imperial College of Science and Technology in London.

Brilliant, well read, and strong-minded, Hopkins thrived in the rich intellectual atmosphere of the postwar British capital. When he was a student in the 1930s, "any physicist worth his salt was a member of the communist party," he recalled.[62] He left the party after the war, but his politics remained leftist, and he avoided the entanglements of Cold War military work. He earned accolades from the optical world by designing the first zoom lens that worked as well as a standard fixed-focus lens. That was no mean feat because the design required elaborate calculations at a time before electronic computers.

Hopkins also developed a taste for fine wines and London social life.[63] At a 1951 dinner party, he met a physician who had just performed what Hopkins called "a particularly distressing endoscopy using the old rigid type of gastroscope." When the physician learned that Hopkins specialized in optics, he asked if more flexible instruments could be made to look into the stomach.[64]

Intrigued, Hopkins first thought of inserting a single glass fiber down the throat, and moving it to look around the stomach. He soon realized that was as impractical as mechanical television cameras that looked through holes in moving disks; the physician could see only if there was a bright lamp in the patient's stomach. Then he decided to try a flexible bundle of fibers. He had every reason to think the idea was new. Baird's and Hansell's patents and Lamm's paper were buried in dusty archives, and van Heel's work initially was classified. The major London newspapers ignored Møller Hansen,[65] although Hopkins might have heard of his work through other channels.

To test his idea, Hopkins got glass fibers made for other purposes; they were only 20 micrometers (0.8 thousandth of an inch) thick, so fine he could hardly see them. His first experiments were the simple ones a curious child will do with an optical fiber today. He made loops in the fiber and pointed one end at a lamp while looking at the other end through a microscope. The fiber tip was bright until he put his hand between the other end and the lamp. The light was still visible after passing through 4 feet (1.2 meters) of fiber. Delighted, Hopkins adjourned with friends to the local pub and celebrated with several pints of beer.[66]

Hopkins knew he needed more time and money to make image-transmitting bundles, so in July 1952 he asked the Royal Society for a grant of £1500 a little over half earmarked to pay a graduate assistant. He got the grant in three months and offered the assistantship to Narinder S. Kapany, an ambitious Indian national who had asked earlier about graduate study. The selection was as much political as academic; Hopkins felt "very much motivated, as indeed were many liberal-minded people at that time, by sympathy for someone who had come from part of Britain's imperial past."[67] It would prove a fateful choice.

Born in 1927 in Moga, Punjab, Kapany still wears the untrimmed beard and turban traditional for followers of the Sikh religion. Raised in a well-educated upper-class family, he grew interested in optics after his father gave him a box camera in high school. He had attended an Indian university and worked in a military optics factory, then studied optics for a year at Imperial College, and worked for a Glasgow optics company. He hoped eventually to return to India to start his own optical business.[68] He jumped at the chance to work with Hopkins.

The two started with 25-micrometer (0.001-inch) glass fibers that had been made for glass cloth. The fibers carried light only 75 centimeters (30 inches), but Hopkins thought that would suffice for a demonstration. Neither he nor Kapany thought of cladding the fibers.

The biggest challenge, to Hopkins' mind, was assembling 10,000 to 20,000 fibers into a bundle with their ends aligned to carry an image.[69] It's the sort of task that sounds mundane until you try it. The idea was simple. First he and Kapany wound a fiber thousands of times around a spool, then clamped the looped fiber at several points and unscrewed one side of the spool to remove the loop. Next they cut the loop at the clamps, yielding bundles four inches (ten centimeters) long, with fibers clamped at the ends but loose in the middle. They checked the bundles by looking through them at a razor blade; if they saw a straight line, the fibers were in the right places. They spent months assembling equipment to wind fibers and getting it to function properly.

In November 1953, they sent a letter to *Nature*,[70] which published it on January 2, 1954, just below van Heel's long-delayed letter. It's not uncommon for journal editors to group papers on the same topic—the mystery is the timing. The journal received van Heel's letter May 21, and normally it would have appeared months before the one from Hopkins and Kapany was received on November 22. Suspicious minds suspect Hopkins was asked to referee van Heel's paper and delayed it until he could submit his own. In fact, Brouwer says that when van Heel asked about the delay, the journal referred him to Hopkins as its letters editor.[71] (*Nature* has no records from that period.)[72] Yet before his death, Hopkins denied responsibility, saying he did not see van Heel's paper until shortly before it appeared, when *Nature* noted the similarities and sent each author copies of the other's paper.[73]

The two *Nature* papers do not reference each other and take rather different approaches. Van Heel concentrated on transparent coatings, mentioning image transmission only in passing. His bundles contained only several hundred fibers, enough to demonstrate the idea without elaborate winding machines. In contrast, Hopkins and Kapany assembled bundles containing many more fibers, but their fibers were unclad.

Neither paper was quite sufficient by itself. Yet taken together, the two papers in one of the world's most widely read scientific journals launched modern fiber optics.

Aftermath

In America, the angry O'Brien realized that van Heel had set the clock running on his patent claims; he had to file within a year of the first publication of the idea. He pressed American Optical lawyers to file. Shortly after the *Nature* paper appeared, van Heel belatedly sent reprints of the Dutch article, which the Americans had never seen. An American Optical engineer translated it, confirming what O'Brien could see from the English abstract and the illustrations, that van Heel had written about clad fibers.[74] The reprints had "*De Ingenieur*, No. 24, 1953" printed on the cover but were undated inside. Someone hand-wrote the publication date on O'Brien's copy: "12/6/53." Thinking the paper had appeared late in the year, the lawyers filed O'Brien's application November 19, 1954, and in due time he received a US patent.[75]

Hopkins consulted a London law firm about patenting image transmission. However, an elderly partner recalled Baird's 1927 patent, which he thought would preempt Hopkins. With little money to waste on a losing battle, Hopkins gave up on patents.[76] He and Kapany worked on fiber bundles through 1954, eventually making a 30-inch (75-centimeter) bundle.[77] They developed faster winding machines, a practical necessity when one bundle included 30 miles (50 kilometers) of fiber. They analyzed light collection and transmission, but never tested claddings.

Even without a patent, Hopkins sought an industrial partner that could offer the resources needed for further development: "glass technologists, precision machinists, and cooperation with a traditional endoscope maker." However, the eminent young professor had no more commercial success than Møller Hansen, and he concluded the companies were "dead from the neck up."[78]

Kapany and Hopkins fell to quarreling even before the ink was dry on Kapany's 1955 dissertation on fiber optics. Hopkins complained that Kapany claimed too much credit for the concept.[79] Kapany admits he did not contribute to the original grant proposal that suggests imaging bundles,[80] although his vita claims he is "widely acknowledged as 'the Father of Fiber Optics.' "[81] Years did not abate the feud. Months before his death in October 1994, a feisty Hopkins told me Kapany "contributed nothing to the brains of the project; he was a pair of hands."[82] For his part, Kapany says the professor was too much a theorist to appreciate his practical skills.[83]

99 Percent Perspiration

The Birth of an Industry (1954–1960)

The definition was good enough to read large print, but the color was green, and light loss was so great as to make long fiber bundles impractical. Nevertheless, it was flexible and did transmit an image, and that was enough to set one dreaming.

—Basil Hirschowitz, recalling his first look into a bundle made by Hopkins and Kapany.[1]

A good idea is not enough to launch a new technology, as the cases of C. W. Hansell, Heinrich Lamm, and Holger Møller Hansen testify. It takes a critical mass of credible evidence to get the idea rolling, careful engineering and a technological infrastructure to make practical devices, and money and marketing savvy to sell them.

The two *Nature* papers provided the critical mass of evidence, along with an important new idea, the transparent cladding. Lamm had hit only the easiest of targets, recording a jagged image of a bright light-bulb filament. Hopkins and Kapany made a better bundle that showed the letters GLAS.[2] Lamm was a medical student writing in an obscure journal; van Heel and Hopkins were prominent scientists writing in one of the world's most widely read journals. Moreover, 1954 was a better time to launch new ideas than 1930, when Europe and America were in economic crisis and the ugly tide of Nazism was rising in Germany.

Yet the *Nature* papers represented only a first step. They didn't mention unpleasant realities such as the rapid degradation of transmission when bundles of fibers were handled. Much more work was needed to translate laboratory demonstrations into practical devices, but the pioneers ran out of steam. Unable to get industrial backing, Hopkins went on to other projects.[3] Van Heel stopped after the Dutch National Defense Organization lost interest,[4] and he could not interest the Dutch giant Philips N. V. in commercial applications.[5] Europe was full of ideas but slow to apply them.

Narinder Kapany, full of enthusiasm, saw better opportunities in America and headed across the Atlantic. Two others also picked up the new idea. One was Basil Hirschowitz, a young surgeon excited by the idea of a flexible gastroscope. The other was an anonymous spook deep in the bowels of the Central Intelligence Agency who thought image scramblers might solve some American security problems.

"Enough to Set One Dreaming"

A native of South Africa born in 1925, Hirschowitz studied gastroenterology in London, then began a fellowship at the University of Michigan in Ann Arbor in mid-1953. Hirschowitz was so intrigued by the *Nature* papers[6] that he arranged a 1954 London vacation that included a visit to Hopkins and Kapany at Imperial College. Looking through one of their bundles, Hirschowitz saw the possibility of truly flexible gastroscopes that could be threaded easily down the throat into the stomach, and of flexible endoscopes that would use the same technology to inspect the intestines and other parts of the body.[7]

Well aware of the drawbacks of semirigid gastroscopes, Hirschowitz wanted an instrument that was thin, flexible, easy to handle, sealed against body fluids, and able to deliver enough light for the physician to see clearly. He knew that he couldn't build it alone. He sold the idea to his Michigan supervisor, Marvin Pollard, and the two started to assemble a team. The physics department pointed Hirschowitz to C. Wilbur "Pete" Peters, a 36-year-old optics professor. Pollard invited Kapany to join them when he finished in London. As a first step in the spring of 1955, Hirschowitz ordered $12 worth of the same glass fibers used by Hopkins and Kapany.

Kapany disrupted their plans when he landed a better offer from Bob Hopkins, O'Brien's successor at the University of Rochester. That job included consulting for Bausch & Lomb, a big Rochester optics company with much more research money than Pollard and Hirschowitz.[8] It was an offer too good to pass up for the ambitious and energetic 28-year-old, recently married and with a new doctorate.

Peters suggested they fill the void by hiring a physics student and recommended Lawrence E. Curtiss. Hirschowitz sat down with the student, explained his ideas, and demonstrated a rigid gastroscope. "That put the fear into me," Curtiss says.[9] The young physician probably did not realize that Peters didn't consider the project very challenging. Hirschowitz definitely did not know that Larry Curtiss was still in his teens and about to start his sophomore year in the fall of 1955. Curtiss had spent the summer after high school building equipment for a small Michigan hospital, developing an interest in medical instruments that left him full of questions in freshman physics class. When Hirschowitz asked for help, Peters remembered the bright kid who kept asking medical questions.

Curtiss recognized an opportunity that went beyond earning much-needed college money. He read the papers by van Heel, Hopkins, and Kapany. It seemed a straightforward matter of following Kapany's detailed recipes for making bundles, once he untangled the fibers, which had become badly snarled en route from England. Curtiss trusted Kapany's instructions so completely that didn't even try to measure fiber characteristics. He was dismayed to find that his first short bundle transmitted almost none of the light that entered it.[10]

The failure of the published recipe gave Curtiss a problem to solve, making the project much more interesting. Kapany's published articles gave no clues; the student had to figure out what had gone wrong for himself. He found that unclad fibers transmitted light better after he cleaned them with chromic acid. However, five percent of the light vanished each time he touched them afterward. That enigma had a deceptively simple explanation. Touching the fibers left behind fingerprint oils, which after drying have a refractive index close to the 1.5 value of the glass in the fibers. From an optical standpoint, each fingerprint roughened the surface of the glass such that total internal reflection did not work in that region. In addition, the fibers Hirschowitz had bought were little more than window glass stretched into fine threads, so they were not very clear and even had the same green tint that's visible at the side of a sheet of window glass.

The Michigan team thought they could solve those problems by drawing their own fibers from clearer glass with a higher refractive index. They found the rods they wanted at the Corning Glass Works, made of optical glass with a refractive index of 1.69. That was enough higher than fingerprint oil that touching the glass would not stop total internal reflection.

With a tiny budget, the little group scrounged equipment to make fibers. Their jury-rigged apparatus cost them less than $250.[11] They mounted the glass rod vertically in a four-inch tubular furnace, heated it, and pulled fiber from its molten bottom. They wound their fibers onto the cardboard cylinders of two-pound boxes of Mother's Oats, which Curtiss picked after rejecting rug rolls. Peters collected the oatmeal in a corrugated box and took it home to feed his children.[12] Fortunately, the kids liked the cereal, though they still joke that eating it was their contribution to fiber optics.[13]

Pulling fibers proved easy, although freshly drawn fiber had to be handled carefully. The first bundle Curtiss made from homemade fibers transmitted images well, as long as the fibers were loose on the table. However, the images faded when he put the fibers inside a protective rubber tube. Kapany had written that there was little crosstalk between fibers, but Curtiss found that light leaked freely between bare fibers when they were packed together. More losses came from surface scratches, where the fibers rubbed each other. Once again the recipe didn't work, and it was back to the drawing board.

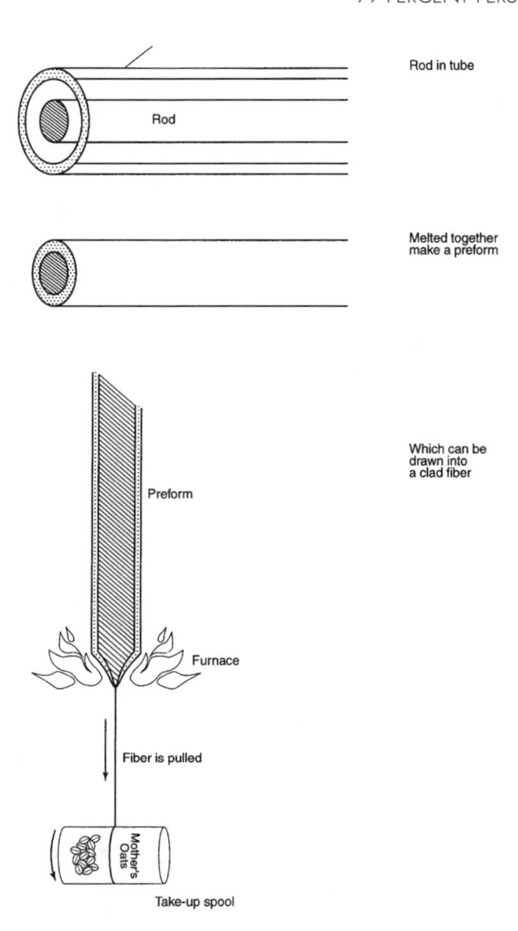

Figure 6-1: Larry Curtiss made a glass-clad fiber by melting a glass tube onto a rod, heating the tip, and stretching a fiber from it.

In the summer of 1956, after finishing his sophomore year, Curtiss took the problem to an informal committee of the physics faculty, a group of professors who played bridge at lunch in the laboratory basement. They unanimously agreed the best solution was to apply a transparent cladding of plastic or lacquer, just as van Heel had suggested. Curtiss tried to convince Peters and another professor that a glass cladding would be better,[14] but the older men doubted it could be applied uniformly, and worried that the glass might crack or craze.

Curtiss wondered if he could solve the problem by putting a tube of low-index glass around a rod of higher-index glass, melting them together, and stretching the assembly into a fiber (figure 6-1). But he spent the rest of the summer and the early fall making bare glass fibers and dipping them in plastic and lacquer. That reduced crosstalk, but only at the cost of reducing light transmission. Nonetheless, they made progress, assembling a three-foot bundle, the longest yet. Peters thought it was good enough to share at the annual meeting of the Optical Society of America, held in October 1956 at Lake Placid, New York. He asked Curtiss to write a brief report, and the two drove in the professor's old Packard to save money.

They barely made it. The Packard blew a universal joint near Utica at midnight, so they had to stop for repairs. When they finally reached Lake Placid, Kapany was speaking eloquently about the simplicity of fiber optics. Curtiss followed with a down-to-earth report that lacquer coating could cure crosstalk, although transmission remained poor. Later at the meeting he met a physicist a decade older who had already spent nearly two years working on similar problems at American Optical.

A Project for the CIA

Brian O'Brien's work on clad fibers should have given American Optical a commanding head start. However, his genius lay in ideas, not administration. American Optical had all the inertia of a big company, and its best engineers were busy with Todd-AO. Image scramblers remained an idea stuck in the piles of paper on O'Brien's desk.

They might have remained buried there if O'Brien had not told the CIA about the potential of image scramblers. The spy agency could use one half of the bundle to make ciphers that only the spy with the matching half could decode. The spy, in turn, could send messages back that only the spy masters could decode. Diplomats could use them to send secret messages back home. Each pair of encoders would have a unique pattern, so loss of one scrambler would not break security for people using others. With the theft of atomic secrets and the Cold War fueling fears of Soviet spies, the CIA made security devices a high priority. Yet American Optical was doing nothing about image scramblers because its best engineers were busy with Todd-AO, and van Heel had leaked the idea in *Nature*: "Coding and decoding of two-dimensional pictures proved to be practicable."[15]

Frustrated CIA officials sought new talent to launch the American Optical project. They found Will Hicks, a 30-year-old physicist from Greenville, South Carolina, studying textiles for the Milliken Research Trust. His background looked good, but the CIA didn't realize that Hicks had neither worked on fibers nor completed a course in optics.[16] In the long run, that lack of background may have been a blessing because it helped him take a fresh approach. Hicks also brought incisive insight, tremendous energy, a quick mind, and a stubborn and independent nature.

The son of a judge, John Wilbur Hicks, Jr., had thought of becoming a pianist but majored in physics at Furman University in Greenville. After working briefly on the Manhattan Project at the end of World War II, he spent four years in graduate school at Berkeley, studying quantum mechanics. He excelled in coursework but never settled down to write a dissertation. Homesick, and married with one small child and another on the way, Hicks returned to South Carolina in 1950. When the CIA contacted him in 1954 he was restless, unhappy with the conservative South, and fighting with the state over taxes.

Hicks arrived at American Optical in September 1954. The first day, O'Brien briefed him on fiber optics and gave him a couple of plastic-clad fibers O'Brien had made in Rochester. O'Brien told Hicks that his assignment was to make and clad fibers and assemble them into an image scrambler. The busy O'Brien then sent Hicks off to work in a building separate from the main research lab and had little more to do with the project.

In theory, Hicks reported to associate research director Steve MacNeille and later to Walt Siegmund, a protégé of O'Brien's from Rochester. In practice, he was largely on his own while everyone else concentrated on Todd-AO, an arrangement well matched to his independent nature. Not one to merely follow the boss's lead, Hicks began by convincing himself that low-index claddings were essential before the Michigan group reached the same conclusion. Then he tackled scrambler development, a job much tougher than merely demonstrating a crude imaging bundle because it essentially required a whole new technology.

The security-conscious CIA used gastroscopes as a "cover" project to conceal work on image scramblers. At a simple level, the two applications share the need for fiber bundles that transmit intact images from end to end. To make an encoder, you scramble the loose fibers in the middle, glue the mixed region solid, then saw through it.

Look closely, however, and you find quite different requirements. A gastroscope must be long, thin, and flexible to examine the stomach. A scrambler should be at least a couple of inches across to encode pictures or signatures, and short, fat, and rigid to handle easily. These differences are crucial and lead to quite different devices. Whatever American Optical thought about the potential of medical instruments, their customer was the CIA, and the spy agency wanted image scramblers. Although both applications required fibers, the overall goals were different, ultimately leading Hicks and American Optical along a path different from the Michigan team.

Hicks first tried drawing fiber from a hole in the bottom of a platinum crucible full of molten glass and then coating it with plastic.[17] The process was quick and fairly easy, but plastic-clad fibers did not transmit light well. When several months of refinements did not solve the problem, Hicks started work on glass cladding, without completely abandoning plastic.[18] His first idea was a logical extension of drawing fibers from a single crucible. He nested two crucibles together, melting low-index glass in the outer one and a high-index glass in the inner one. Then he pulled fibers through concentric holes at the bottoms of the crucibles, so glass in the outer crucible formed a cladding around a core of glass from the inner crucible.[19] The technique worked but did not yield good fibers.

Hicks took a break in October to attend the Lake Placid meeting. He met Peters and Curtiss and was impressed by the progress they were making on a minimal budget. Their discussions led Hicks to visit Ann Arbor some weeks later, but both sides knew a race was on and guarded their secrets.

Taking a Chance on Glass Tubes

The poor results with lacquer left Curtiss itching to try making a fiber from a glass tube collapsed onto a glass rod. After one or two preliminary tests, he finally got his chance on December 8, 1956, when the professors were away at a conference. He bought tubes of soft glass, with a refractive index lower than the Corning rods, from the chemistry supply office. Then he put a rod of the high-index Corning glass inside a tube, melted the two together in a furnace, and drew a fiber from it.

"It was probably the most exciting day of my life," he recalls. He walked down the hall, pulling the fiber. "I was 40 feet down the hall and I could still see the glow of the furnace. I knew it was good."[20] His glass-clad fiber transmitted light far better than any other fiber ever made. The college junior had broken through his professors' conceptual logjam.

In part, Curtiss was lucky, as he readily admits today. But his was the luck that comes to the intelligent and observant. His lacquer experiments made it clear that plastic claddings would not work well. It seemed logical to try glass cladding instead. Like any science student of the time, Curtiss had seen plenty of glass tubing. He had stretched rods into fibers. Why not slip a rod into a tube, melt the two together, and see what happened when he stretched the composite into a fiber?

Later research identified the big problem with plastic claddings. Tiny inhomogeneities about the size of a wavelength of visible light, 0.0005 millimeter (500 nanometers), form as a liquid plastic film solidifies. They make the glass-plastic boundary irregular, so it scatters light instead of guiding it along the core. These losses limit transmission efficiency. Glass has its own inhomogeneities, but drawing it into a fiber stretches them to many wavelengths long. They become so large that light doesn't "see" them at the core-cladding boundary. The same effect happens in plastic, but you can't stretch a plastic-clad glass rod into a fiber, because the two materials melt at much different temperatures.

The Michigan team also did not initially realize the importance of rod preparation. Their search for rods of high-index glass happened upon some that Corning had fire polished to form a very clean, smooth surface. American Optical took the tried and true approach of mechanically grinding and polishing their rods down to the right shape. Mechanical polishing produces exceptionally smooth optical surfaces but leaves behind residual surface contamination that can ruin the core-cladding layer.[21] Fire polishing removes contaminants, but nobody knew that in 1956.

Once Curtiss showed Peters and Hirschowitz his clear glass-clad fibers, the project went into high gear. They fused tubes and rods together, mounted them upright in a four-inch-long furnace, and drew fibers from the molten bottom. Curtiss built a system that measured fiber diameter as it was drawn, without touching the glass. They controlled the rate of fiber drawing by hand, averaging about five miles (eight kilometers) an hour. Like Hopkins and Kapany, they made bundles with ends aligned in the same position by looping the fibers, fixing one part of the loop in place, and then cutting carefully through the fixed spot. They needed 25 miles (40 kilometers) of good fiber to make a single bundle of 40,000 fibers.[22] The freshly drawn fibers were only about 0.04 millimeter thick, so thin that plastic had to be coated on top of their glass cladding before they could be handled. Hirschowitz and Peters paid the physics shop $800 to $900 to build a machine to wind fiber off the oatmeal boxes into bundles, but almost everything else was homemade.

Unlike O'Brien, Hirschowitz wasted no time filing for patents; his first application, covering the gastroscope, was filed December 28, 1956, with rights shared with Peters and Curtiss.[23] Curtiss applied for a patent on glass-clad fibers on May 6, 1957, assigning part interests to Hirschowitz and Peters.[24]

The glass-clad fiber was the last piece needed to build a gastroscope. Hirschowitz knew exactly what he wanted, and it took the three men just two months to assemble the first working model. The tip that went inside the stomach included a little illuminating lamp and optics to collect and focus light onto the end of the bundle in the stomach. Hirschowitz himself swallowed it first.[25] A few days later, February 18, 1957, he tested it on his first patient, a small woman who had an ulcer. The instrument went down successfully, and the moment seemed triumphant—until Marvin Pollard arrived.

Academia has its pecking orders, and Marvin Pollard had thought he was at the top. He had not been deeply involved in development; Curtiss hadn't realized he was involved at all until Pollard invited him to a reception. However, as the senior researcher, Pollard wanted to try the new instrument first. Furious that Hirschowitz hadn't waited for him, he confiscated the new endoscope and locked it in his safe, so no one else could use it.[26] Tensions between Pollard and Hirschowitz escalated rapidly when Pollard demanded a share of patent rights.[27] He never got them, but he later blocked Hirschowitz from getting a permanent job at Michigan.[28]

Curtiss, who had thought the project was over, had to make a second gastroscope 5/16 inch (0.8 centimeter) thick and a meter long for Hirschowitz to show at a meeting of the American Gastroscopic Society in Colorado Springs.[29] A May snowstorm disrupted transportation, so fewer than 40 people attended his 10-minute talk, but those who attended were impressed. Kapany was there and claimed Bausch and Lomb had almost completed an instrument, but it never appeared.[30] Rudolf Schindler, by then a grand old man of gastroscopy, recalled that Heinrich Lamm had the same idea but thought Lamm had failed to assemble a bundle of fibers.[31] A few people asked for a demonstration, so Hirschowitz took them to the nearest phone booth and showed them fine print in the telephone directory.

The whole project cost about $5500 over two years, nearly $4000 of which went to pay Curtiss. American Optical, the CIA, and Bausch & Lomb probably spent far more.

A Problem with Image Scramblers

When progress stalled at American Optical in the fall of 1956, the company started looking for alternatives. Lee Upton, an AO glass specialist, had worked in a plant that collapsed glass tubes onto electrical components to insulate them, and he suggested adapting that idea to making fibers. In October, Steve MacNeille brought in an outside consultant, a retired MIT professor named Frederick H. Norton. Norton first proposed coating bare fibers with molten glass, but later also suggested a rod-in-tube approach.[32]

Norton also borrowed an idea from ancient makers of miniature glass portraits. They stacked rods of colored glass in the desired pattern, heated the assembly until the edges of the rods melted together, and stretched the bundle into a single fused cylinder. Slicing the fused rod like a salami yielded disks bearing portraits or abstract patterns. The method is called millefiori, for "thousand flowers," and is still used today to make paperweights. Norton realized that many clad fibers could be assembled together in the same way and stretched into a rigid "multifiber." This avoided the need to draw many fine, separate fibers and assemble them into a bundle, because a single operation stretched all the fibers at once. The trade-off was the loss of flexibility; the many fibers emerged fused together in a solid rod. That rigidity made them useless for gastroscopes but was fine for image scramblers, which were much fatter and thus needed many more fibers.

It took Hicks time to assemble the equipment he needed to test the new ideas. The big company's advantage in resources was more than offset by its ponderous bureaucracy. By November, Hicks had found thin-walled tubing but not rods of high-index glass suitable for the core of the fiber. American Optical had the right kind of glass, but it came in rectangular blocks, and Hicks couldn't persuade anyone at AO to grind them down to the cylindrical shape that he needed. He finally ordered polished cylinders from another company. While he waited for the rods to arrive, he hand-built a glass-melting furnace in the AO shop, because he couldn't get the company to pay its machinists to do the job for him. It was January 1957 before Hicks had the equipment to try Upton's approach to drawing glass-clad fiber.

By this time, development of the Todd-AO optical system was winding down, and American Optical was shifting some of those engineers to the fiber project. However, they and Hicks faced tough problems debugging fiber production. Contaminants from mechanical polishing remained at the core-cladding interface, causing troublesome losses that made the glass-clad AO fibers poorer than those Curtiss drew from fire-polished rods. These problems helped keep plastic-clad fibers alive at American Optical; although they transmitted less light than good glass-clad fibers, loss was not as critical for short image scramblers as it was for longer gastroscopes.

Hicks's group made a demonstration scrambler by looping single fibers many times around a drum, gluing one region, and sawing through that spot. After they showed that the flexible bundle could transmit an image, they scrambled the loose fibers in the middle, glued them, and sawed through that region to make an encoder/decoder pair. Then they took photos of the scrambled and unscrambled images, packed them up with the bundles, and sent them to the CIA.

Hicks, feeling satisfied, took his first vacation in three years at American Optical. As his body relaxed on the sand at Myrtle Beach, his mind began to analyze the coding theory behind the image scrambler, and Hicks came to an unpleasant realization. Using the same scrambler many times could "wear out" the code, so someone who intercepted enough scrambled images could decode them by comparing the patterns. "You wear it out in a hurry, too," recalls Hicks, because the fixed fibers made each device scramble the picture in exactly the same way each time it was used. He figured breaking the code would take only 18 images scrambled by the same device and access to a "fairly large computer"—by 1957 standards.[33] That meant image scramblers could not offer the security the CIA wanted.

Most people try to avoid concluding their own work is futile, and very few willingly reveal that unpleasant truth. Not Will Hicks. He wrote to the CIA, explaining why image scramblers would not work and suggesting they instead use one-shot keys, which are simpler but unbreakable. "The people I was working with weren't very knowledgeable about coding theory, and it was news to them. It may have been to everybody at the CIA,"[34] Hicks says. But he convinced them, and they canceled the project.

American Optical management was rather less pleased with Hicks's insight and considered firing him. However, cooler heads prevailed. He still had the "cover" project of the endoscope and was already working on another military fiber-optic program. "I was sort of a pain in the ass when I was young," Hicks says. "Maybe I am still."

The Odyssey of Narinder Kapany

Narinder Kapany was far from idle at the University of Rochester. He built some devices, did some contract research, and spoke and wrote prolifically. By 1957, when he moved to the Illinois Institute of Technology Research Institute, he was the best-known spokesman for fiber optics. Self-confident and secure in his Sikh traditions, the solidly built Kapany was a striking and exotic figure in white-bread America. Nearly six feet tall, he was well dressed in western suits but his turbaned head and bearded face stood out in a crowd. Charming and articulate, he made the case for fiber optics with a voice carrying hints of his years in India and England.

His energy, enthusiasm, and dash of flamboyance left a vivid impression. While attending a faculty Halloween party in Rochester, he volunteered to answer the door, knowing that to Americans he was already in costume.[35] He called sending an image through a knotted bundle of optical fibers his "Indian optical rope trick."[36] He also wrote a series of sober, scholarly papers in the *Journal of the Optical Society of America*[37] that outlined the basic principles of fiber optics. As Curtiss discovered, Kapany's papers were not infallible—among other things, he was slow to realize the need for a cladding—but they were the best introduction to a promising new field.

In fact, they were the only readily available introduction because nobody else wrote much. O'Brien was too busy with new ideas.[38] H. H. Hopkins and van Heel abandoned fiber optics soon after their early papers. Hicks wrote little but patent applications, and much of his early military work was classified until the late 1960s.[39] Curtiss found himself in a bitter patent battle that lasted until 1971, and his lawyers discouraged him from writing about technical details.[40] Hirschowitz wrote about building and using medical instruments, not about fiber properties.

From 1955 to 1965, Kapany was the lead author on 46 scientific papers and coauthor of 10 more—an average of over five a year. That represented a staggering 30 percent of all the papers published on fiber optics during those years, including reports on medical treatment.[41] It was Kapany who wrote a 1960 *Scientific American* cover story on fiber optics,[42] and the first book on fiber optics,[43] based partly on his earlier papers.[44]

Kapany's writings spread the gospel of fiber optics, a highly visible role that cast him as a pioneer in the field. However, his technical innovations are widely regarded as less crucial than those of men like Curtiss and Hicks. Through 1969, he collected a respectable 10 fiber patents, but Hicks earned 22.[45] His greatest gift may have been for promotion, though he dislikes the word. "I was just doing what came naturally, which is to go and get the resources to work on the crazy ideas I come up with, and talk about it in conferences and publish."[46]

Finding a Market for Gastroscopes

By the spring of 1957, Hirschowitz, Peters, and Curtiss had a working gastroscope and patent applications in the works. Hirschowitz, the salesman of the group, went hunting a company to make commercial instruments. American Optical turned him down because the instrument division didn't think they could sell enough endoscopes.[47] American Cystoscope Makers Inc. eventually licensed the technology on the condition that the Michigan group help them start production. As the expert fiber maker, Curtiss found himself shuttling back and forth to New York to lecture American Cystoscope production engineers while still an undergraduate.[48]

Building machinery for industrial production was far more difficult than jury-rigging equipment at the university. American Cystoscope could not borrow glass samples and wind fibers on oatmeal boxes. To mass-produce bundles reliably, they had to buy special machinery, find reliable sources of the right kind of glass, and assure quality of their product.[49] That took time. Curtiss got into graduate school at Harvard, then took off a semester to help start production. The interruption became a career, and he never returned to Harvard.[50]

Hirschowitz tested the first commercial gastroscope in 1960 at the University of Alabama Hospital in Birmingham, where he had moved after leaving Michigan. Other doctors quickly adopted them.[51] They were a dramatic advance, the first instruments that allowed doctors to see inside the stomach with little risk to the patient. Only a handful of specialists mastered the delicate techniques of using semirigid lensed gastroscopes; many more physicians could use thin, flexible fiberscopes. By the late 1960s, they almost totally replaced lensed gastroscopes, quickly becoming important new tools in treating many internal conditions. Only recently have tiny electronic cameras begun to replace flexible fiber bundles.

Fiber Optics Brighten Military Images

Shortly before the CIA gave up on image scramblers, an old Berkeley friend at the MIT Lincoln Laboratory near Boston asked Hicks for help. Bill Gardner was working on image intensifiers, tubes that sense weak visible or infrared light and turn it into brighter images, so soldiers and pilots can see better. The Pentagon thought that using one tube to amplify the output of another should yield even brighter images. However, the output faces of the tubes were curved so sharply that ordinary optics couldn't focus light from one tube onto another. Gardner asked Hicks if fiber optics could do the job.[52]

The tube's input was a flat screen, which released electrons when light hit it. An electric field inside the tube accelerated the electrons so that they hit a phosphor screen like a black-and-white television, producing a brighter image. That screen had to be curved outward to collect electrons in the right places to replicate the image accurately. Standard optics could not focus light from the curved phosphor screen onto the flat input face of another tube without losing light and distorting the picture.

Gardner and Hicks decided the solution was to mount a short, fat bundle of parallel fibers between the tubes. They ground out one side so that it fit perfectly over the curved end of the tube. The other end was flat. The array of short fiber segments, called a "faceplate," carried light straight from the curved face to the corresponding point on the flat face of the next tube.[53] The idea was a logical extension of Baird's 1927 television patent. (A mineral known as ulexite forms similar structures naturally—see box, page 74.)

It was also a natural application for the fused multifiber technology American Optical had started developing for image scramblers. Faceplates sealed the end of the tube, so they had to be rigid and vacuum tight. The early tubes had screens a couple of inches across, and the faceplates were disks, which could be sliced from a fused bundle of multifibers like a salami—or millefiori paperweights.[54]

American Optical charged ahead, fueled by ample Pentagon funding and its new fiber-optic skills. AO engineers mastered the practical techniques of making large bundles, drawing them into large multifibers, and slicing the bundles into faceplates. As they made progress, the strong-willed Hicks began to clash with straight-laced company management. Annoyed when a guard insisted he sign in with an excuse for arriving late at 8:30 one morning, Hicks wrote: "I was on the potty. Sorry." Hicks had worked past midnight the night before, but research director Steve MacNeille had to save him from being sacked by an offended company official.[55]

Ironically, the final blow came after Hicks had done what he thought American Optical wanted—developed the fiber-optic faceplate into a viable product. He and MacNeille visited division managers but couldn't interest them in producing faceplates. American Optical remained a stodgy century-old maker of spectacles and microscopes, disconnected from its new research lab and unwilling to venture into a new field.

Furious that AO didn't care about the technology he had spent four years developing, Hicks quit in 1958, telling MacNeille he was taking the technology with him. "I walked out the front door at noon with everybody watching," he recalls. With friends from American Optical, he immediately started a company called Mosaic Fabrications.[56] Military customers for faceplates soon arrived, starting with an Army engineer who heard Hicks give a talk at an Optical Society meeting. In the classic style of small companies, he started the company in his barn but soon moved to a building at the nearby Southbridge Lumber Company. With no money for new equipment, he scrounged junked car parts to build fiber drawing machines. Soon he was making prototypes, and military agencies were coming back for more.

Full of energy and ideas, Hicks proved a hard-driving hands-on entrepreneur who pushed himself and the technology relentlessly. In no position to turn down business, Hicks took a Navy order for a 12-inch faceplate when the standard width was one inch—a daring leap, but one that he made work.[57] He battled AO management over patent rights[58] but collaborated with company scientists on research. He was tough on competitors but generous to employees. Once the company was doing well, Hicks handed out raises to everyone—and when another local business owner complained he was driving wages up, Hicks defiantly handed out a second raise. He ran things himself, building Mosaic Fabrications into a thriving business as a military contractor and developing a loyal following among his employees with his unconventional style.

Patent Problems at American Optical

Things did not go as well at American Optical. O'Brien retired at 60 in early 1958 but kept very busy as a consultant to industry, government agencies, and the National Research Council.[59] Mike Todd died March 22, 1958, in a plane crash, sealing the fate of Todd-AO wide-screen films as an artistic success but commerial failure.[60] Walt Siegmund, a gentlemanly O'Brien protégé from Rochester and Todd-AO veteran, took over fiber optics, but muddled corporate management gave little support to the new technology.

The most daunting blow may have been the collapse of American Optical's vaunted patent position. O'Brien and American Optical lawyers thought they had an airtight case for priority on the crucial concept of the clad fiber. Hansell's 1930 patent covered imaging, but by the time O'Brien's patent was issued on March 4, 1958,[61] it was clear that imaging needed clad fibers. O'Brien had licensed American Optical, so the company turned its lawyers loose on potential competitors. American Cystoscope Makers was at the top of the list.

An American Cystoscope lawyer did some legwork and found a copy of the issue of *De Ingenieur* containing van Heel's paper. That was something no one at American Optical had seen. Van Heel had sent American Optical only a reprint, with "De Ingenieur, No. 24, 1953" printed on the cover and the handwritten date "12/6/53."[62] The sharp-eyed lawyer American Cystoscope saw the crucial publication date on the magazine cover: June 12, 1953. He must have chortled with glee as he copied it to send to American Optical.

It is not enough for an inventor to show that he or she was the first to have an idea. The inventor must apply for a American patent no more than a year after the idea is first described in public anywhere in the world. The American Optical legal department had interpreted the handwritten date in American style as December 6, 1953, which gave them plenty of time to file their application. They didn't remember that Europeans write dates differently, with the date before the month. When the lawyers filed the patent application November 19, 1954, they had missed the proper deadline by over five months. "We knew the game was up" when the copy of the cover arrived, recalls Siegmund.[63] An angry Brian O'Brien felt cheated.

American Optical did not give up easily. With its star patent worthless, the company pressed an application Norton had filed on glass cladding. However, Curtiss had filed his application earlier in 1957. A determined attack by a horde of American Optical lawyers made the Curtiss patent one of the most litigated in history, but it survived to issue in 1971—a delay that ironically increased royalty payments.[64]

More Uses for Fibers

While the lawyers battled, the young field grew. The Michigan group, American Cystoscope Makers, Mosaic Fabrications, and American Optical concentrated on making practical instruments. Both Mosaic Fabrications and American Optical found new uses for fused fiber optics. They made fused bundles in which a special glass filled the space between fibers; etching away the filler glass left loose fibers. They twisted bundles of hot glass fibers in the middle, so when they cooled the image appeared upside down. They interleaved sheets of fibers into Y-shaped bundles that could combine or split images. They tapered bundles so that they shrank or magnified images, a wonder Dave Garroway demonstrated on his nationwide television show, *Today*.[65]

Some applications were quite specific. Lured to IBM by a handsome salary offer after earning America's first doctorate in fiber optics from Rochester,[66] Bob Potter put fibers to work reading the punched paper cards that symbolized computers in the 1960s. Early card readers used an array of 12 small and short-lived "grain of wheat" bulbs, one to illuminate each row of holes in the moving card. Potter replaced them with a fiber bundle that collected light from one long-lived bulb and split into 12 smaller bundles, one going to each hole. Before it became the first fiber-optic system to march into obsolescence, it earned American Optical some money on its fiber-optic investment.

Flexible bundles soon spread beyond medicine. American Optical developed bundles 9 and 15 feet long for NASA testing of the Saturn boosters used in the Apollo program. An American security agency investigated flexible bundles for surveillance.[67]

Ulexite: A natural fiber bundle

Nature makes its own bundles of optical fibers: a mineral called ulexite. It's a complex boron compound formed where mineral-rich lakes evaporate, found in Boron, California. As the mineral crystallizes, long, thin crystals grow parallel to each other and eventually merge into a solid block. Slice the block perpendicular to the filaments and polish the faces smooth, and it acts like a fiber-optic faceplate. Toy and rock shops sometimes sell it as "television stone."

Although ulexite was discovered around 1850, nobody took a close look at its optical properties until roughly a century later. It inspired no fiber-optic pioneers; some had never heard of it until I mentioned it. Only in 1963, several years after the faceplate was invented, did Bob Potter recognize ulexite as a natural fiber bundle.

Parallel filamentary crystals in ulexite guide light along their lengths like fibers in a faceplate. (Courtesy Dan Garlick, from G. Donald Garlick and W. Barclay Kamp, "The strange optical properties of ulexite," *Journal of Geological Education 39*, pp. 398–402, 1991)

Other developers tried less serious uses like decorative lamps, where a bundle of plastic fibers splays out sparkling light at their tips. Will Hicks threaded plastic fibers through a plastic Christmas tree, trying to sell them as decorations at a trade show in New York. When a potential customer visiting his hotel room asked if the fiber was fireproof, Hicks lit a match to see—and ignited both the tree and the fibers, getting himself tossed out of the hotel.

Nobody gave much thought to communicating through optical fibers, although the invention of the laser in 1960 brought optical communications into the spotlight. It was intuitively obvious to any physicist worth his salt that no solid material could be as clear as air, and communications should go through the clearest material possible. Hicks made some rough calculations, but the results were discouraging; other frontiers seemed much more promising. He didn't realize until later that he had made a big mistake.[68]

7

A Vision of the Future

Communicating with Light (1880–1960)

> In the region that includes the visual spectrum [and] the near infra-red, there is about nine hundred million megahertz of bandwidth waiting to be used. This is enough for seven thousand million speech channels of present-day standards, using a digital method such as PCM [pulse code modulation]. . . . It would be enough too for the PCM transmission of one and a half million individually used TV channels, with color and stereo-type 3-D features, of standards that the future will require. If the traffic loaded it at all fully, even one percent of this bandwidth per conductor could easily make all lower frequency methods that are now foreseeable as obsolete eventually as the stagecoach.
>
> —Alec Harley Reeves, 1969[1]

At first glance, communicating through optical fibers seems simpler than imaging. You only need a single fiber to carry a light signal between two points. Yet anyone who looked closely at the idea in the late 1950s saw a much different picture.

The first and most obvious problem was transparency. Larry Curtiss, Will Hicks, and the engineers who worked with them had made tremendous progress, but their clearest bundles carried images only a few yards or meters. Six feet of fiber is more than enough to look into the stomach, but it may not reach from your desk to the wall jack for a telephone line. No one knew how to make fibers carry light over the much greater distances needed for communication lines.

A second problem was the tremendous success of radio and electronics technology. Wires and radio links reached around the globe; satellites were on the horizon. Optical communication through the air had been tried in the nineteenth century but lost the race to the electrical telegraph and telephone. Radio transmission was pushing to higher frequencies; semiconductors were opening new possibilities in electronics. Optics seemed a backwater in comparison.

Moreover, communications engineers knew their task had become far more complex than carrying voices from one room to the next, as Alexander Graham Bell had done in 1876. Telephone signals traveled circuitous paths, following wires from one home to a switching center, which routed them to another switch, and another, until they finally reached their destination. The system was as vast and complex as the network of tiny streams that ultimately combine into the mighty Mississippi River. The switches represented another crucial level of complexity, because they had to work together to make connections between any pair of telephones. A whole infrastructure of electrical, mechanical, and electronic technology supported the telephone system and seemed destined to dominate in the future.

You had to be a little crazy to challenge that conventional wisdom, and Alec Harley Reeves had a dash of the requisite madness about him. He was an idealistic and eccentric British bachelor whose hobby was investigating the paranormal. Yet he was not just another dotty, aging gentleman. The cognoscenti of communications knew him as a genuine visionary, an intuitive engineer who had seen the potential of digital electronics in the 1930s. In the 1950s, Reeves started looking to light. Initially, fiber optics were not part of the picture.

An Array of Pipes and Switches

The job of the telephone network is to carry signals from your phone to any other phone. In recent years, the network has expanded to carry many signals besides voices, and mobile as well as fixed phones, but the principles remain the same. The network that does the job is, in essence, an array of pipes and switches. The pipes carry the signals from point to point; the switches direct the signals. You need both pipes and switches to have a working phone system. Either one or both can be bottlenecks that limit how well the system operates.

When the phone system began, telephone companies ran wires from every phone to a central "office" where people did the switching needed to complete calls. Operators sat at switchboards with arrays of holes, one for each phone, and people signaled them when they wanted to place a call. The operator responded by connecting an instrument to the hole leading to the caller's phone, and asking for the phone number being called. Then the operator plugged the proper wires into the corresponding holes on the switchboard and listened to see if the person on the other end picked up the phone. This worked for a town with only a few phones, but it didn't scale very well. Think of it as a manual system, like making calculations by writing numbers on pieces of paper.

As the number of phones grew, phone companies added more operators and switchboards. They also set up extra offices or switching centers to serve different parts of larger cities. To complete calls from one part of town to another, the operator would patch a call through to a second central office, where another operator would make the connection. The same principle worked for long distance.

The more phones, the more complex the network grew and the more operators were needed. Eventually, telephone engineers started to replace operators with mechanical switches that moved in response to electrical signals, making and breaking connections. A simple switch might rotate to one of ten positions (the digits on a dial) and stay in that position while the next switch in line rotated to the proper position. One of the most popular switches was designed by a Kansas City undertaker, Almon B. Strowger, who wanted to automate the phone system because he feared that operators were routing business elsewhere.[2] Dial phones automated the system, the click of each digit of rotation telling the mechanical switch to take another step. Think of mechanical switches as adding machines, clanking, chattering, and clunking away to route phone signals.

Expansion of the phone system meant more people could be calling from one town to the next at the same time. That required ever bigger "pipes" to carry more and more signals between local switching offices. The more traffic, the bigger the pipe was needed. Switches were expensive, so they were used only where traffic was heavy and the task not too complex. All long-distance calls had to be placed through operators until long-distance direct dialing was introduced in the 1950s,[3] and for many years afterward smaller towns still routed long-distance calls through operators.

At the same time, phone companies were developing all-electronic switches that routed signals without mechanical motion. Mechanical switches had simply reached their own limits; solid ranks of them filled brick buildings in town after town across the country, and they couldn't keep up with system growth. The search for all-electronic switches led Bell Laboratories to develop the transistor, which not only could amplify signals, but also could turn them "off" and "on" in response to a control input. That let transistor circuits serve the same function as electromechanical switches. Put enough circuits together and you have a special-purpose computer that decodes the numbers you dial and routes your phone signals accordingly. These electronic switches are indeed computers, far more powerful than the electromechanical switches that were the adding machines of the telephone world.

The Quest for Capacity

Both the switches and the pipes determine the capacity of a telephone system. Reeves focused on developing a new family of pipes with higher capacity. Engineers had been heading in the same direction for decades, although they didn't always express their goals in those terms. Two trends pushed the demand for higher capacity. One was the sheer volume of traffic; the other was the shift toward carrying signals that carried more information and thus needed more capacity per signal. The voices carried on a telephone line represented more information than the dots and dashes that represented letters in the telegrapher's Morse code. The tiny black-and-white pictures John Logie Baird and Francis Jenkins sent over their first television transmitters carried more information than voices. Standard color television channels carry more information than those first television transmissions.

Even back in the Victorian era, engineers tried to pack as much information as they could onto a single communications line. The first electrical telegraphs could carry only one set of dots and dashes along their wires at a time. As telegraph traffic increased, engineers realized it would be much cheaper to send multiple signals down the wire than to string more wires. A host of inventors turned to the task of multiplexing telegraph signals; among them was Alexander Graham Bell, who extended his ideas to sending voices along wires and made the telephone.[4] As telephone service spread and forests of overhead wires threatened to darken cities, engineers invented ways to send multiple conversations through one set of wires.

Along the way, engineers developed a framework that helped them understand the process of communication, and how to extend their technology. A simple telegraph works by letting current flow through a wire for short (dot) and long (dash) periods. Thus, the telegraph key modulates the current. Step back, and you see a more general process—a signal that carries information modulates a steady "carrier" that can be an electrical current, an audio tone, a radio wave, or a beam of light (figure 7-1). In a telephone, a voice modulates the electrical current, which varies at the same frequency as the voice.[5]

At first, the carrier was a steady electric current. Then engineers realized they also could modulate the intensity of a wave. An oscillator generated a particular frequency, while a separate signal modulated the strength of the carrier at that frequency. This is the basis of AM (amplitude-modulated) radio transmission. Later, engineers discovered they could modulate the frequency of a radio signal, used in FM (frequency-modulated) radio and television broadcasting.

The transmission capacity or "bandwidth" of a system depends on the frequency of the carrier; the higher the frequency, the greater its bandwidth. Since radio transmission began, engineers have moved to higher and higher frequencies to increase transmission bandwidth. They started at frequencies of thousands or tens of thousands of hertz (cycles per second), then moved

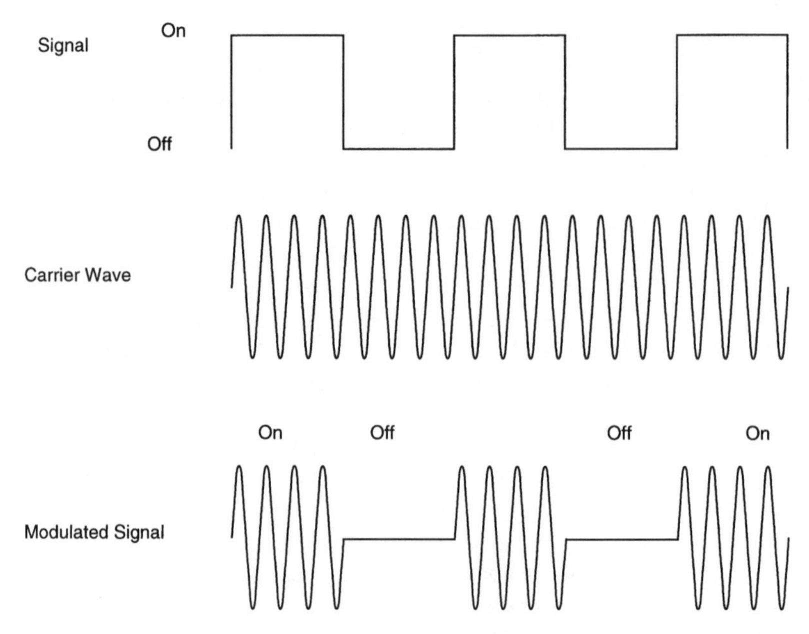

Figure 7-1: Modulation of light waves.

to millions of hertz, megahertz. By World War II, they were testing microwave frequencies, at billions of hertz or, equivalently, gigahertz.

As Alec Reeves watched the radio spectrum grow crowded in the 1950s, he considered the possibility of moving to even higher frequencies. He knew that radio waves and microwaves are part of the broad spectrum of electromagnetic waves. All are essentially the same phenomenon, waves composed of oscillating electric and magnetic fields.[6] Their properties differ with their frequency. Move beyond the microwave spectrum and you reach millimeter waves, infrared radiation, and visible light; ultraviolet light, X rays, and gamma rays have even higher frequencies. For visible light, the frequencies approach 900 trillion hertz, or 900 million megahertz in Reeves's terms. He estimated maximum possible bandwidth as roughly the carrier frequency, and that meant visible light had a truly staggering capacity.

The years around 1960 were the heyday of technological optimism, when the space frontier seemed within reach, nuclear power promised a glittering future, and pollution was unknown. Any down-to-earth engineer could have done the same calculations, but it took a visionary like Reeves to believe them.

Optical Telegraphs and Photophones

When Reeves turned to light, he was putting a new spin on an old idea. The first telegraphs were optical, part of a system developed in France at the end of the eighteenth century and copied by many other countries but now largely forgotten. The optical telegraph was essentially a series of relay towers, where operators peered through telescopes to spot semaphore signals from a tower on one side, then displayed the same signals for the next operator to read. The signal schemes were ingenious, but the technology was crude—men looked through telescopes and pulled ropes to set flaps and flags.[7] Labor-intensive, expensive to run, and unusable at night or in bad weather, the optical telegraph was replaced by the electrical telegraph. (Semaphores and signal flags survived to send messages between ships at sea.) Electricity could send telegraph signals farther through wires than people could see through the air.

The telephone followed over wires, but after the dust and the patent lawsuits had settled, Alexander Graham Bell wasn't quite satisfied with the results. He wanted to send signals without wires, and at the time the only wireless transmission he could imagine was by light. On an earlier trip to England, Bell had seen that exposure to light changed how much current could pass through the element selenium. In early 1880, he realized that selenium might detect changes in a beam of light modulated by voices. He doubted the idea would prove practical but was nonetheless intrigued.[8] Thomas Edison also envisioned a wireless optical telephone about the same time but never built one, perhaps because he was more practical.[9]

Within a month, Bell demonstrated his "photophone," which operated by focusing sunlight onto the surface of a flat mirror that vibrated when moved by sound waves from voices (figure 7-2). The vibrations modulated the amount of sunlight reflected onto a selenium cell in a telephone receiver circuit, reproducing the voice. The excited Bell wrote: "I have heard articulate speech produced by sunlight! I have heard a ray of the sun laugh and cough and sing. . . . I have been able to hear a shadow, and I have even perceived by ear the passage of a cloud across the sun's disk."[10] He dispatched his first photophone to the Smithsonian Institution in a sealed box, and packed off another several weeks later after reaching a distance of 700 feet (213 meters). He proudly described his invention to the then-young American Association for the Advancement of Science at the end of August.[11]

Yet the photophone came to naught. Wires were messy, but they carried signals in fair weather or foul, day or night, farther than Bell could send light. He talked some about the idea in later years but never worked seriously on it again, and it was soon forgotten.[12] In the decades that followed, the only serious work on optical communications was by military agencies worried about enemy eavesdropping. Those systems never got very far.[13]

The Triumph of Radio

It was radio that filled the need for wireless transmission and it did the job so well that dictionaries still define wireless as another word for radio. The first radio signals were sparks spanning so many frequencies that they blocked each other, sometimes with disastrous consequences for ships frantically tele-

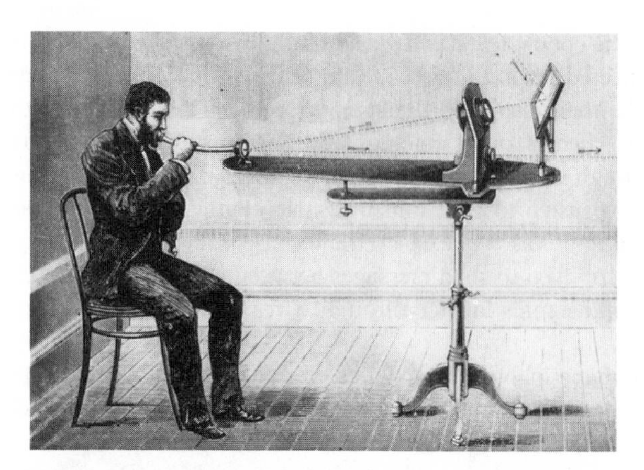

Figure 7-2: Alexander Graham Bell's photophone modulated reflected sunlight with sound vibrations, then detected the sound by using the light to illuminate a piece of selenium. It carried voices without wires, but not very far. (Courtesy Lucent Technologies)

graphing "SOS" for help. Progress came with the advent of electronic circuits tuned to oscillate at specific frequencies so that several transmitters could share the spectrum.

Radio began at low frequencies, then moved relentlessly higher. In the early 1920s, Hansell helped build the first radio transmitter to send voices reliably across the Atlantic.[14] It oscillated 57,000 times a second—57 kilohertz in radio parlance—generating what were called "long waves" because each one stretched 5.26 kilometers (3.27 miles). At night, both the ionosphere and the ocean reflect long waves, so they could carry whole speeches across the Atlantic, something not possible with earlier transmitters.[15]

Vacuum tubes soon made it possible to generate higher frequencies reflected better by the ionosphere. Soon after Hansell established the RCA Rocky

Point Laboratory in 1925, he built a 20-megahertz (million-cycle) transmitter emitting "short waves" 15 meters (50 feet) long. Not only did the signals reach South America in daytime, but the short waves did not require the gigantic antennas needed for long waves, so the short-wave transmitter cost a mere $15,000, compared to $1.5 million for a long-wave transmitter that didn't work as well.[16] International broadcasters and amateur radio operators still transmit short-wave signals around the world. Higher frequencies followed as electronic circuits improved, and radio broadcasts claimed chunks of the spectrum.

As electronic circuits improved, telephone engineers learned how to multiplex many voices, using each one to modulate a different frequency, and sending them all through the same wires. Multiplexed 24-phone channels require 24 times the bandwidth of a standard phone line, but it's much cheaper than stringing 24 separate pairs of wires. Further improvements in electronics allowed multiplexing hundreds of telephone channels. National telecommunications networks spread around the world, often carrying the voices for radio networks as well as telephones.

Increasing demand, and prospects for future television systems, pushed engineers to ever higher frequencies. Above about 10 megahertz, the only reliable radio transmission is in the line of sight, so Hansell's team at Rocky Point developed chains of radio towers to relay experimental television signals. They started at 80 megahertz and by 1939 had reached 500 megahertz.[17] Although they built the system for television experiments, the same equipment could carry radio and telephone signals.

Radio frequencies passed a billion hertz (a gigahertz) with the development of microwave radar during World War II. However, ordinary vacuum tubes were not fast enough to amplify gigahertz signals. Reaching those frequencies required complex new tubes, called klystrons, magnetrons, and traveling-wave tubes, that were much bulkier and costlier than ordinary thumb-sized vacuum tubes.

The continued spread of telephones and the advent of commercial television broadcasting put new demands on the postwar communications network. After the war, chains of microwave relay towers operating at a few gigahertz (billion hertz) sprouted across Europe and America. They were the biggest information pipelines money could buy in the 1950s, but the towers could be no more than about 50 miles (80 kilometers) apart or the Earth's curvature would block the microwave beam.

To bridge the Atlantic, an international consortium of public and private telephone companies turned to coaxial cable, a copper wire separated from a metal sheath by a plastic insulator, like modern television cables. Submarine telegraph cables had been in service across the Atlantic since the 1860s, but they did not need amplifiers. Telephone cables did, and only in the 1950s were electronic amplifiers up to the job of working on the ocean floor for the 25-year period needed to make the cables economical to operate. The consortium spent $37 million on undersea equipment to replace radio telephones dating from the 1920s and 1930s. AT&T supplied the cable from Scotland to

Newfoundland, while Standard Telephones and Cables built the shorter segment from Newfoundland to Nova Scotia. The new cable, called TAT-1, could carry 36 simultaneous conversations across the Atlantic when it started service in 1956. Work started almost immediately on a second transatlantic cable, linking Newfoundland to France.[18]

The growth of television and telephone traffic on land threatened to choke microwave relay towers in densely populated areas. Coaxial cables were an alternative where there wasn't room for microwave towers, but their capacity also was limited. Color television was coming; phone companies expected video telephones to follow. Communication networks needed a new generation of technology with even higher capacity. Most engineers thought the next logical step was higher microwave frequencies. Initially Alec Reeves was among them.

A Man Ahead of His Time

Born March 10, 1902, in Redhill, Surrey, Reeves began his career as electronics was starting its spectacular rise after World War I. Like Hansell, Reeves spent virtually his whole career at a single large communications company, a typical pattern for successful engineers of their generation. During that career, Reeves accumulated over a hundred patents.

Some engineers ponder with mathematical formulas, but Reeves had an intuitive feel for electronics. He sometimes built complex mechanical contrivances to demonstrate his inventions. His vision was not unerring and often ahead of its time, but the International Telegraph and Telephone Corporation recognized his gift and put him in charge of exploratory research at its Standard Telecommunication Laboratories north of London. His work was not day-to-day paperwork and administration, but developing ideas and programs. His influence was strongest behind the scenes, where one colleague recalls "he could always get someone to pull strings."[19]

Reeves had earned that influence. Working at ITT's Paris laboratory in 1937, he had devised an elegant and powerful technique to keep noise from building up in a long telephone line. He proposed measuring signal strength at regular intervals—eight thousand times a second—and converting that measured value to a number between 0 and 31, depending on its strength. Those numbers could be transmitted in digital form, as a series of five binary pulses separated by 1/40,000th of a second, and converted back into speech at the other end.[20]

The idea grew from a crucial insight. Electronic noise blends with signals transmitted in the continuously varying analog format of speech so the two can't be separated, and the noise becomes part of the signal. Add too much noise, and you can longer understand the words. However, electronics can recognize a series of digital pulses in a background sea of noise. Convert the pulses carrying the digitized sound back into an analog signal, and the noise is gone.

Engineers call that technique pulse code modulation, and today it's the standard way to transmit digitized voice and video signals around the world. Yet when Reeves filed his patent application, the state of the telephone art was big black Bakelite phones with rotary dials, connected by clattering mechanical switches and banks of switchboards into which operators plugged wires by hand. The first transistor was a decade in the future; the vacuum-tube electronics of the time were too slow to make Reeves's idea work. ITT never collected a penny in royalties on his patent, but they recognized his genius.

Reeves's colleagues fondly remember him as unfailingly open and honest, even "saintly."[21] He served as a scoutmaster, worked for charities, and tried to help delinquent boys. He also was eccentric. He worked from midday to 3 A.M., often in a spare bedroom that served as a home laboratory, and would call co-workers at 2 A.M. to share his bright ideas. He sometimes appeared at work wearing a tie in place of a belt, and a crocodile clip—used to clamp electronic wires together temporarily—as a tie clasp.[22] In his spare time, he experimented with the paranormal, trying to understand dowsing, mental telepathy, and psychokinesis.[23] He was not a true believer, but his restless mind liked to explore unconventional ideas.

Invading German troops chased Reeves from Paris in 1940. Back in England he was slow to embrace war work, but eventually devised the Oboe radar system, the most precise tool for guiding Allied bombers to their targets.[24] After the war, ITT moved him to its new Standard Telecommunication Laboratories, a subsidiary of Standard Telephones and Cables. At STL, Reeves developed a family of semiconductor devices, later eclipsed by transistors,[25] and remained heavily involved in military systems. He also looked to the future of civilian communications.

The Pipe Dream

By 1950, the telecommunications establishment thought the next logical step was to use microwave frequencies above 10 gigahertz, called millimeter waves because their waves are millimeters long. Wires were the local streets serving home and office phones; coaxial cables and microwaves were the main roads linking local telephone switching offices in urban areas. But long-distance traffic was growing steadily, and the industry wanted to replace its aging microwave relays with a higher-capacity intercity backbone network, like America wanted high-speed interstate highways to carry trucks and cars across the country. Multiplying the carrier frequency by a factor of 10 promised the extra communications capacity, as well as beams that could be focused more tightly toward relay receivers.

Unfortunately, the shorter the wavelength, the more often weather gets in the way. Radio waves an inch or more long don't see clouds, rain, and fog, but water droplets can block millimeter waves. That wouldn't work for phone companies; their customers don't want to wait until the rain stops to call long distance. They decided to shield millimeter waves inside hollow pipes called waveguides.

The workings of waveguides are more subtle than those of William Wheeler's light pipes or early optical fibers. Waveguides started as the sort of abstract problem that intrigues theoretical physicists facile with advanced calculus. They wondered what would happen to an electromagnetic wave inside various structures, such as long tubes made either of electrically conductive materials or of nonconductive insulators.

Waveguide behavior depends on "boundary conditions"—how the walls affect the electric and magnetic fields that make up radio waves, light, and other electromagnetic waves. Conductive metal walls reflect electromagnetic waves, so metal tubes guide waves along their lengths. Grind through the mathematics, and you find that waveguides don't work for wavelengths longer than a particular cutoff value. In essence, those waves don't fit inside, although the details are more complicated and depend on the waveguide shape. The minimum wavelength is half the wider dimension of rectangular waveguides, and a little longer for round ones. Filling the waveguide with plastic or something else more substantial than air increases the cutoff wavelength further.

This restriction meant that waveguides were strictly of academic interest in the early days of radio. A 100-megahertz signal has waves three meters long, so it would require an impractical 1.5-meter (5-foot) waveguide. Only when engineers pushed frequencies to several gigahertz, where wavelengths are ten centimeters (four inches) or less, did waveguides become practical. The technology spread quickly with the development of radar during World War II, and with postwar advances in microwave communications, making waveguides seem attractive for the new generation of high-capacity long-distance systems.

In America, Bell Telephone Laboratories settled on circular hollow waveguides with inner diameter of five centimeters (two inches) to carry signals at 60 gigahertz, with wavelength of five millimeters (0.2 inch). In 1950, management put Stewart E. Miller in charge of millimeter waveguide development.[26] In England, Harold E. M. Barlow, a professor at University College London, proposed a slightly different circular millimeter waveguide.[27] The British Post Office, which ran the country's phone system, began work at its Dollis Hill Research Laboratory in London. Standard Telecommunication Laboratories quickly followed, the first British company in the field.

The technical challenges didn't daunt mid-century providers of telephone service. They were government or private monopolies, and regulations assured that customers or the government would pay the bill. AT&T led the way in America, generously funding basic research at handsome Bell Labs facilities scattered about suburban New Jersey. In 1956, Bell Labs put the new technology to the test at its Holmdel development lab, burying 3.2 kilometers (2 miles) of a millimeter waveguide made by embedding a tightly wound coil of copper wire in protective plastic. The experiment confirmed one growing concern—signals leaked from any kinks or bends, even small ones caused by uneven settling of the soil. Like a high-speed railroad line or interstate highway, it required broad, sweeping curves. That would make installation costly, but AT&T could accept that. Waveguides promised tremendous capacity, and they were to run mostly between cities, not within them.

After two years of tests, Bell Labs settled on a design for 50.8 millimeter (2-inch) waveguides each carrying 80,000 conversations at frequencies between 35 and 75 gigahertz. The signals would be digitized and transmitted by pulse-code modulation, as Reeves had proposed 21 years earlier.[28] The overall data rate would be a then-staggering 160 million bits per second.

Meanwhile, Bell Labs was also pursuing another long-distance alternative, the communications satellite. Arthur C. Clarke, a British engineer and writer, had come up with the idea during World War II as an alternative to radio relays and coaxial cables. Both of those systems required chains of repeaters to span long distances. However, Clarke realized that a satellite with an orbit lasting exactly one day would stay continually over the same place on the equator, so a transmitter on board could relay signals between any two points on its side of the Earth.[29] John R. Pierce, a top Bell Labs communications engineer who also worked on millimeter waveguides, picked up on the idea in the 1950s and pushed it as the Space Age emerged. Like Clarke, Pierce had published science-fiction stories, but Pierce was primarily an engineer and saw the practical potential of satellite communications, which became the first important civilian use of space technology.[30]

Standard Telecommunication Labs concentrated on millimeter waveguides, but Reeves was not impressed by early trials. Congested Britain didn't have room to bury pipes with sweeping curves, small irregularities caused disturbing losses, and no technology was available for the amplifiers needed to compensate for the inevitable losses. He didn't like the costly, brute-force approach, so he considered a bolder alternative—moving all the way to light. The gap between microwave and optical frequencies is a factor of 100,000. That seemed overwhelming to most, but Reeves realized the difference was the same as the gap from long waves to microwaves he had seen crossed in his 35 years of engineering. He had a hunch light would work better, and he was a man who listened to his hunches because they often were right.[31] Reeves hoped that mental telepathy might be the ultimate communications technology,[32] but he didn't know how to tame that, so light would have to suffice.

A Pioneering Effort

Reeves knew little about light until STL landed a military contract in the field in 1952. The contract required a working knowledge of optics, so he hired Murray Ramsay, a young physics graduate from University College London, who had been a scout in Reeves's troop before the war. The ever-curious Reeves pumped the young Ramsay for information and pondered the prospects for optical communications in his smoke-filled office at STL. He tested ways to modulate and guide light from special high-performance lamps.[33] About 1958, he assembled a small team to study optical communications, including Ramsay, an older engineer named Charles C. Eaglesfield, and four others who reported to Len Lewin, a senior manager.[34] Reeves monitored their work and added his own ideas, although he had many other projects to distract him.

The millimeter waveguide project continued under Lewin, developing a 7-centimeter (2.8-inch) waveguide Barlow had proposed. In 1958, Lewin put that project under Antoni E. Karbowiak, a microwave engineer who had earned a doctorate under Barlow. Born in 1923 in Poland, Karbowiak fought with British troops during World War II, earning British residence and an education.[35] Quiet and reserved, he combined a mastery of mathematical waveguide analysis and a fertile imagination with a solid physical intuition missing in many theorists.[36] His grasp of advanced mathematics complemented nicely Reeves's less mathematical intuition, enthusiasm, and drive.

Working in the same department as the millimeter waveguide project, the little optics team began thinking of optical waveguides. That was an innovation. Since the days of Alexander Graham Bell, most people had automatically assumed optical communication would go through open air. Yet living near London, notorious for its murky smogs, the STL team needed only look out their windows to see the problems of sending light through the atmosphere.

A few people had had similar ideas before, but none had gotten far. Both Bell Labs[37] and RCA[38] had patented schemes for sending light signals through transparent rods or hollow pipes, but neither did anything with the idea.[39] At the end of the war, R. V. L. Hartley, a Bell Labs scientist, concluded that transparent rods did not transmit light well enough for communications, and that hollow reflective metal pipes were too sensitive to bends.[40] Those were reasonable conclusions in 1945, but times were changing.

Light Pipes

At STL, Eaglesfield proposed a disarmingly simple idea: an optical "pipeline" of one-inch steel pipe coated inside with silver, the most reflective metal available. "It is a little strange that this subject has received apparently no published treatment," he wrote,[41] evidently unaware of the patent William Wheeler had received 80 years earlier.

On paper the idea looked good. If light passing down the pipe spread out at an angle of no more than half a degree, in theory only about 0.05 percent would be lost at each reflection. Eaglesfield calculated the loss in terms of decibels, a logarithmic scale handy for measuring loss because you can add the decibel losses of two segments to get total loss, or multiply the transmission distance by the loss per unit length to get total loss. The lower the loss in decibels, the better the transmission line (see box, pages 115–116). Eaglesfield calculated that loss should be 2.5 decibels per mile, meaning 56 percent of the light that entered a one-mile length would emerge from the other end. (In metric terms, loss was 1.6 decibels per kilometer, and 70 percent would emerge from a one-kilometer length.) That was about 10 times better than the theoretical loss of Wheeler-style light pipes under the same conditions.[42] Eaglesfield estimated his optical pipeline could carry light around very gradual bends, with a radius of about half a mile (0.8 kilometer).

To test the idea, STL assembled 35 four-foot (1.2-meter) segments, coated inside with epoxy resin to provide a smooth base for the silver film. After considerable trouble joining the pipes, STL technicians stretched them along eight concrete posts sunk deep into the ground to give a sound footing. The assembly was within 1/16 inch (1.6 millimeters) of being perfectly straight, but the results were disappointing. Eaglesfield had predicted 97 percent of the light should emerge after a 276-foot (84-meter) round trip, but the measured amount was under 9 percent, corresponding to a loss of more than 200 decibels a mile.[43] That meant that only 10^{-20} of the input light would have emerged from a mile-long pipe. Eaglesfield complained that the measurements didn't do his idea justice, and a few years later Czech engineers did somewhat better.[44] However, Reeves held out little hope and went looking for other ideas.

One came from Ramsay, who suggested arranging a series of lenses along the inside of a pipe such that each one focused an image of the previous lens onto the next lens (figure 7-3). Such a "confocal lens waveguide" could relay light along the pipe such that none was lost by hitting the sides.[45]

The little team also began looking at optical analogs of a less common kind of microwave waveguide made out of a nonconductive material like glass or plastic. The edges of such "dielectric" materials also can guide electromagnetic waves. The materials absorb microwaves, so thick rods that guide microwaves inside themselves do not work well. However, in the late 1940s, engineers at RCA Laboratories in Princeton found they could do much better with plastic rods thinner than about a quarter of the microwave wavelength.[46] In that case, most of the microwaves travel along the *outside* of the waveguide, not inside where the material can absorb it. That means that thin plastic waveguides have very low loss, but only if they are perfectly straight; like many other waveguides, they radiate energy at bends. This was a serious practical problem at microwave frequencies, but it didn't keep STL engineers from considering making dielectric waveguides of transparent materials for optical communications.

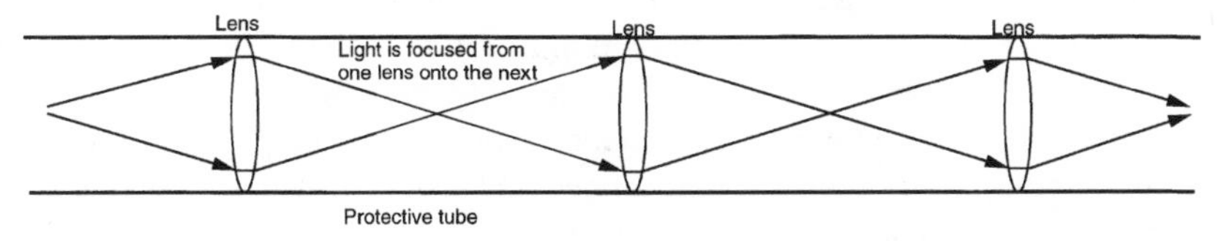

Figure 7-3: A confocal waveguide guided light from lens to lens without hitting the walls of the pipe.

Fibers as Dielectric Waveguides

To a theorist, an optical fiber is a dielectric waveguide for light. The process classical optics sees as total internal reflection is the optical equivalent of the process that guides microwaves along the inside of a thick plastic rod. Brian O'Brien probably was the first to recognize an optical fiber as a waveguide, but he never settled down to document the idea. In practice, it didn't matter much as long as the fiber core was much bigger than the wavelength of light. The traditional optical view of total internal reflection works, and the concept is simpler.

Differences arise when fiber cores are shrunk close to the wavelength of light, restricting the number of modes, or paths the light can follow through the core. That didn't happen until the late 1950s, when Will Hicks wondered how fine he could stretch optical fibers in a fused bundle. It was a natural experiment to try, and one with practical import because the core size limits resolution of a fiber bundle (bundled fibers use very thin claddings). The smaller the cores, the finer the details you can see. As Hicks shrank the cores, he saw a strange phenomenon: Geometric patterns and different colors began to appear in individual fiber cores. He eventually decided it must be a waveguide effect but didn't settle down to document it before he quit American Optical.

The topic was still a hot one at American Optical in early 1959, when Steve MacNeille, Walt Siegmund, and Lewis Hyde interviewed Elias Snitzer over lunch. Siegmund pulled out a photograph and asked the young physicist if he could explain it. Snitzer asked what it was and grew excited after Siegmund said it was very fine fibers. "That's waveguide modes in the visible region of the spectrum. I don't believe anybody's ever seen that before!" His response passed the test, helping convince MacNeille to hire him despite a blot on his record. The Lowell Institute of Technology had recently fired him because he refused to cooperate with an investigation of left-wing student politics by the House un-American Activities Committee.[47]

Snitzer recognized the phenomenon because he had worked on microwave systems. It appears when core diameters approach the wavelength of light, leaving only a few modes, which form curved geometric patterns inside the fiber cores. The patterns and colors varied from fiber to fiber because slight differences in core shape and diameter gave them different mode patterns. Shrink the fibers far enough, and only one mode remains. Once he started work at American Optical, Snitzer teamed with Hicks to experiment on fiber modes, analyze their structure in detail, and report the results.[48]

American Optical was not in the communication business, but Snitzer and Hicks knew microwave waveguides were used in telecommunications. Before they published anything, they applied for a patent, suggesting that the ability to transmit light in "separate well-defined and readily detectable and distinguishable electromagnetic modes . . . may be used advantageously in various ways" to transmit "data, information, signals and the like."[49] They didn't worry too much about how far their fibers transmitted light; their main concern was controlling the mode structure, because that was important in microwaves.

Word about fiber modes spread quickly. Narinder Kapany duplicated the experiment, and *Scientific American* used his photograph of the effect on its November 1960 cover.[50] However, the idea of fiber optic communications stalled in America. Kapany's article didn't mention communications. Hicks had a business to run; he turned to other ideas after a rough calculation indicated atoms would scatter too much light for practical communications even if the glass was perfectly clear.[51] American Optical lacked the clean room needed to make ultrapure materials, and Snitzer's attention was diverted to a hot new idea—the laser. That left the matter of fiber-optic waveguide communications to Alec Reeves and the others at STL.

8

The Laser Stimulates the Emission of New Ideas

(1960–1969)

Usable communication channels in the electromagnetic spectrum may be extended by the development of an experimental optical-frequency amplifier announced by Hughes Aircraft a few days ago.

—*Electronics* magazine, July 22, 1960[1]

Eli Snitzer was not the only person distracted by the laser. Its invention was big news in 1960. Military planners and science-fiction fans saw it as the ray gun of their dreams, but engineers and physicists recognized it as the first optical oscillator, and the first practical source of coherent light.[2] They immediately recognized its development as a milestone on the road to optical communications, although it was far from clear where the road was going.

Radio engineers use oscillators to generate the pure carrier frequencies that are modulated to transmit radio signals. Radio oscillators drive antennas so they radiate coherent radio waves, which share the same frequency and stay in phase with one another, like a troop of identical soldiers marching in step on parade. That frequency must be pure so that radio receivers can essentially cancel it out to recover the transmitted signal.

When Reeves started looking at optical communications, there were no practical sources of coherent light. Light bulbs, stars, and virtually all other common sources emit light spanning much of the spectrum that spreads unsynchronized across the universe like a crowd leaving a baseball stadium after the game is over. You can switch a light bulb off and on, but it doesn't generate a very "clean" signal. To realize the tremendous theoretical capacity of optical communications, you need an oscillator that generates light. Back in 1951, Bell Labs had concluded that the only way optics could match the capacity of the millimeter waveguide was with a coherent light source.[3] With no coherent light in sight, Bell Labs pursued the millimeter waveguide. The emergence of the laser got Bell Labs and many others to take optical communications seriously.

First-Generation Lasers

The laser was years in the making, and like many other new ideas of the time, it grew from microwave research. Charles H. Townes, then a physics professor at Columbia University, took the first step in 1951 when he realized how to make a new type of coherent microwave oscillator. He called it the "maser," for microwave amplification by the stimulated emission of radiation.[4] His idea was to collect a group of molecules excited so that they possessed extra energy, and stimulate them to emit that extra energy in the form of microwaves. Three years later, he had a working maser.

Other masers followed, and Townes realized he could expand the principle to make an optical oscillator. He teamed with his brother-in-law Arthur L. Schawlow,[5] then working at Bell Labs, to work out the details. Their theoretical proposal[6] started a race to build the device, later christened the laser. On May 16, 1960, Theodore Maiman won the race by firing pulses of red light from a small ruby cylinder at Hughes Research Laboratories in Malibu, California.[7] After the editors of *Physical Review Letters* summarily rejected Maiman's paper reporting his discovery,[8] he fired a 300-word letter to *Nature*,[9] and Hughes called a press conference on July 7. Some observers were skeptical, but Schawlow soon duplicated Maiman's feat from newspaper accounts. The laser era was off and running.

Electronics magazine put communications at the top of its list of potential uses for the new invention.[10] Laser beams were focused much more tightly than microwaves (a consequence of their shorter wavelength), so they looked promising for sending signals long distances through space or the atmosphere. Rudolf Kompfner, head of transmission research and director of Bell's Crawford Hill Laboratory, was quick to see the possibilities. In October, *Electronics* showed two Bell Labs scientists firing their new ruby laser through 25 miles (40 kilometers) of clear air.[11] In reality, the laser fired only one pulse at a time, so it could not send useful information, but the experiment showed that the potential was there. So was the demand for communications, as the microwave spectrum and existing cables were filling. Millimeter waveguides were set to be the next technological generation, but lasers might follow them on the ground, and perhaps replace microwave links in space. Looking farther into the future, Townes suggested interstellar communications might be possible with highly directional, high-power laser beams.[12]

First, however, scientists needed better lasers.

Maiman used ruby because he understood its properties, but it is not an ideal laser material. It only fires pulses, and converts only a small fraction of the input energy—supplied by a flashlamp—into laser energy. Bell Labs wanted a laser that oscillated continuously, emitting a steady light beam that could be modulated with a signal like a radio carrier frequency. Bell had a young Iranian-born physicist hard at work toward that goal, Ali Javan, who had studied under Townes at Columbia.

Javan was trying to generate a laser beam by passing an electric current through a gas. Others used the vapors of alkali metals like sodium and potassium as the gas, but Javan picked the rare gases helium and neon, which are simpler to study and much easier to handle. He filled glass tubes with helium, to capture energy from electrons passing through the gas, and a dash of neon, which borrowed energy from the helium and turned it into light. He mounted highly reflective mirrors on both ends, with one allowing a small fraction of light to escape. On a snowy Monday afternoon, December 12, 1960, he was elated when his helium-neon laser emitted an invisible infrared beam at 1.15 micrometers.[13] It was the first laser to emit a continuous beam, and the first laser to operate in a gas. Management, which was growing impatient, thought it was about time.

The helium-neon laser proved as important as Javan hoped; it soon became the standard laboratory laser and remains the most common gas laser. However, it did need some modifications. Other Bell Labs scientists developed a version that emitted at 633 nanometers in the red part of the spectrum. That was much better for communication experiments, because the beam was visible as well as stable, and very coherent. Output powers were milliwatts to tens of milliwatts, fine for research; external devices could modulate the beam by changing their transparency over time. Few developers thought the helium-neon laser was ideal for communications, but they didn't have ideal millimeter-wave sources either, and they thought laser technology had plenty of time to grow.

If they had an ideal laser, it was the solid-state semiconductor type, first demonstrated in the fall of 1962. The transistor age was on a roll; semiconductors were swiftly replacing vacuum tubes in electronic circuits. But doubts remained about how far semiconductor technology could go, and progress on semiconductor lasers soon stalled, leaving the best devices able to operate for only a short time at the −196°C (−321°F) temperature of liquid nitrogen. A few were used in experiments, but helium-neon lasers remained far more practical.

Clouds in the Picture

Communications engineers started sending laser beams through the atmosphere as soon as they got their hands on lasers. Air looked like an ideal transmission medium; the eye could see it was clear, and it carried radio and microwave signals easily.

Experiments soon revealed it wasn't quite as simple as pointing the pencil-thin laser beam at a distant target. A couple of Bell Labs engineers hauled an early ruby laser to the top of a microwave tower at Murray Hill, New Jersey, and aimed it at the Holmdel lab 25 miles (40 kilometers) away, where a third engineer watched a movie screen for signs of the red pulses. They hooked up phone lines, and the two in the tower phoned their partner in Holmdel each time they fired the laser. "He didn't see the pulse very often," recalls one.[14]

Helium-neon lasers followed as soon as they were available. Bell put one on the roof of its main development building in Holmdel and another atop the smaller and more research-oriented Crawford Hill Lab 1.6 miles (2.6 kilometers) away. In clear weather, the red laser spot spread as wide as a dining-room table on its journey and glowed like a fireplace.[15] Yet thick New Jersey fogs blocked the beam when they rolled in from the shore.[16] So did rain, snow, sleet, and haze. It should not have been a surprise, but no one had thought about it. Reference books gave no warning because scientists always had measured air transmission in clear weather. No one had systematically studied how weather affected light transmission, much less laser beams, and most Americans had yet to recognize air pollution.

Poor transmission in bad weather wasn't a showstopper for all laser communications. Some companies merely wanted to send laser beams between buildings on opposite sides of a street; their signals didn't have far to go, and they could wait if they had to. Nor did it discourage NASA or the Air Force from thinking of laser communications above the atmosphere. They had money and energy to burn in the salad days of the space race, and hoped tightly focused laser beams would be both more efficient and more secure than microwaves. (They eventually proved too narrow to hit distant receivers reliably.)

However, unreliable transmission was a big issue for phone companies that made network reliability a matter of pride—especially when convincing regulators to approve expensive new projects. AT&T wanted optical communication systems to be out of service no more than one hour per year.[17] Early tests showed that would not be easy in open air. Fog, rain, or snow could attenuate a laser beam by more than a factor of one million over the 2.6 kilometers between Holmdel and Crawford Hill.[18]

British engineers, accustomed to murky air, were quicker to recognize the problem. One military engineer bluntly told a 1964 conference: "The atmosphere is completely inimical to laser transmission systems."[19]

Bell Labs didn't give up as easily on air. As soon as they got the first high-power lasers, Bell researchers used them to burn holes through fog, but new fog filled the holes as fast as the laser beam opened them.[20] Fortunately, the telephone monopoly had vast resources and applied some of them to an alternative, optical counterparts of microwave and millimeter waveguides.

Waveguides for Light

American research on optical waveguides began at the Army Electronics Command in Fort Monmouth, New Jersey, about a half-hour drive from Holmdel. It was the brainchild of Georg Goubau, a former engineering professor at the University of Jena in Germany, who came to America after World War II with German rocket scientists as part of Operation Paper Clip.[21]

Goubau began with what he called a "beam waveguide," in which uniformly spaced lenses or iris-like openings guided a beam of coherent microwaves along a pipe. It was a subtle innovation that worked differently than the usual waveguide. When people started talking about lasers in the late 1950s, Goubau realized his concept also should work for coherent light. He first tested it with millimeter waves in 1958, before any lasers were available.[22] When the laser arrived, he shifted his attention to the different scale of optical wavelengths. (The version with lenses was equivalent to Ramsay's confocal lens waveguide in England.)

For his optical experiments, he built a beam waveguide inside six-inch (15-centimeter) aluminum irrigation pipe that hung above ground. It stretched 970 meters, over half a mile, "the total length . . . of available real estate."[23] Ten internal lenses relayed light from one lens to the next inside the guide, but the beam did not follow the straight path it was supposed to. On clear days, sunlight heated the pipe, creating a temperature gradient that bent the beam off course. Army engineers tried mounting a smaller plastic pipe inside the metal pipe, but eventually despaired, "it still appears questionable whether operation during sunlight is possible or whether further shielding of the light channel is required."[24] Over the next couple of years, they managed to get up to 80 percent of the optical signal through the pipe by pumping air out of the tube and carefully aligning lenses to within 0.1 millimeter.[25] However, they couldn't maintain that demanding alignment very long.[26]

Meanwhile, the idea migrated down the road to Crawford Hill, where it caught the eager eye of Kompfner. Born in Austria in 1909, Kompfner was one of those rare engineers gifted with both technological vision and management skills. Originally trained as an architect, Kompfner emigrated to England in 1934 and worked there until interned as an enemy alien at the outbreak of World War II. The pragmatic British moved him to the other side of the barbed wire when he showed his engineering skills, and he justified their deed by inventing the traveling-wave tube, an important microwave amplifier. He came to Bell in 1951, where he founded the Crawford Hill lab—using his architectural training to design the building.[27] Dynamic and outgoing, Kompfner was a charismatic figure who freely scattered the seeds of new ideas.[28]

He had divided transmission research at Crawford Hill between two groups, each of about 30 people. Leroy Tillotson headed one on atmospheric transmission, which tested laser links through the air as well as microwave relays. Stew Miller headed guided-wave research and had been developing millimeter waveguides since 1950. Kompfner added optical waveguides to his charter.

It was a logical choice on a couple of levels. There are inherent similarities in the physics of millimeter and optical waveguides, so skills should be transferable. Moreover, millimeter waveguide development had reached an awkward stage in late 1962. Bell had made reasonable progress on the waveguides themselves, but they suffered some losses, so signals would have to be amplified as they crossed the wide open spaces of America. Because millimeter waveguides were expensive to build and install, they would have to carry a tremendous signal volume, so the amplifiers would have to handle signals with a bandwidth of 11 gigahertz. That was beyond the state of the art for amplifiers. Special vacuum tubes could reach that frequency, but they didn't last long enough for practical use. Solid-state semiconductor electronics were much more reliable, but they had yet to come close to the required frequency. Miller's group had done their job; somebody else would have to develop the amplifiers. While AT&T waited for semiconductor researchers to develop fast electronics, they shifted the millimeter waveguide project to another group,[29] and shifted Miller's group to optical waveguides in 1963.

Like most electrical engineers at the time, Miller had little experience with light. Nearly a decade younger than Kompfner and much more reserved, Miller had a solid record of technical innovation in radio transmission, waveguides, and coaxial cables. He joined Bell Labs after receiving a master's degree from MIT in 1941, but never finished the doctorate that was a hallmark of the research staff. Adept at working within the AT&T corporate bureaucracy, he rose up the management ladder, but retained a keen eye for the technology.

Like Goubau, Miller's group used their mastery of arcane mathematics to develop theoretical models of optical waveguides. Bell Labs soon took the lead in the field; unlike the Army lab, they had a clear application for optical guides. They also hired Detlef Gloge, a young West German who as a graduate student tested similar lens waveguides in old heating tunnels under the University of Braunschweig.[30]

A major worry with confocal waveguides was the fact that every lens surface inevitably reflects some light. Sit in a lighted room at night and you can see reflections on the dark windows. Clean, uncoated glass reflects 4 percent of the light that hits the surface from air, but coatings can reduce that reflection if they have a refractive index between that of air and glass. Kompfner had to make some very optimistic assumptions when he predicted that a confocal-lens transmission line could carry signals 650 kilometers (400 miles) before they had to be amplified.[31] To test those predictions Bell Labs buried a waveguide with six lenses spaced along 840 meters (2750 feet). Gloge bounced a laser beam back and forth 150 times through it and was happy to find that the beam quality remained good, and the loss was only about 9 percent (0.4 decibel) per kilometer.[32] However, meeting Kompfner's goal required reducing loss below 0.08 decibel (1.8 percent) per kilometer.

On paper, confocal waveguides offered tremendous capacity. External modulators could vary the strength of the laser beam at high speeds, although the technology was not well developed. In addition, their large diameters held the potential of carrying many separate beams. Gloge calculated a 20-centimeter (8-inch) guide could carry some 300 separate beams.[33]

Unfortunately, there were a plethora of practical problems. Temperature variations, ground vibrations, and mechanical instabilities perturbed the beam even when the waveguide was buried to isolate it from the environment, so low loss was very hard to maintain. Bends are inevitable on any communications route, but they presented serious problems because bending the waveguide required moving the lenses much closer together than straight segments. The more lenses, the more their reflection losses accumulated, even when reduced to about 0.5% per surface. The bends had to be gradual; it took about ¾ kilometer (2500 feet) of lensed waveguide to turn 90 degrees, and even then only 1/1000th of the light got through, a loss of 30 decibels. Engineers hoped to reduce that loss by replacing lenses with pairs of focusing mirrors like those used in some periscopes, but the mirrors posed other practical problems.[34]

Gas Lenses

Meanwhile, Kompfner went looking for help on the problem of surface reflection. His typical approach was to scatter ideas among other Bell Labs scientists, trying to catalyze them to make innovations. He mentioned the problem to managers at the Murray Hill lab, whose mission was basic research. One of them mentioned the idea to a young scientist working for him as they stood talking in the hall. Remembering how mirages form, Dwight Berreman suggested making lenses of air.[35]

Temperature gradients in air bend light to form mirages because of a simple principle of physics. Heat a gas and it expands, thinning out; cool it, and it becomes denser. The refractive index of air changes with its density, so heating and cooling change its refractive power. You can see the effects if you look over a stretch of asphalt road on a hot sunny day, or across the hood of a hot car. Rising pockets of hot and cool air bend light from distant objects back and forth, so they look rippled. Berreman thought the same effect could make air in a tube focus light like a lens. Heating the walls of the tube would warm gas near the walls, making it expand, while the gas in the middle stayed cool. The resulting gradient in density would cause a gradient in refractive power, focusing light toward the middle of the tube like a lens. Unlike a lens, it would have no surface to cause reflection losses (figure 8-1).

Berreman's department was not working on optical communications, but Murray Hill let its scientists play with new ideas. He built a couple of short gas lenses, satisfied himself that they worked, and wrote up the results before going on to other projects.[36]

At Crawford Hill, Stew Miller embraced gas lenses wholeheartedly, as a welcome way to circumvent troublesome surface reflection. In fact, Miller liked them considerably more than Berreman did after giving his idea a more careful second look. Berreman found that slight fluctuations tended to make

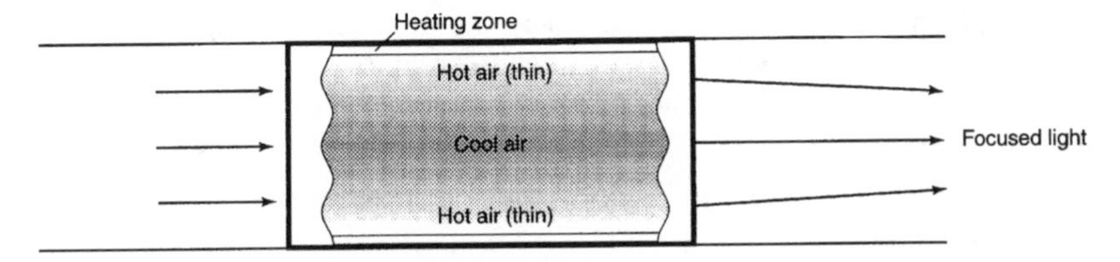

Figure 8-1: Heating the sides of a hollow tube thins the air along the sides, so its refractive index is lower than that of the cooler gas in the middle. This refractive-index difference makes a gas lens focus light.

the beam wander off the center of the gas lens, and once it slipped off center, it kept on going. "The only way to recenter it was by brute force," he recalls, so he wrote off the gas lens waveguide as "a dead duck."[37] He predicted long chains of glass lenses would suffer similar centering problems. An interactive computer system that sensed beam motion and moved the beam back on target could solve the beam wander problem, but that seemed a forbidding task at a time when a computer filled a whole room. Berreman wrote a paper detailing the problem, but Miller would hear none of it. He sat on the negative results until one of his own analytical wizards came up with essentially the same answer.[38]

Miller put a new man on gas lenses, Peter Kaiser, who took a new approach. Kaiser blew cool gas into a hot tube about 6 inches (15 centimeters) in diameter, so the tube heated gas near the walls to focus light. To bend light around troublesome curves, he placed vents 30 inches (75 centimeters) apart. Adding more vents focused light more sharply. He placed about 80 gas lenses in a demonstration waveguide that ran 200 feet (60 meters) down a Bell Labs corridor, and found loss was too low to measure. However, the apparatus was elaborate, and he had to use argon rather than air.[39]

Those problems did not discourage Miller and Kompfner, who could see a role for laser communications in the future that AT&T could carefully plan as a regulated monopoly. Its Picturephone video-telephone was set to debut in 1970, and AT&T had planned its evolving network around the new service, even designating the # key on push-button phones to signal video calls.[40] The company expected Picturephone to spread steadily but not spectacularly, reaching 100,000 sets in 1975 and a million in 1980. Millimeter waveguides would provide the extra long-distance capacity to handle the early years of that growth, but AT&T expected to need the tremendous capacity of optical waveguides would be needed once Picturephones became commonplace, probably after 1990.

Bell expected either gas or glass lens waveguides to be so elaborate and expensive that they would have to carry at least a million telephone circuits to be economical.[41] That was not a showstopper for the world's biggest and richest telephone company. Thanks to regulations that assured the company a return on its investment, AT&T had ample money to spend on both approaches, as well as on other research even less likely to generate near-term profits. Nor were research managers deterred by technical difficulties; millimeter waveguides had come a long way in the past dozen years. "Today there are probably more physicists and engineers working on the problem of adapting the laser for use in communication than in any other single project in the field of laser applications," Miller wrote in the January 1966 *Scientific American*.[42]

Checking Long Shots

Bell Labs also searched for long-shot alternatives to confocal or gas-lens optical waveguides. One idea was making hollow waveguides from nonconductive dielectrics or reflective metals. The numbers looked good for 0.25-millimeter tubes if they were perfectly straight—a third of the light would be lost in a kilometer-long metal tube, and just a little more would be lost in a dielectric. However, bending caused serious problems. Loss of a metal guide doubled if it was curved over 48 meters (157 feet), comparable to a freeway off-ramp. The dielectric guide was much worse, with loss doubling for a 6-mile (10-kilometer) bend.[43] Those numbers, and the difficulty of making thin, perfect tubes, stopped that line of research.

Optical fiber was hard to ignore. Kapany's *Scientific American* cover story on fiber optics appeared as Javan closed in on the helium-neon laser. Bell had its own resident fiber expert, Jeofry Courtney-Pratt; he didn't work in communications, but when he saw a 1961 paper[44] that described a fiber as a waveguide, he passed it along to Miller.[45] Miller was intrigued by the idea of optical analogs of solid plastic microwave waveguides.

However, the reality was daunting. Kompfner cut to the heart of the issue by asking a simple question: How clear are the best glasses? Calls to glass manufacturers and trips to the reference library yielded similarly discouraging answers. The clearest glasses had attenuation of at least one decibel per meter. That meant that 20 percent of the light entering a fiber was lost going the width of a desk. That was adequate for an endoscope, which need only reach into the stomach. It was hopeless for communications. Go 4 meters (13 feet), the width of a typical room, and you lose 60 percent of the light. Go a hundred meters—the length of a football field—and only one ten-billionth (10^{-10}) of the light remains. It was no wonder Kompfner said "forget it."[46]

The door at Bell Labs stayed closed for years. When Kompfner outlined optical communications research in 1965, he shrugged off optical fibers because "numerous serious problems" remained unsolved.[47] In mid-1966, Miller and Roy Tillotson reviewed optical communications for the technical journal *Applied Optics*, but said nary a word about fibers.[48]

They saw the millimeter waveguide as the next generation of communications technology. By 1966, Bell Labs had buried several miles of experimental millimeter waveguide at Holmdel and was designing solid-state repeaters. A single two-inch waveguide was designed to transmit 50 channels each carrying 281 megabits, roughly a total of 15,000 million bits per second. The millimeter waveguide was "at least four years from commercial service," Miller declared in 1966. Yet he added, "No new inventions or fundamental advances in techniques or materials are needed to make it a technically practical system."[49]

A Growing Fiber Industry

The fathers of fiber-optic imaging virtually ignored laser communications. They had their hands full with building an industry around other fiber-optic applications.

Brilliant, hardworking, and strong willed, Will Hicks built Mosaic Fabrications into a thriving business making fiber-optic faceplates for military image intensifiers. He was a leader, determined to steer his own course, with the rare gift of technological charisma that dazzled those around him. Demand rose as the technology improved and the Vietnam war escalated. The wily Hicks drove off potentially troublesome competitors like the Corning Glass Works.[50]

By the mid-1960s, Mosaic occupied a handsome new building paid for by hefty profits from military contracts. Everything was going fine until the Pentagon decided Mosaic was doing too well at government expense, and took action to recover the excess profits. Hicks had put most of the profits back into the business. Lacking cash to repay the government, he sold the company for several million dollars to Bendix, a big aerospace firm, and split the proceeds with stockholders. Bendix signed him to a five-year contract, but he lasted less than a year, unwilling "to put up with somebody else's nonsense"[51] after years of running his own business. Restless, Hicks bought a restaurant and got involved in civil rights projects. His children lectured him on the evils of the Vietnam war, and he grew angry with himself for ignoring how the military used his faceplates. Today we'd call it his mid-life crisis.

American Cystoscope Manufacturers started selling fiber-optic endoscopes in 1960 and soon had a hit on its hands. The company initially expected to sell some 2000 instruments over the 17-year life of Hirschowitz's patent. They quickly sold 2000 a year, as younger doctors turned en masse to flexible fiberscopes. Larry Curtiss developed instruments in which fibers carried light into the body from an external bulb, giving physicians 10 times more light and greatly easing prostate surgery.[52] His one-semester leave from graduate school at Harvard became permanent.

Fiber-optic fever spread through the medical community, and word of the new invention eventually reached a middle-aged surgeon in Harlingen, Texas, a small city near the southern tip of the state. Heinrich Lamm, who had settled there in the 1930s, dug into his files for the tattered preprint of the paper he had written 30 years earlier and translated it into English for those who didn't read German. "I share the fate of many who had a good idea and could not carry it to fruition," he wrote to the author of an article in the *Texas State Journal of Medicine.* "I am nearly certain there is no previous published report of an image transmitted by a bent fiber bundle."[53]

Word must also have reached C. W. Hansell, who moved to RCA Laboratories in Princeton after the company shut the Rocky Point Lab in 1958, and retired in 1963 to Florida. Yet no one remembers him mentioning fiber optics before he died in 1967.[54] Perhaps he felt no need to boast after collecting over 300 other American patents, and helping develop FM radio, color television broadcasting, radar systems, microwave relays, and high-speed aircraft communications.[55] Perhaps he had forgotten an idea that seemed to go nowhere nearly four decades earlier.

The fiber-optic illuminator became standard equipment in the card readers that processed the punched "IBM cards" ubiquitous in computer centers through the 1960s. American Optical landed the contract to make them, generating healthy profits that kept its fiber-optic division going. Eli Snitzer and some of the company's other physicists turned to lasers, starting a glass laser group.

In California, the energetic Narinder Kapany pushed Optics Technology into laser development as well as fiber optics and contract research. Polished, charming, and articulate, Kapany has the charisma of a scholarly businessman. With turbaned head and bearded face over a well-tailored suit, he cut a striking figure in the board room. The company built one of the first lasers used in eye surgery[56] and became one of the first to mass-produce helium-neon gas lasers.[57] Full of ideas, Kapany led the company to reach sales of $2.4 million and profit of $140,000 in 1968. However, the company spread itself too thin—although never into communications—and soon began losing money.[58]

Laser research boomed although cynics called the laser "a solution looking for a problem." Communications was only one small area of laser research, and optical communications usually meant sending laser beams through air or space. With the space race in the headlines, Bell Labs' work on optical waveguides drew little attention. America led the world in developing lasers and optical fibers, but it virtually ignored the idea of fiber-optic communications.

9

"The Only Thing Left Is Optical Fibers"

(1960–1966)

> It may well be that what appears an impractical approach today may prove to be a success tomorrow, as a result of developments in materials and/or techniques. . . . We point out clear advantages of some methods of [light] guiding, but the ultimate choice will be made in years to come against the background of materials technology.
>
> —Antoni E. Karbowiak, in a 1964 paper that concluded "of all the [optical] guides to date, the fiber guide appears to hold most promise."[1]

Lacking the immense resources of AT&T, Standard Telecommunication Laboratories looked at many ideas, investigated some, and seriously pursued only the few most promising concepts. Periodically, research managers faced the tough job of weeding out the losers. In the early 1960s, STL essentially abandoned the millimeter waveguide.

It was a sensible commercial decision, although it must not have been a pleasant one for men like Toni Karbowiak, who had invested a decade in the technology. Expensive millimeter waveguides made sense only for high-capacity, long-distance "trunk" communications between far-flung population centers. American needed them because it sprawled across a continent. Britain had different telecommunication needs; its cities spanned an island, and it had little room for systems that couldn't turn tight corners. STL decided it couldn't sell enough millimeter waveguides to justify the high development costs, particularly since its military clients didn't want bulky, delicate, and costly plumbing.

The optical communication project had problems of its own, with the lack of a good transmission medium at the top of the list. British engineers quickly ruled out sending laser beams through the air, although Karbowiak suggested relaying laser beams between tethered balloons floating high above the clouds.[2]

Hollow optical waveguides didn't look much better. Reflective light pipes had not lived up to Charles Eaglesfield's optimistic predictions.[3] Demonstrations of confocal waveguides with glass lenses worked well only briefly in the middle of the night, when thermal fluctuations were at their smallest.[4] No one at STL had thought of gas lenses, but they wouldn't have done much good anyway. In principle, Karbowiak wrote, hollow optical waveguides might be "capable of attenuation as low as one decibel per mile, but the engineering difficulties associated with beam structures are likely to render this scheme commercially impractical."[5]

However, laser communications was still young, and Alec Reeves, Len Lewin, and the others were not about to give up easily. They sat down and mulled the prospects. Both light pipes and confocal waveguides required manufacturing tolerances that in the early 1960s were closer to completely impossible than merely damnably difficult. Like the millimeter waveguide, they offered nothing to ITT's major military customers. "We had to conclude that none of these were likely to finish up with something practical for high-capacity, long-distance communications," recalls Karbowiak. Their none-too-optimistic conclusion was: "The only thing left is optical fibers."[6]

The Troublesome Matter of Modes

With the millimeter waveguide gone, Karbowiak turned more to optics, although the group still reported to Lewin. Karbowiak believed most problems of millimeter waveguides and hollow optical waveguides arose from how they guided waves. His main concern was the troublesome matter of modes. The rectangular waveguides that worked well for microwaves transmit only a single mode because they are less than half a wavelength across. Millimeter waveguides are many wavelengths across, so they carry many modes. Light pipes and confocal lens waveguides were thousands of wavelengths of light across, so light could travel in a tremendous number of modes.

Mathematically, a single-mode waveguide is an ideal and relatively simple system for a master of electromagnetic theory like Karbowiak. Mathematical simplicity was important when computers were room-sized giants just moving out of the vacuum tube era. More crucially, single-mode transmission was physically simple as well, so the signal traveled in the same predictable way through the entire waveguide.

Multimode transmission is messy, and the more modes, the messier the transmission becomes. Waves in different modes can travel at slightly different speeds, and those small differences build up over long distances. Fire an in-

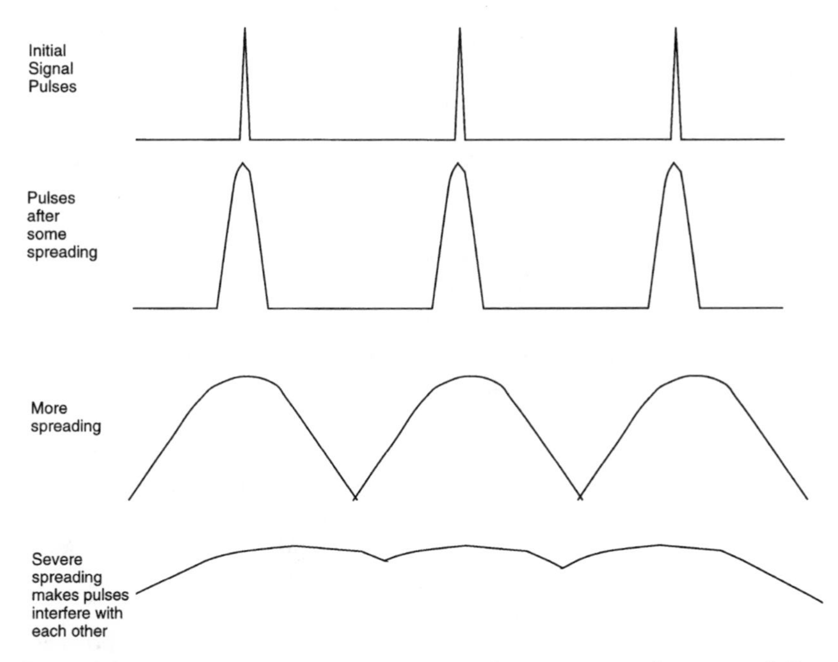

Figure 9-1: Pulses stretch out as they travel along a waveguide or optical fiber because of differences in mode and wavelength. The longer the distance, the more the pulses stretch, until they overlap and become indistinguishable.

stantaneous pulse down a multimode waveguide, and it stretches because some modes move faster than others. If successive pulses spread too much, they can interfere with each other, limiting how fast signals can travel down the waveguide (figure 9-1).

There are other problems as well. Different modes may experience different amounts of loss, or interfere with each other, changing beam intensity inside the waveguide. To complicate things further, bends in multimode waveguides can shift waves from one mode into another. The results were noise and interference, traditional enemies of electronics engineers. With all these complications, engineers could not calculate exactly how multimode waveguides would behave; they had to make approximations and hope they guessed right.

The developers of millimeter waveguides had accepted those limitations because they considered single-mode waveguides impractical. A single-mode guide could be only half a wavelength across, and at a frequency of 50 gigahertz that was only 3 millimeters (under ⅛ inch). Not only were small guides hard to make, but reducing diameter increased the loss, making long lengths impractical. Bell Labs had weighed the trade-offs in settling on a five-centimeter (two-inch) waveguide.

Karbowiak also weighed the trade-offs and looked for fresh ideas. He asked a young Chinese-born engineer, Charles K. Kao, to calculate the properties of multimode millimeter waveguides using a novel theory, but Kao found little that was promising.[7] As Karbowiak looked at the problems of confocal optical waveguides and Eaglesfield's light pipes, he realized many of them arose from multimode operation. He decided the solution might lay in a single-mode waveguide for light.

The starting point for Karbowiak was the single-mode dielectric waveguide, a thin plastic rod that guides microwaves along its surface. It had found a few uses in microwave systems, where hollow metal waveguides usually were more practical. But he thought the balance might shift in favor of a thin, nonconducting waveguide at optical wavelengths. In theory, moving from microwaves to light was as simple as dividing all the dimensions by 100,000, the difference between microwave and optical wavelengths. It's considerably more complex in practice, of course, but the theoretical simplicity was alluring for an engineer at home with equations.

An obvious problem was the tiny dimensions required for waveguides to carry visible light, which has a wavelength under one micrometer (0.001 millimeter). A simple optical version of a single-mode dielectric waveguide would have to be even smaller, too small to see and too fine to handle. Most people would have given up at that point; some very bright people at Bell Labs did. But Toni Karbowiak had some ideas of how to make single-mode waveguides, and he asked Kao and another young engineer just two years out of school, George A. Hockham, to evaluate them. It wasn't a big project; STL paid for it out of internal funds set aside for such efforts, betting on Reeves's ability to pick winners. The theory was straightforward; the difficulty was making a practical single-mode optical waveguide.

Karbowiak, Kao, and Hockham were not alone in looking at optical analogs of the dielectric microwave guide. The same idea occurred to Jean-Claude Simon and Eric Spitz at the central research laboratory of CSF,[8] the French equivalent of RCA in the Paris suburb of Corbeville. Spitz, an engineer in his early thirties who headed the microwave lab, was intrigued by the coherence of laser light.[9] He and Simon, who directed the whole lab, thought of extending microwave concepts to light.

The British and the French both looked first at the simplest type of dielectric waveguide, a thin rod suspended in air, like an unclad optical fiber. Most energy travels along the outside of the rod, avoiding absorption by the material, an effect that Karbowiak had exploited in making microwave devices for satellites. Both groups estimated the transmission loss should be attractively low.[10]

There was, however, a serious practical problem—those single-mode waveguides were only a fifth of a wavelength thick. That was fine for ½-inch (1.25-centimeter) microwaves; the waveguide could be a 3/32-inch (2.5-millimeter) polystyrene rod.[11] However, at optical wavelengths it led to impossibly small fiber diameters of only 0.1 to 0.2 micrometers—4 to 8 millionths of an inch.[12] The developers needed something that behaved like a dielectric waveguide but was large enough to handle.

Fortunately, microwave dielectric waveguides can take more complex forms. The crucial requirements were that the material surrounding the central dielectric could not conduct electricity, and that (for light) its refractive index be smaller than that of the central dielectric.[13] Karbowiak realized, as Brian O'Brien had a decade earlier, that the surrounding material did not have to be air. Any kind of cladding or coating would make a minuscule optical waveguide easier to handle. In theory, an infinitely thick cladding should behave the same way as one just a few wavelengths thick, so the cladding could be as thick as the designer wanted—thick enough to ease handling but thin enough to remain flexible.

More subtle, and it turned out more important, surrounding the waveguide with another material changed the diameter needed for single-mode operation. The critical number is the difference in refractive index between the waveguide (or fiber core) and the surrounding cladding. The larger the difference, the smaller the core must be to transmit only a single mode. For glass in air, the difference is 0.5, so an unclad glass fiber must be no larger than 0.1 to 0.2 micrometer to transmit light in a single mode. Anything larger operates multimode. However, the smaller the difference between the refractive indexes of the core and the surrounding material, the larger the diameter for single-mode transmission. Apply a cladding with a refractive index just one percent lower than the core, and the core or central waveguide layer can transmit single-mode light even if it is several micrometers thick. That's still small, but it's getting into the realm of feasibility, especially because the surrounding cladding can be many times thicker.

Increasing the size of the waveguide offered a crucial benefit for optical communications. Directing light into a fiber core is like threading a needle; the bigger the target, the easier it is. In the 1960s, no one knew how to focus light onto a spot much smaller than its wavelength (it's still very difficult). You couldn't get a useful amount of light into a bare fiber waveguide 0.1 to 0.2 micrometers wide. However, with great care you could aim a laser beam into a core several micrometers across in a clad fiber. Thus, adding a cladding put single-mode optical communications into the realm of possibility.

Eli Snitzer had formulated the same rules earlier at American Optical, but he had come to single-mode fibers from a different approach. Imaging fibers typically have large cores surrounded by thin cladding layers, so that they can transmit the brightest image possible. Karbowiak had a different goal—transmitting light in a single mode for communications—and he envisioned a different structure, with a tiny core surrounded by a thick cladding.

While the cladding solved some problems, it added another: Light had to travel in the transparent material, instead of in the air. (In fact, cladding changes the properties of a single-mode waveguide such that most light travels in its core, rather than along the surface.) This threatened to raise transmission loss tremendously. Even the clearest solids available in the early 1960s absorbed too much light for a practical optical waveguide, Karbowiak concluded in an internal report. Yet other possibilities looked even worse. "It would be unwise to dismiss any of the proposed means of [optical] communications as too impractical or too costly," he told a London meeting on laser applications in September 1964. "Nonetheless . . . of all the guides known to-date the fiber guide appears to hold most promise if due to advances in materials technology it becomes possible to manufacture cladded fibers having effective loss . . . about two orders of magnitude better than at present."[14]

In France, Spitz asked the French glass manufacturer Saint Gobain to make glass cylinders with thin inner cores, which could be drawn down into fibers. Then he assigned further research to Alain Werts, who had just started at CSF after finishing his undergraduate degree.[15] Spitz and Simon also studied ways to suspend thin unclad fibers in air.[16] That would not improve light collection, but it would make the filaments easier to handle.

A Search for New Waveguides

While Toni Karbowiak knew ultraclear glass could cure the problems of clad fiber waveguides, he was not a materials specialist and had no idea how to make it. He did know waveguide theory, and he applied that expertise to inventing a new type that could guide light along its surface with low loss. A crucial problem was suspending it without obstructing the surface wave.

He devised a simple and elegant alternative to hard-to-handle fine filaments: a flat waveguide a fraction of a wavelength thick but many wavelengths wide. His theoretical analysis showed the thin film ribbon could guide light along the middle of its surface in a single mode, although not in the same mode as a cylindrical fiber. It could be a centimeter or more wide, large enough to collect light from a focused laser beam. Because light traveled along the middle, a frame supporting the edges would not affect the surface wave. He predicted its loss should be no more than a few decibels per kilometer—so roughly half the signal that entered the waveguide would remain after one kilometer.

Nothing confined the light in the plane of the thin film, but twisting the waveguide in a spiral pattern should avoid "any noticeable loss of energy," Karbowiak wrote after filing a patent application in April 1964.[17] He hoped to reach attenuation of a few decibels per kilometer at infrared wavelengths of 1 to 10 micrometers.[18] After finishing his theoretical work, Karbowiak asked Kao and Hockham to make and test samples. It became their top priority. Number two on the list was finding low-loss materials to clad fiber waveguides. That looked like a long shot, because Toni Karbowiak, like Rudy Kompfner, had already ascertained that the clearest optical glass on the market was far too lossy for the job.

An Unexpected Offer

As Kao and Hockham struggled with the tough problems of making thin-film optical waveguides in late 1964, opportunity knocked unexpectedly for Karbowiak. With a doctorate and some three dozen articles published in scholarly journals after a decade at STL, he was a technical heavyweight at 41. The University of New South Wales thought he would make an ideal chair for its department of electrical engineering.

Karbowiak had talked about academic posts before, but his comfortable job at STL paid more than an ordinary professorship. He hesitated when the Australians asked, saying he didn't know much about the country. The vice chancellor responded by sending first-class tickets for Karbowiak and his family to visit Australia. The university wined and dined him, promising him money to continue his research in optics and other areas.[19]

It was an opportunity too good to resist. While STL had abandoned millimeter waveguides, other leading communications labs had not—and Karbowiak had invested years in that technology, becoming a recognized expert and writing a book that was nearing publication.[20] In late 1964, it was far from obvious to him that STL was on the verge of an optical breakthrough. Academia was a big step up the technical prestige ladder, and the university chair paid well. It offered him more freedom to investigate new ideas than he could have at a company with its own product agendas. Karbowiak started packing, much to the surprise of the young men working for him.[21]

A Problem of Materials

Kao inherited management of the little optical waveguide program. He was young to manage a group, but the group was tiny—only Hockham reported to him.

Born November 4, 1933, in Shanghai, Charles Kuen Kao was the son of a judge who tried to raise his family in traditional Chinese style. That was a difficult task in unsettled times. The Japanese army lurked ominously in nearby Manchuria from 1932 until it attacked the French concession in Shanghai on December 7, 1941, the same day Japanese planes bombed Pearl Harbor. The Kao family survived the war and in 1948 fled by boat to British-ruled Hong Kong ahead of the communist takeover of the mainland.

Like other Chinese children in British schools, the young Kao Kuen took an English name—Charles—as he learned the language.[22] It was his third language, after Chinese and French, but he learned to speak it clearly with only a trace of accent. Chemistry was the first science to interest him, but by the end of elementary school he turned to electronics and communications, building standard electronics projects like crystal radio sets. No Chinese colleges offered electronics when he graduated from high school, so in 1952 he left for England, graduating from the University of London in 1957 in electrical engineering.

He stayed in England to work for Standard Telephones and Cables, comfortable in the country's cosmopolitan culture. He courted and married a young STC computer engineer, born in England of Chinese parents. Ambition ran strong in the boyish-faced Kao, and he grew frustrated by the limitations of current telecommunications technology. In 1960 he resolved to return to school but got a better offer from the company's research division—a chance to earn an "industrial" doctorate while working on practical problems at STL. The combination of work and school was demanding, but the salary beat starving as a full-time graduate student.[23] When Karbowiak left, Kao was finishing his thesis for millimeter waveguide pioneer Harold Barlow. He also was busy at home helping raise two small children; his wife had continued working after they were born, a rarity in the early 1960s. While the 26-year-old Hockham raced motorcycles, the slightly older Kao had little time for outside recreation.

Both young engineers had mastered waveguide theory; their other skills were complementary. Kao had a good physical intuition and a knack for assembling components into working systems. Hockham's gift was mathematical analysis of how waveguides radiate energy, an arcane art of the utmost importance for transmission lines and antennas. Antennas are supposed to radiate energy; transmission lines are not. Hockham's job was to make sure waveguides didn't act like antennas.[24]

Before Karbowiak left for Australia, Hockham tested a larger polyethylene film model scaled to work with eight-millimeter microwaves. Karbowiak found the results "most encouraging, showing small attenuation, good field confinement, and ability to negotiate bends and twists."[25]

Optical thin-film waveguides proved more troublesome. Light waves are over 10,000 times shorter than eight-millimeter waveguides, so Kao and Hockham had to shrink the guide dramatically. To make films thin enough, they dissolved plastic in a solvent which evaporated readily at room temperature, then dropped the solution gently onto water. The solvent evaporated, leaving behind an extremely thin film that they had to gently lift off the water. After a series of experiments, they finally made films less than half a micrometer thick, so thin that they were iridescent, like an oil slick, because light waves interfered inside them.

Karbowiak's theory said the delicate films should carry light in just one mode. Kao and Hockham played with the films, aiming the red beam from a helium-neon laser into the thin guide. When they put a film guide on a curved support, light spread over the walls of their laboratory, leaking prodigiously from the bent waveguide. "It was a spectacular sight, and we took a photo to record this event," recalls Kao.[26] "And that was the end of the surface waveguide, because there was no way you could use it."[27] He and Hockham had done their job, testing their former superior's idea. With Karbowiak gone, no one remained to advocate and refine the thin-film guide. They turned to the idea Kao considered more promising—clad optical fibers (figure 9-2).

Seeking Clearer Fibers

Theory clearly showed that a cladding would keep light from leaking out at bends in a single-mode waveguide, solving the problem that killed the thin-film guide. That left the question of material transparency.

Kao and Hockham began analyzing requirements for optical waveguides months before Karbowiak left. They targeted needs of the British Post Office that were quite different than AT&T's plans for high-capacity systems to cross

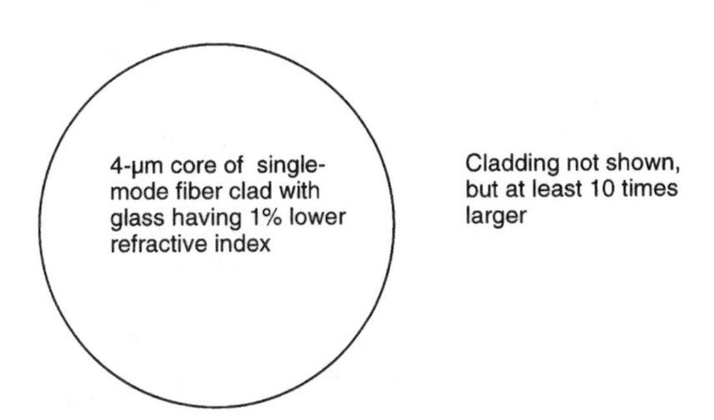

Figure 9-2: A clad single-mode optical fiber is gigantic compared to an unclad single-mode waveguide and the thin-film optical waveguide. This picture shows only the comparatively small single-mode core. One wave of visible green light is about 0.5 micrometer (μm) long.

the wide open spaces of America. The Post Office wanted better technology to send signals between local switching centers that typically were a few miles apart. They wanted something easy and inexpensive to install in heavily developed areas, not high-priced huge-capacity systems to span vast distances. The goal was local arteries for communication traffic, not long-distance superhighways.

Starting from the Post Office wish list, Kao and Hockham calculated their targets. They knew how much power a laser transmitter could generate and how weak a signal an optical sensor could detect. From that, they calculated loss allowable between transmitter and receiver. Dividing that number by the distance gave attenuation, the loss per mile or kilometer. The answer came to 20 decibels per kilometer, so one percent of the light entering a waveguide should remain after traveling a kilometer. (That is equivalent to 32 decibels per mile, so under 0.1 percent of the light would remain after a mile). That was a challenging target, because the best fibers available reduced light intensity by 20 decibels over a distance of just 20 meters (66 feet).

Kao and Hockham faced two crucial questions: Was any material clear enough to meet their target of 20 decibels per kilometer loss? Would the tiny fluctuations in dimensions that are inevitable in any real waveguide make light leak out as if the fibers were miniature antennas?

The two bounced ideas back and forth and loosely divided the problems. Hockham, the antenna expert, concentrated on waveguide irregularities that could scatter light out of a fiber. The geometry of a clad optical fiber looked good for a waveguide. Light spread out in the plane of a thin-film guide, but the cylindrical cladding should confine light entirely within the fiber. However, Hockham knew this simple model made some unrealistic assumptions—that the waveguide had no discontinuities, no bends, and no changes in diameter. Experience with the millimeter waveguide warned that such assumptions could gloss over serious difficulties, so Hockham took the harder course of calculating what would happen in real fibers with the inevitable minor imperfections.

He expected core diameter would be hard to control, so he calculated the effects of fluctuations along the fiber. He worried that such variations could shift light into different modes and cause some light to leak out of the fiber. The numbers were encouraging; it looked like the effects should be small. But he didn't stop with theory. To test the predictions, he built model waveguides, scaled up to carry microwave signals. "There's nothing magic about it," says Hockham; the larger-scale models were easier to test than fibers. He still has some of the curious-looking copper tubes, which vary in diameter along their length and have disk-shaped metal fins along their sides. The experiments confirmed his theoretical predictions, and the project became the core of Hockham's doctoral thesis.[28]

The Materials Problem

Kao concentrated on the transparency of the most common optical material, glass. He and Hockham searched painstakingly through the scientific literature but found very little information. They visited glass specialists and found very little more. Mostly, they learned how little people knew. Glass specialists blamed impurities for the residual absorption in the clearest optical glasses, but no one knew what set the fundamental limits on glass clarity. There was good reason for that lack of knowledge. Before the invention of the fiber-optic endoscope, nobody had any reason to send light through more than a few inches of glass.

The state of the art was not encouraging. The best imaging fibers had losses of about one decibel per meter, or three decibels every 3 meters (10 feet), so half the light was lost in going the distance across a small room. Go another three meters and half the remaining light was gone. After 20 meters (67 feet), such fibers soaked up 99 percent of the input light, as much as Kao had allocated for one kilometer (3300 feet) in a communication system. Go 100 meters, the length of a football field, and one ten-billionth of the light remained. It was no wonder Rudy Kompfner gave up hope.

Charles Kao did not. Materials science is often empirical; specialists make measurements first, then try to explain them. Kao came from a different field, electromagnetic theory, where elegant formulas precisely predict what experiments should measure. You can calculate the behavior of a waveguide from fundamental laws of physics, but not the transparency of glass. Trained in a field where fundamental limits were known, Kao asked about the fundamental limits on glass transmission. Three different factors enter into the equations. One is surface reflection, which wasn't a major concern because it happens only on the ends of fibers. A second is scattering of light by atoms in the glass, which sends it in some direction other than down the fiber. A third is absorption of light by atoms in the material.

Early on, Eaglesfield asked about scattering and came back with an encouraging estimate that it was less than five decibels per kilometer for quartz.[29] Later Kao found a formula for light scattering derived several years earlier by Robert D. Maurer of the Corning Glass Works.[30] When he and Hockham plugged in the numbers, they got an estimate even more to their liking—one decibel per kilometer at a wavelength of one micrometer. That implied scattering should not be a big problem for communications.

That left the issue of light absorption. Kao recalls, "I was seeking the answer to the question, 'What are the loss mechanisms and can these mechanisms be totally removed?' It appeared that no one had really asked this question before."[31] Where others had asked for the best existing glass, Kao sought the fundamental limit. The experts didn't have a ready answer. When he pressed them, they blamed most absorption on impurities.

The stuff we call glass is a mixture of things. The basic raw ingredient is sand, the debris left after the weather wears down rocks until only the hardest crystals remain—grains of quartz, which chemically is silicon dioxide, also called silica. To melt sand at reasonable temperatures, glass makers add soda, potash, and lime. They add other compounds to make special glasses for purposes from optical instruments to fine crystal ware. Add cobalt and the glass turns a rich dark blue; other metals give other tints.

Traditional glasses are not chemically pure, but they are adequate for their usual jobs. Small dashes of impurities don't absorb enough light to notice in the thickness of a sheet of window glass or a camera lens. However, the absorption becomes noticeable if the light has to go a long distance through a fiber. Iron, copper, and some other elements soak up light, darkening the glass.

How clear would glass be if you removed all the impurities? Many experts were only guardedly optimistic. They weren't sure because they hadn't measured absorption in extremely pure glasses. They weren't sure how pure glass could be made. They simply didn't have the answers to Kao's questions.[32] Yet they also had no showstoppers, and Kao heard some encouraging words. Professor Rawson of the Sheffield Institute of Glass Technology said he was convinced that removing impurities could reduce absorption below the target level of 20 decibels per kilometer.[33] If all went well, that meant fiber optic communications might be possible.

Putting the Pieces Together

Encouraged, Kao and Hockham drew a few fibers and tested them at STL. The fibers were lossy, but those with cores smaller than four micrometers transmitted the red light from a helium-neon laser in a single mode. They experimented with semiconductor lasers and white light. They tested Hockham's microwave guides and analyzed the results. They convinced themselves that fiber-optic communications could work. But like Heinrich Lamm, they knew they could not develop a whole new technology by themselves. They needed to interest others. Convinced they had "enough evidence to commit our findings to paper,"[34] Kao and Hockham sent an article to the *Proceedings of the Institution of Electrical Engineers* in November 1965.

Their analysis was careful, but their conclusions were daring. The decibel scale (see box, pages 115–116) understates the immense gap between the best existing fibers and their goal because it's logarithmic. In 1965, the best fibers had attenuation of 1000 decibels per kilometer; their goal was 20 decibels per kilometer. Twenty decibels is a factor of 100; lose 20 decibels and you have one percent of the original light left. A thousand decibels is 10^{100}; lose a thousand decibels and you have only $1/10^{100}$ of the original light. Actually, you have no light, because you lost it all long ago. That drop in intensity is worse than starting with the mass of the whole universe and ending up with one atom.

That didn't discourage the editors, who probably had seen crazier schemes. They asked for revisions, which Kao and Hockham completed in February. The final version runs eight printed pages, packed with equations and charts and thick with electronic jargon.[35] The details were central to convincing their fellow engineers that their new communications medium could offer huge transmission capacity at low cost.

Published papers mark milestones, but they take months to reach print. Kao wasn't about to wait; he launched the proposal with a January 27, 1966, talk at the London headquarters of the Institution of Electrical Engineers, which counts Sir Francis Bolton, the impresario of illuminated fountains, as one of its founders. STL management thought it worthy of a press release, which announced: "Short-distance experimental runs of these optical waveguides have been operated successfully. They have exhibited an information-carrying capacity of one gigacycle, which is equivalent to about 200 television channels or over 200,000 telephone channels."[36] The release soberly summarized the state of the art, but closed with what must have seemed a wildly optimistic prediction: "When these methods are perfected, it will be possible to transmit very large quantities of information (telephone, television, data, etc.) between say, the Americas and Europe, along a single undersea cable." That was impossible with hollow structures like millimeter waveguide or light pipes.

The British magazine *Wireless World* allocated the bottom part of one page to Kao's talk in March.[37] A small American newsletter named *Laser Focus* also took note.[38] Yet otherwise it sank without a trace. Scans through a sampling of major American science and technology magazines, the *New York Times* index, and the *Reader's Guide to Periodical Literature* show nary a word about Kao's proposal.[39] If the editors noticed it at all, they probably dismissed it as just another crazy scheme, hardly likely to be practical in the twentieth century. After all, in 1966 satellites were the future of telecommunications. The role of light would be to travel through confocal waveguides or gas lenses as was spelled out in the lead article of the January *Scientific American*[40]—then the semipopular journal of record for American science—by Stew Miller of the prestigious Bell Labs. Responsible journalists knew they should trust such authoritative sources. When the Kao and Hockham paper finally appeared, it also made no discernible mark even in the technical press.

No Silver Medals for Invention

The French followed the same trail after Spitz talked the French Ministry of Defense out of a small grant, but they lagged behind STL. Werts modeled light transmission along a fiber "surface" with 1.4-millimeter microwaves, the shortest wavelength available. Then he calculated how light should travel through a clad single-mode optical fiber. Only after Werts started did Spitz discover Snitzer's careful analysis of fiber modes. Werts had to make his own laser for fiber-transmission experiments, buying mirrors from America and filling a glass tube with helium and neon.

The Decibel Score

Decibels are very handy units for engineers evaluating signal strength but can be quite confusing to other people. If you want to avoid complexity, you can think of decibels as a way of keeping score. In fiber optics, they usually measure how much of the signal is lost when being transmitted. Lower loss is better, so lower decibel numbers are better. Engineers usually measure transmission loss per kilometer of fiber, so the standard units are decibels per kilometer. Metric units are standard for the research and development community, but you needn't worry too much about them. The relative scores for fibers are the same whether the loss is measured per kilometer, per foot, per mile, or per light year. (The numbers, however, are quite different. Kilometers are used here because they're the most common scale.)

Strictly speaking decibels measure the ratio of output power to input power on a logarithmic scale. The formula used for fiber optics is

$$\text{decibels} = 10 \log \left(\frac{\text{power out}}{\text{power in}}\right).$$

For optical fibers, the number is negative because power output is less than input, but the sign is usually ignored. (The only way to have higher output is in an optical amplifier.)

Decibels greatly simplify engineering calculations because you can find total loss by multiplying fiber loss in decibels per kilometer by the length of a fiber. For example, 50 kilometers of 0.5 decibel/kilometer fiber has a 25-decibel loss. However, because the scale is logarithmic, it's easy to underestimate the impact of loss measured in decibels. Loss of 10 decibels means one-tenth of the signal remains. Loss of 20 decibels means only one percent remains, while loss of 100 decibels means only 10^{-10} (0.000,000,000,1) remains. The table below gives some examples, with physical analogies that may be helpful.

Table 9-1 What Decibel Losses Mean in Optical Fibers and Elsewhere

Loss in decibels	Fraction of power remaining	Physical analog
0.2	0.954992586	Loss undetectable by the eye
0.5	0.891250938	Light lost reflecting from aluminum
1	0.794328235	
10	0.1	Reflection from a dark surface
20	0.01	One percent of light remains
30	0.001	One part per thousand
40	0.0001	
50	0.00001	
60	0.000001	One part per million
100	10^{-10}	One part in 10 billion
200	10^{-20}	
300	10^{-30}	One atom in a ton of matter
400	10^{-40}	
500	10^{-50}	
600	10^{-60}	Less than one atom in the sun
700	10^{-70}	
800	10^{-80}	One atom in the visible universe
900	10^{-90}	
1000	10^{-100}	

Werts and Spitz measured some glass losses and borrowed other material data from Kao.[41] They also borrowed Hockham's data on the effects of waveguide irregularities. Werts had to stretch the fibers they got from Saint Gobain to shrink the cores so they carried just one mode. "The only originality of my work," he recalls, was showing that fiber properties could limit light transmission to a single mode.[42] (Snitzer and others had observed modes in bundled fibers but had not systematically studied individual fibers.)

No one considered the fiber project particularly important. Werts took a while to get around to writing the paper, and when he did both Simon and Spitz declined to be listed as co-authors. It appeared in the French-language journal *L'Onde Electronique*[43] just two months after the Kao and Hockham paper was published in England. There the matter stopped. The military money was gone, and no one at CSF was willing to bet his reputation on fiber-optic communications. "I didn't really consider at that time it was an important paper," recalls Spitz. He had ample reasons to be skeptical. French glass specialists told him that it was impossible to remove the iron impurities Kao considered the main cause of glass absorption. Other projects looked more promising; his group turned to storing data in optical form, an idea that led to today's audio compact discs and computer CD-ROMs.[44]

Daniel Colladon, the Swiss engineer and physicist who first reported light guiding in jets of water. (Courtesy French Academy of Sciences)

Jacques Babinet, French physicist who guided light in bent glass rods. (Courtesy French Academy of Sciences)

Clarence W. Hansell, the prolifically inventive electronic engineer who founded RCA's Rocky Point Laboratory and patented fiber-optic imaging, holding a vacuum tube that was one of his inventions (Courtesy Robert and Patricia Sisler)

Heinrich Lamm, the German Jewish medical student who sent the first image through a bundle of glass fibers in 1930. (Courtesy Michael Lamm)

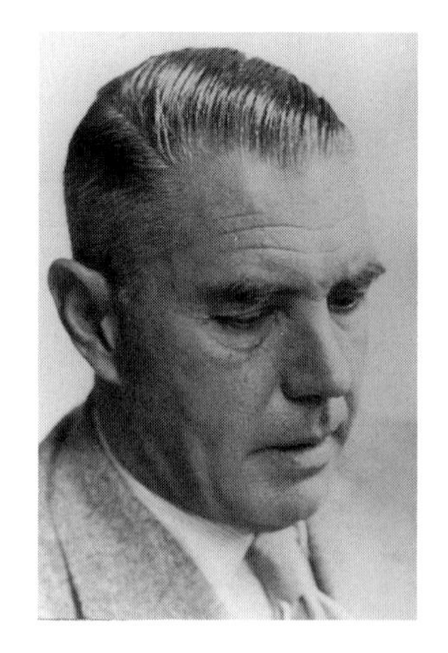

Abraham C. S. van Heel, Dutch physicist who made the first bundles of clad fibers for image transmission. (Courtesy H. P. van Heel)

Holger Møller Hansen, Danish inventor who independently thought of image transmission through fiber-optic bundles, shown in his home workshop. (Courtesy Møller Hansen)

Harold H. Hopkins, English physicist and optical designer who also invented imaging fiber bundles. (Courtesy Kelvin Hopkins)

Basil Hirschowitz, the young gastroenterologist with a vision of a flexible fiber-optic endoscope to look into the stomach. (Courtesy Basil Hirschowitz)

Will Hicks as a young fiber-optics entrepreneur, captured at a trade show wearing an uncharacteristic white shirt and tie. (Courtesy Walt Siegmund)

Developers of the Todd-AO wide-screen movie system, which distracted Brian O'Brien and American Optical from fiber optics. From left, Bob Surtees (director of photography for *Okalahoma!*), producer Arthur Hornblow, Mike Todd, director Fred Zinnemann, O'Brien, and composer Oscar Hammerstein. (Courtesy Brian O'Brien, Jr.)

Larry Curtiss with the equipment he used to make the first glass-clad fibers as a junior at the University of Michigan. (Courtesy Bentley Historical Library, University of Michigan; collection of University of Michigan News and Information Services)

The first glass-clad optical fibers were wound onto oatmeal boxes by Larry Curtiss at University of Michigan (Courtesy Corning Glass Museum)

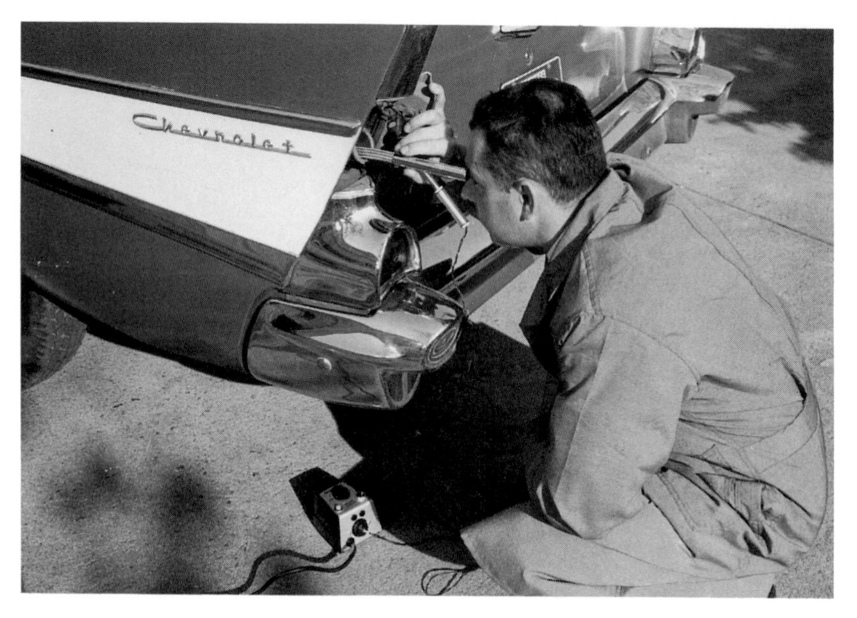

Wil Bazinet of American Optical tests an early fiberscope by examining the inside of a gas tank; he was Will Hicks's first assistant. (Courtesy Walt Siegmund)

Alec Reeves, the visionary British engineer who invented the standard method for digital communications and encouraged development of fiber-optic communications at Standard Telecommunication Labs. (Courtesy Nortel)

Stewart E. Miller, who headed guided-wave transmission research for over 30 years at Bell Labs in Crawford Hill. (Courtesy Lucent Technologies)

Charles Kuen Kao, whose theoretical analysis, experiments, and advocacy lauched fiber-optic communications, first at Standard Telecommunication Labs and later worldwide. (Courtesy Nortel)

George Hockham, with the metal waveguides he studied to understand how small internal variations might cause losses in optical fibers. (Courtesy Nortel)

F. F. Roberts, the crusty visionary who launched the British Post Office's fiber-optics program. (Courtesy John Midwinter)

Donal Keck, Robert Maurer, and Peter Schultz (left to right) pose at Corning after making the first low-loss fibers. (Courtesy Corning Inc.)

Russian scientists savor their success in making the first room-temperature semiconductor laser. From lower right, clockwise: Zhores I. Alferov, Vladimir I. Korol'kov, Dmitriy Z. Garbuzov, Vyacheslav M. Adreev, and Dmitriy N. Tret'yakov. (Courtesy Zhores I. Alferov)

Izuo Hayashi and Mort Panish discuss structure of their room-temperature diode laser at Bell Labs. (Courtesy Lucent Technologies)

Barney DeLoach shows off a million-hour laser, the result of years of development at Bell Labs. (Courtesy Lucent Technologies)

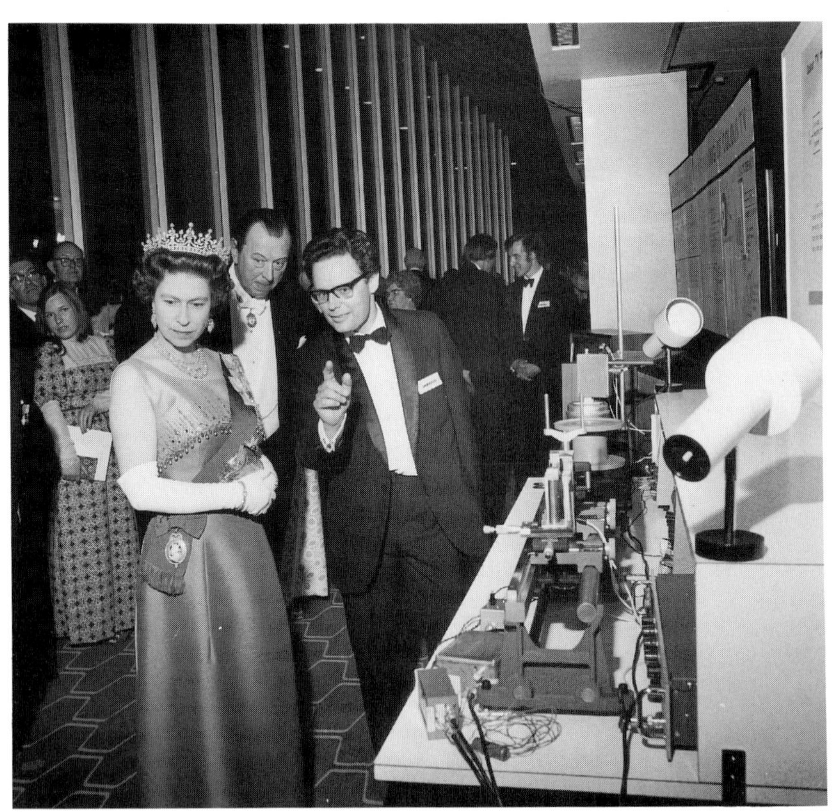

Murray Ramsay demonstrates fiber optics to Queen Elizabeth in May 1971. The receiver is at the near end; the fiber is coiled on a spool on the optical bench beside Ramsay. The laser sits in a Dewar of liquid nitrogen behind it. Lord Nelson, chairman of the Institution of Electrical Engineers, is looking over the Queen's shoulder, and Ramsay appears to be pointing at the photographer. (Courtesy Murray Ramsay)

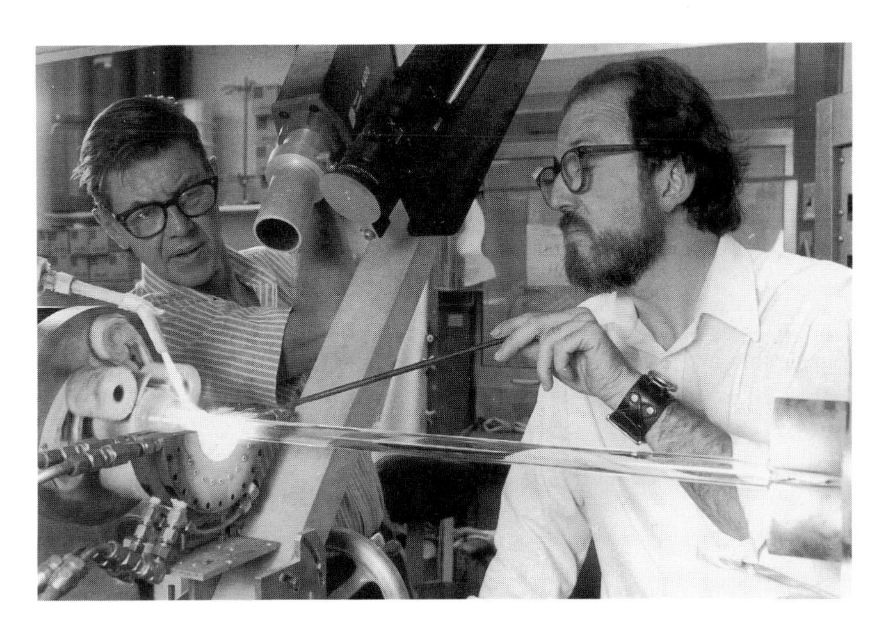

John MacChesney (left) and Paul O'Connor make a preform using modified chemical vapor deposition at Bell Labs; one preform yielded about 15 kilometers of fiber. (Courtesy Lucent Technologies)

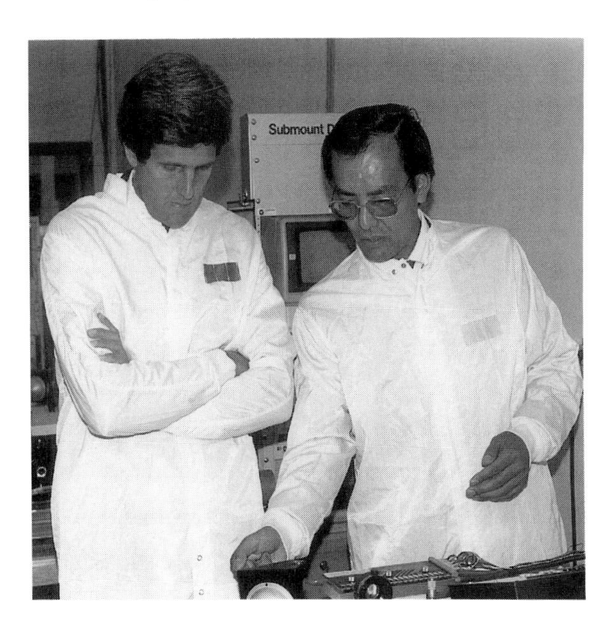

J. Jim Hsieh shows long-wavelength semiconductor-laser manufacture at Lasertron to Massachusetts Senator John Kerry. (Courtesy Lasertron)

Masahara Horiguchi, NTT engineer who opened the long-wavelength windows for fiber communication. (Courtesy M. Horiguchi)

Mashahiro Kawahata at a control console for Japan's Hi-OVIS system, the first project to bring fiber-optic communications to homes. (Courtesy M. Kawahata)

10

Trying to Sell a Dream

(1965–1970)

> If you really look at it, I was trying to sell a dream. . . . There was very little I could put in concrete to tell these people it was really real.
>
> —Charles K. Kao[1]

Charles Kao saw the future of communications and it was fiber optics. Fibers would simply and elegantly avoid the problems that plagued millimeter waveguides and hollow optical waveguides. All that he needed to reach that future was glass as clear as air. Will Hicks, Toni Karbowiak, Rudy Kompfner, and Stew Miller, men of no mean imagination and intelligence, had turned back at the sheer challenge of making so perfectly transparent a solid. Only Charles Kao had the vision and daring to charge full-speed ahead.

He could see no fundamental barrier blocking his goal of clear glass fibers. He had the good fortune to enlist the visionary support of Alec Reeves, who could plead his case to top management. Kao had not invested any crucial part of his ego in millimeter waveguides, gas lenses, or confocal lens waveguides. He could see that thin, flexible fibers not only offered high transmission capacity, but also would be simpler and far cheaper to install than delicate bulky pipes that required underground burial.

Fiber optic communication was a simple and elegant idea, tiny flexible "pipes" compared to the brute-force technology of thick millimeter waveguides that had to be laid absolutely straight. In his twenties, Kao had seen the transistor revolution sweep through electronics, with compact solid-state devices making bulky vacuum tubes obsolete. He could envision optical fibers as another technological revolution, doing the same thing to millimeter waveguides. It was the chance of a lifetime. Ambitious young engineers dream not just of making something work, but of making it work so well that it brings them fame and fortune, putting them in the league of people like Reeves, or perhaps even Thomas Edison and Alexander Graham Bell.

The dream is not easy to achieve. Kao could not be sure in 1966 that glass could ever be as clear as he needed. Other engineers were skeptical. "We were talking about a system concept which required a light source which at that time was working intermittently in liquid nitrogen, and an order of magnitude improvement in fiber that was so far out that people could not believe it was an attainable goal," Kao told me on the phone. He had to sell them on the dream. "I sometimes say I must be a very good salesman."[2] His words sped from Hong Kong to Boston through 8000 miles of fiber-optic cable so clear I could catch the faint touch of China beneath his British-English accent.

A Customer at the Post Office

The January 1966 talk in London opened Kao's sales campaign. His dream caught the imagination of a crucial member of the audience, Robert Williamson White, head of a waveguide development section at the British Post Office Research Station at Dollis Hill in London, the British counterpart of Bell Labs. The Post Office had about a hundred people working on the millimeter waveguide,[3] but White was growing skeptical about its prospects and called it "the pipe dream."[4]

What bothered White was the mundane matter of installation. Millimeter waveguides had to be buried five to six feet underground and run in straight lines. That was going to be expensive. The cost might be justifiable to provide the very high transmission capacity needed for the "trunk" lines between major cities, especially where they ran through open countryside. However, the Post Office also wanted to improve the much larger "junction network" linking local telephone switching centers. Like local roads feeding into an expressway, there are many miles of junction network for each mile of trunk. Connections typically span only a few miles, but they thread through existing cities and towns. Millimeter waveguides were as ill-matched to winding along the tangled streets of London as rigid metal pipes were to looking down a patient's throat into his stomach. Post Office engineers wanted a cable they could snake through utility ducts already buried underground. Thin and flexible, optical fibers sounded worth investigating.

Technology led the way at Dollis Hill. Managers felt "if the technology can deliver it, let's do it and see how the customer reacts,"[5] says Jack Tillman, a former deputy director for research. That attitude led the British lab to study telephone access to remote computer systems as far back as 1960.[6] That interest led to a commercial service called Prestel, which floundered in the

1980s,[7] but anticipated the Internet and the World Wide Web. The Post Office watched Bell Labs, but steered its own course, avoiding some Bell projects including Picturephone[8] and confocal optical waveguides.

Kao's idea struck a responsive chord at Dollis Hill. John Bray, a radio engineer recently named to head the lab, was intrigued when IEE asked him to review Kao and Hockham's paper before publication.[9] He and Tillman asked Frederick Francis Roberts,[10] an engineering manager under Tillman, to investigate. It was a logical choice; Roberts had a penchant for new ideas,[11] and his main project was stalled.[12]

Born of British parents in France in the spring of 1917, F. F. Roberts was a career civil service engineer. Reserved by nature, he was rather stiff and formal even for his generation of Englishmen. Tillman, who had worked with him on radar during World War II, called him Frank, but most people knew him by his initials, and even as the Beatles made London swing in the late 1960s, his subordinates addressed him as "Sir."[13]

Intense, curious, and dedicated to his job, Roberts had the relentless drive of a classic "Type A" personality. He took a formal, rigorous approach to problems, analyzing them systematically. That approach served him well in 1951, when he was the first to demonstrate an important microwave effect.[14] He climbed the management ladder, and by the 1960s had 50 to 80 people working for him.[15] A stickler for doing everything properly, from filing trip reports to calculating the properties of materials, he drove himself as intensely as his staff, unable to relax on the job. Yet the crusty bureaucrat had his own gift. "Roberts was a difficult, cantankerous old sod, but he was a man of vision,"[16] says Charles Sandbank, who became Kao's division manager after Len Lewin left for a professorship.

As the operator of the British telephone system, the Post Office used telecommunications equipment, but it didn't manufacture any. Roberts's goal was to see if the new technology could meet Post Office needs, and to encourage its development. He was not a glass expert, but he knew what questions had to be asked and answered. If there were problems, F. F. Roberts would find them.

The transparency of glass was crucial, so he set George Newns to work purifying glass at Dollis Hill. He supported Kao's research. Roberts also turned to experts outside the Post Office, asking questions and trying to stimulate interest. He visited big British glass companies, but they showed little interest. The hard-driving Roberts pumped every possible contact, including a roving scientist from the Corning Glass Works who in 1966 visited Dollis Hill fishing for new uses for glass. William Shaver was a man of easy enthusiasm for new ideas, but he didn't have anything to offer Roberts off the top of his head.[17] Back at Corning, he mentioned the idea of fiber communications to Bill Armistead, the corporate research director. Armistead agreed it was promising, and they passed the idea along to Robert Maurer, who managed a small glass research group had earlier studied light scattering in glass. Maurer sat down to think about the possibilities.

Military Support

Kao made another early convert in Don Williams of the Royal Signals Research and Development Establishment in Christchurch. Like the Pentagon, the British Ministry of Defense was investing heavily in new technology, seeking new ideas from the likes of Alec Reeves. Military electronics were big business for ITT in both Britain and America in the 1960s. Millimeter waveguides and hollow light pipes were far too cumbersome for military use, but small flexible fibers might replace the thick copper cables that weighed down portable communications systems. Fibers also had another attraction at a time when military planners worried about fighting nuclear wars. Nuclear blasts produce a strong burst of electromagnetic waves that can induce strong current pulses in metal wires, and those pulses can fry delicate electronics. Fibers are immune to that effect because they don't conduct electricity. That made fibers promising for airplane and ship communication systems that had to withstand nuclear effects. Because those systems don't have to carry signals very far, military systems could function with fibers less transparent than needed for civilian telephone networks. That was an important boost, because it opened a market for fiber communications even if loss could not be reduced to 20 decibels per kilometer.

Williams gave STL a small research grant. The military connection gave Kao access to sophisticated equipment able to measure important quantities, such as low levels of impurities in glass.[18] Williams also funded William Alec Gambling, an electronics professor at the University of Southampton. Gambling had thought of fiber-optic communications as far back as 1964[19] but had never done as much analysis as Kao and Hockham. His group concentrated on large-core fibers because the short, cheap, low-bandwidth systems wanted by the Ministry of Defense did not require single-mode transmission.

The Traveling Fiber Salesman

Support from Roberts and Williams was a step in the right direction, but only a step. Charles Kao went on the road to sell others on the case for fiber optics. He had made his first trip to America while working on his fiber proposal. In early 1966, he took a second trip to talk with American experts in fiber fabrication, optical glass, and lasers. He visited Eli Snitzer at American Optical, glass expert Norbert Kreidl at Rutgers University in New Jersey, the optics giant Bausch & Lomb in Rochester, and Kapany's Optics Technology in California.

His stop at Bell Labs was disappointing. Miller's group at Crawford Hill was less then enthusiastic. "They have requested the materials people to investigate into the possible means of obtaining the very-low-loss materials required. That was as much as they would do to influence the work of other departments. They now just have to wait and see. If the low-loss material was forthcoming, then they would be very interested in looking into the fiber waveguide as a possible optical wave guiding medium,"[20] Kao wrote when he returned home. Without solid evidence of more transparent glass, Bell would stay with gas lenses and confocal waveguides. Yet Kao felt a blessing from Bell Labs was critical. When a younger engineer asked why he wanted to invite a tough competitor into the field, Kao replied, "The thing will only take off if we get them into it."[21]

Kao and Roberts visited Spitz and Werts at CSF,[22] and Kao toured German labs. Many listened to his sales pitch, but initially few bought it.

An Invitation to Japan

In late 1966, Kao's fiber-optic campaign yielded him invitations to speak at Tohoku University and Nippon Telegraph and Telephone in Japan. Japanese engineers had also been thinking about fiber communications and wanted to hear what Kao had to say.

Engineering professor Zen-ichi Kiyasu grew interested in optical communications after leaving NTT and joining the university faculty, but he could not see much future for hollow optical waveguides. In 1964, he told another Tohoku professor, Jun-ichi Nishizawa, that he had not heard any proposals for optical communications that would be reliable enough for practical use. A couple of days later, Nishizawa suggested using optical fibers, evidently inspired by fiber-optic endoscopes.[23]

The fibers used in endoscopes have large cores and thin claddings, so they transmit many modes, like millimeter waveguides. The two professors quickly realized that was a problem, but instead of turning to single-mode fibers, they invented a new way to guide light along a fiber. Imaging fibers rely on total internal reflection at a sharp boundary between two materials. Specialists call them "step-index" fibers because the refractive index changes abruptly at the boundary between the light-guiding core and the cladding. Kiyasu and Nishizawa proposed grading the refractive index so that it changes gradually from core into cladding. Instead of reflecting light abruptly from a sharp boundary, a graded-index fiber bends it back gradually. You can visualize the light rays as following a wavy path, rather than the zigzag path defined by total internal reflection in a large-core step index fiber. The Japanese hoped this would reduce losses caused by imperfections in the core-cladding boundary, a problem with some early fibers. Nishizawa filed for a patent in November 1964 and was later surprised to learn that Stew Miller at Bell Labs had filed for a patent on a similar idea in February.[24]

High glass loss stalled the Tohoku group, who like most communications engineers knew little about glass. The Japanese listened carefully to Kao and took him to the Japanese equivalent of Bell Labs, the NTT Electrical Communication Laboratory, which gave Kao a sample of its own experimental single-mode fiber.[25]

A Budgetary Windfall

As Charles Kao preached the gospel of fiber optics, Dollis Hill blundered into a bit of budgetary good fortune. American management consultants told the Post Office that its telephone division wasn't spending enough money on research. That may seem preposterous in an era of corporate downsizing, but the 1960s were flush with technological optimism. Management allocated an extra £12 million for research. "Goodness knows how we are going to spend it all," Tillman confessed to Richard Dyott when he started work at Dollis Hill on March 1, 1967.[26]

The extra money gave Bray and Tillman the luxury of investing in wild schemes, and the realists at Dollis Hill counted fiber optics among the wildest. It would be nice to have a flexible waveguide to thread through convoluted urban underground ducts, but it was not a pressing need. Tower-to-tower microwave transmission worked well and avoided messy construction in developed areas. Engineers saw millimeter waveguides as "the next logical step from microwave towers for long-distance stuff."[27] Fiber-optic communications was a long shot, worth a small bet from the suddenly flush research budget because it might solve some annoying problems.

Roberts gathered a small team to work on fibers and glass, and they collected all the optical equipment they could find. He put George Newns in charge of developing ultrapure glasses with low loss. Dyott headed a group devoted to making fibers and studying their properties. Hugh Daglish was to develop optical techniques to measure fiber properties, a critical concern for a stickler like Roberts. Most of the team came from electronics or millimeter waveguide development and were surprised to be assigned to the unfamiliar world of light.[28]

They started with minimal resources. Dyott sealed a thin rod of tungsten glass inside a thick Pyrex tube, which had a lower refractive index, to make a preform that could be stretched into single-mode fiber. He had no fiber-drawing equipment, but his lab was long and narrow, so Dyott improvised. "We heated up the preform at one end of the lab and Jacqueline Viveash, who by chance had a pair of tennis shoes handy, raced to the other end carrying the end of the preform in a pair of tongs to the cry of 'Run, Jacquie, run.' " he recalls.[29] Jacquie's dash yielded a fiber that carried a single mode at the red helium-neon wavelength. Dyott was pleasantly surprised to find that the fiber collected more than half the laser light focused into it through a microscope objective. Unfortunately, the light didn't go very far. The loss was 30 decibels per meter—so high that less than 0.1 percent of the light that entered the fiber on one side of a desk would emerge on the other.

Dyott also borrowed time on a university computer to solve the complex equations that Snitzer had formulated for single-mode fiber. It was not an easy task because input had to be submitted to the computer center punched on five-hole paper tape, and a single mispunched hole could stop the program, yielding only a cryptic error message hours later. Yet the results helped explain how pulses spread along fibers.

Meanwhile, Roberts pushed ahead, forming a consortium to develop low-loss fibers, including the Post Office, STL, the Scottish optics company Barr & Stroud, and British Titan Products, which made titanium and ultrapure materials. Several days before the consortium's first meeting on July 26, 1967, Roberts returned from a short business trip to France and started driving toward the south of England for a short holiday. On route, he suffered a severe heart attack in the car. Realizing what was happening, he turned around and drove himself to the hospital.

The seizure was a shock to everyone; just under six feet tall, Roberts was thin, active, and had appeared fit. It was November before he returned to work. He had to climb five flights of stairs to reach his office, which may have contributed to a second heart attack in January 1968.[30] A bypass operation followed, but the diligent Roberts insisted on receiving progress reports, which his wife delivered to his hospital bed.

Pushing Hard at STL

Charles Kao sought clearer, purer glass at STL, driving the project "every way that he could. He believed in it, and that's what it took," recalls Martin Chown, who worked for him.[31] Management was largely skeptical with the crucial exceptions of those in the line of command above Kao[31]—Sandbank, Reeves, and Jock Marsh, STL's managing director. Marsh didn't come down hard when Kao exceeded his budgets for travel and experiments.

Contracts from Roberts at the Post Office and Williams at the Ministry of Defense provided vital support. ITT's contribution was modest. Sandbank penciled in $34,000 for fiber research in one early year. "It was very difficult to get it through," he recalls.[32] Reeves's backing saved the day, because the head of ITT laboratories had worked for him back in 1940.[33]

The fiber communications project grew slowly. Optics were incidental to Hockham, who left to form his own antenna technology group in mid-1967.[34] One early addition was Richard Epworth—a young engineer who had read about Kao's ideas in *Wireless World*. Epworth modified a laser transmitter to send video signals through fibers instead of the air. Realizing that individual single-mode fibers would not collect much light, he blew 70 segments of the clearest fiber he could get into a 20-meter tube. The fibers had loss of about one decibel per meter, a total of 20 decibels along their length, but they carried the video signal the full length of the tube.[35] Students do better today in science-fair projects, but in 1967 it was an important feasibility demonstration.

Kao soon realized that such demonstrations were not enough. Skeptics focused their criticism on the key issue of glass transparency. Except for a few small military systems, fiber communications would never work without low-loss fibers. He decided the best way to answer the critics was to roll up his sleeves and measure the best glasses he could find.

It's a big challenge to measure a little attenuation because it's a small difference between two large numbers. One percent accuracy may sound fine, but it's useless if you want to measure the difference between 99 and 99.5 microwatts. Doing better is a challenge for the best of laboratory wizards, and that was not Charles Kao's specialty.

Today, engineers measure attenuation in the clearest fibers by passing light through one kilometer, or ten, or a hundred. That builds up loss until it's high enough to measure easily. Kao didn't have that luxury; no one knew how to draw good fibers from the clearest glasses. The best he could do was work with samples of bulk glass about 30 centimeters (a foot) long.

Nobody had ever tried to measure such low losses before, so he had to develop a new technique. His first study showed glass could be made clearer than standard imaging fibers, but could not measure loss as low as 20 decibels per kilometer.[36] That required him and Mervin W. Jones to devise an even more sensitive instrument, which compared light passing through two glass rods, one 20 centimeters (8 inches) longer than the other. Accurate comparison is a demanding task, but it let them cancel the effects of surface reflection and concentrate on loss in the bulk glass.

Kao sought the purest glass available for his measurements. Ordinary optical glass would not do because the raw materials that go into it are riddled with impurities. Instead, he studied fused silica, a synthetic glass that is essentially pure silicon dioxide (SiO_2), with less than one part per million of the troublesome iron impurities.[37] At first their results looked too good to be true. The samples seemed to be perfectly clear, with no attenuation at all. They knew that meant that the real attenuation was somewhere within their margin of error, which allowed loss between 4 and −4 decibels per kilometer. They spent months analyzing the experiments, to make sure they had everything right, and to be sure their calculations did not yield negative loss (which would imply their measurements were wrong because the glass would have to generate light). In 1969 they finally reported loss of five decibels per kilometer,[38] with the lower limit of their error margin close to zero.[39]

Difficult and elegant, those measurements opened the eyes of skeptics around the world. Before them, Kao had only a handful of believers because ultraclear glass existed only on paper. The measurements demonstrated he was right; extremely pure glass could be clearer than anyone else had imagined. "Kao gave everybody a jolt," recalls Dave Pearson of Bell Labs. "That was the first practical measurement which said, hey, you're not just whistling Dixie."[40]

The Search for Clear Fibers

The measurements were a milestone far from the finish line. They showed that pure glass could be extremely clear; they did not show how to make ultratransparent optical fibers.

A key step was to get rid of the impurities, but how to do that was far from obvious. Kao had measured fused silica, which most major laboratories—STL, the British Post Office, Bell Labs, and the Japanese—considered an impractical material. Its melting point is over 1600°C (2900°F),[41] far higher than other glasses. Virtually no one had furnaces hot enough to soften it for drawing into fibers. More troublesome optically, its refractive index is 1.46, the lowest of any standard glass. That was a big problem because an optical fiber requires a cladding with refractive index lower than the core, but nobody knew how to make glass with lower index than fused silica. STL engineers groped in vain for ideas, even considering cladding pure silica with ice.[42] The difficulties were so large and so obvious that virtually everyone crossed pure fused silica off their list of potential fiber materials.

The consensus was that the best way to make fibers was to purify other glasses, in which other oxide compounds such as phosphates, borates, soda, and lime are added to impure silica. The glass industry had generations of experience with multicomponent glasses. Blending other materials into silica reduced its melting point to reasonable temperatures and gave control over the refractive index. Purity was a problem, but progress was being made in purifying raw materials. It seemed just a straightforward matter of slogging slowly forward to purer and clearer glass.

Dyott's group at the Post Office refined their methods of pulling fibers from preforms made by sealing a core-glass rod inside a tube, but they weren't satisfied. They had problems keeping rod and tube surfaces clean, and wanted to pull fibers continuously, without replacing preforms. Dyott started looking at a process used to make other glass fibers, pulling them from a hole in the bottom of a crucible filled with thick molten glass. It yielded a continuous fiber as long as fresh material was fed into the crucible, but a simple crucible could not make the clad fiber needed for communications. The technical director of a glass company outside London[43] suggested Dyott try a double crucible, with core glass melted in an inner crucible which sat in the middle of an outer crucible filled with molten cladding glass. The core glass emerged from a central hole at the bottom of the inner crucible; cladding glass emerged from a concentric ring around that hole.

Dyott liked the idea, a variation on a scheme invented in the 1930s to make insulating glass fibers.[44] He didn't know that Will Hicks had briefly experimented with a similar approach a decade earlier,[45] but that probably would not have mattered. Dyott thought industrial production would require a process that could draw fibers continuously. If the core and cladding glass were both molten as they emerged from the nozzle, but cooled quickly enough that they didn't mix, he expected them to form a smooth core-cladding interface.

While the double-crucible concept was simple, if was far from obvious how to calculate exactly how the process worked. That didn't stop Dyott, who like Reeves devised intuitive models to test new ideas. He decided he could test the double-crucible process with sugar. Although the choice sounded unlikely, it made eminent sense. Like glass, sugar melts to a thick liquid that can be drawn into fibers (cotton candy), but sugar melts at 107°C (225°F), so it can be handled in brass crucibles instead of the high-temperature materials needed for glass. Dyott dyed the inner core red and started drawing sugar fibers. By adjusting composition to control the refractive index of sugar, he could draw single-mode sugar fibers. The fibers were totally unsuitable for telecommunications, but they helped Dyott understand the physics of making fiber.

Roberts knew nothing about the experiments until January 1969, when he asked Dyott about fiber-drawing progress as they shared a train compartment riding to Sheffield. "I mentioned that we were using sugar, and he blew his top," says Dyott. The precise manager considered the experiments ridiculous; to him, the only proper way to examine a process was by writing and solving formal mathematical equations that gave exact numerical results.

Furious that Dyott could not be bothered to do the requisite calculations, Roberts transferred fiber drawing to Daglish. He dutifully set to work with pencil and paper, telling Dyott, "If Sir wants it calculated, Sir will have it calculated." However, the calculations were beyond Daglish as well, and work stalled until an exasperated Roberts transferred Daglish completely out of fiber optics. George Newns inherited the double-crucible project, and he eventually resorted to experiments with a core of molasses and a cladding of sugar syrup. The two liquids did not solidify, but they yielded enough information to design a platinum double-crucible system for drawing glass fibers, and Newns got away with it.[46]

Newns ordered the purest available raw materials and started drawing fibers at Dollis Hill, but industrial northwest London was a poor environment to use ultrapure materials. He was one of the first people that the Post Office moved to its newly built research laboratories in Martlesham Heath, near Ipswich, some 100 kilometers (60 miles) northeast of London. There he found that platinum particles from the crucibles contaminated the glass, so he switched to crucibles of fused silica, which remains hard at temperatures much higher than the 1000 to 1200°C (1800 to 2100°F) melting temperatures of the compound glasses he was using.[47]

Bell Labs Wakes Up

Kao's careful measurements of fused silica forced Bell Labs to take fibers more seriously. At Murray Hill, the center of Bell's materials research, Dave Pearson put his low-level study of optical fibers on the front burner, adding several people,[48] including two from Miller's group at Crawford Hill.

Like the British Post Office team, Pearson's group saw little hope for fused silica and concentrated on multicomponent glasses. They had no glass-making facilities, so they had outside contractors make glass from the purest available raw materials. They measured light transmission in bulk glass and in fibers. They tested fibers made from rod-in-tube preforms, and assembled their own double-crucible apparatus.[49] They confirmed that impurity absorption was the big problem, and it proved as hard to reduce at Bell Labs as elsewhere. In the spring of 1970, the lowest total loss they measured in fibers was about 0.7 decibel per meter, still far too high for communications.[50]

The main thrust at Bell Labs remained the millimeter waveguide. Field trials were scheduled for 1973, Kompfner reported in March, 1970. He considered laser communications to be another technological generation in the future, but he said some 100 Bell Labs engineers were already working on it, mostly on underground systems using confocal waveguides or gas lenses.[51] Fiber remained a tiny effort.

A New Type of Fiber in Japan

Meanwhile, the Japanese had tackled another problem: getting light into fibers. Squeezing a laser beam into the core of a single-mode fiber required alignment accuracy of about a micrometer. In 1967, that was an extremely difficult task for a specialist in a fully equipped optics laboratory; it seemed an inconceivable task for a technician in a manhole or on a telephone pole. That worried Shojiro Kawakami at Tohoku University, and in the spring of 1967 he suggested the problem could be eased by switching to a new type of graded-index fiber.

Standard large-core fibers were not attractive for communications because they suffered from pulse spreading. Whether you consider the light as rays bouncing around in the core or as modes confined by a waveguide, light could follow many paths through a large-core fiber. Each path took a different time to travel, so the further the light pulse traveled, the more it spread. The more the pulses spread, the longer you had to wait between pulses to keep them from interfering from each other. That reduced transmission capacity.

Graded-index fibers work differently because the refractive index varies with the distance from the center of the fiber. The original reason for the design was to avoid losses at the core-cladding boundary. However, Kawakami realized that careful choice of the refractive-index gradient also could do something else because the speed of light in glass depends on its refractive index. The higher the refractive index, the slower light travels. Thus, light in the high-index center of the fiber lags behind light farther out, where the refractive index is lower. The more the light travels in the outer core, the higher its average velocity, offsetting the delay it suffers from having to travel a greater distance. Kawakami spent a week or two grinding through the numbers to convince himself that this change in speed could compensate for pulse spreading. Complete compensation was impossible, but proper tailoring of the refractive index gradient could reduce pulse spreading by a factor of 100 to 1000. He was thrilled; the effect meant that graded-index fiber would have almost as much transmission capacity as single-mode fiber over distances of a few miles.[52]

Japanese companies started working intently on graded-index fibers. Attenuation remained a problem, but Nippon Sheet Glass claimed loss as low as 100 decibels per kilometer by the end of 1969.[53] They couldn't do that well consistently; their typical value was 200 decibels per kilometer. Nonetheless, it was progress.

Building Real Systems

Standard Telecommunication Labs worked hard on demonstrations as well as on improving fibers. Kao asked Chown, a radio engineer, to build a demonstration system. Chown decided to build a digital repeater, a device to detect a weak optical signal, convert the light into electrical form, and amplify the electronic signal to drive a laser transmitter. Repeaters are vital for long systems, and they demonstrate the essential components of a communication system—a transmitter and receiver. Chown designed his to operate at 75 million bits per second.[54]

Light detectors and electronic amplifiers were easy to build. The laser source was a much tougher problem. Everyone wanted to use semiconductor lasers, but they had to be kept chilled in liquid nitrogen lest they burn out. Chown had to build a miniature transmission line to deliver short but powerful electrical pulses to a cooled laser chip the size of a grain of salt packaged in a metal case about the size of a pencil eraser. To collect the light, he drilled a hole in the window in the case through which the light normally emerged, stuck the fiber through, and glued it to the laser.[55]

Chown proudly demonstrated the repeater at a 1969 exhibition run by the Physical Society. Alec Reeves was not impressed; his mind had once more drifted to the future. He thought it awkward to convert an optical signal into electronic form, then back into light again. Reeves wanted a purely optical amplifier, which could amplify a light signal without first converting it into electronic form.[56] However, Kao was delighted, considering the demonstration as a sign that fiber communication was "ready to be moved into the development phase." At the very least, it convinced "industrial leaders, responsible government officials, and technocrats alike . . . to recommend further investment."[57]

Investment was on Kao's mind because his growing fiber-optic program kept running over budget. When the time came for his annual budgetary pilgrimage to ITT headquarters in New York, Sandbank suggested a severalfold increase in what had been a nominal budget for fiber research. Although he could report progress, he hesitated to ask for too much money. When his turn came at the meeting, he outlined his plans for trying to make low-loss fibers. Al Cookson, corporate vice president of engineering, asked what would happen if the initial approach did not work.

Sandbank said he would try another one.

The American asked, "in series or in parallel"—electronic jargon for one at a time or all at once.

"In series," replied Sandbank cautiously.

"Why not in parallel?" Cookson asked.

"That will cost a lot of money," replied a startled Sandbank.

"Yes, but it will speed up the program," was the response.

That night, Sandbank revised his plan, doubling or tripling the budget, but fully expecting the number to be cut. He was amazed the next day when Cookson said: "That looks like a good program, Charlie."

Later Sandbank learned the impetus for the expansion came from the very top. Harold Geneen, the legendary businessman who built ITT into a global conglomerate, had read an article predicting fiber optics would become the most important advance in telecommunications since the transistor. Noting the development came from "STL in Harlow," Geneen asked, "Is that our STL in Harlow?" Told it was, he asked pointedly, "Well, how is this optical fiber program?" That greased the wheels for a budget increase.[58]

Signs of Slow Progress

The technological boom of the 1960s, fueled most visibly by the space race and less obviously by Cold War military programs, began to wind down after Neil Armstrong walked on the moon in the summer of 1969. The sustaining forces of money, mind-set, and momentum began to dissipate. The space program lost direction. America's love affair with technology faded as the curtains were stripped away from the ugly realities of pollution and the machinery of nuclear Armageddon. The economy sputtered and slowed.

Telecommunications had it easy in comparison. Slower growth left more time for semiconductors to approach the frequencies needed to build millimeter waveguides linking the world's great cities in the 1970s. AT&T had high hopes to revive demand with its Picturephone system, a highlight of the 1964 World's Fair in New York. Bell Labs was working on laser communications to deliver a new generation of even higher capacity systems in the 1980s and beyond. Miller and Kompfner still expected the beams to go through gas lenses or confocal lens waveguides, but fiber optics had joined the race as a dark horse.

Charles Kao had eloquently sold his dream. His success came partly from offering the attractive prospect of a flexible waveguide, and partly from his experimental *tour de force* in measuring the incredible clarity of fused silica. A small cadre of other key scientists and engineers around the world were convinced that fiber loss could be reduced. While the Post Office, STL, Pearson at Bell Labs, and the Japanese stood in the forefront, others labored quietly, including Bob Maurer, who had built a little team of young scientists at Corning.

However, by the spring of 1970 others were coming to suspect fiber attenuation was inevitably high. Theorists suggested minor irregularities in the core-cladding interface might scatter light into modes that leaked out of the cladding.[59] Miller emphasized gas lenses and hollow waveguides when he reviewed optical communications for the prestigious *Science* magazine. With typical managerial caution, he wrote: "Glass fibers are expected to be useful over distances of tens of meters, certainly, and over distances of kilometers if the material purification work proves fruitful."[60] To most people involved, that was a big "if."

11

Breakthrough

The Clearest Glass in the World (1966–1972)

If you do something different from what everybody else is doing, you've got two advantages. One is that you may succeed where they fail, of course, but even if you fail you will gather information that they don't gather. [That] will give you some insights into what might follow

—Robert Maurer[1]

We never told anyone where we were, and I had always assumed that everybody was coming along about the same way we were.

—Robert Maurer[2]

From his first-floor office at the Corning Sullivan Park Research Center, Bob Maurer could see the clear and compelling logic in making fibers from compound glasses. The materials were well known, and the technology for handling them was well developed. Removing the impurities that absorbed light seemed a straightforward matter of purifying raw materials. The approach was so logical that everyone was doing it, including companies with resources far greater than the Corning Glass Works. It was a race Corning seemed doomed to lose.

Never one to pick the beaten path, Maurer looked for alternatives. His native caution warned him to be careful in picking long shots. He wanted to take advantage of Corning's extensive expertise in glass technology. Fused silica stared him in the face. Everyone else had scorned it, and that appealed to his contrarian nature.

Others avoided fused silica because its refractive index was too low for fiber cores, and its melting temperature too hot for their furnaces. Maurer worried about these difficulties; he was always a worrier. But he knew that fused silica was the purest glass in the world, and that Corning had three decades of experience working with it. After some preliminary research, he decided the long shot was a sensible bet.

A Company Built on Glass

Corning is a company built on glass, with its roots in the nineteenth-century advances that made glass a cheap, widely used material, and its headquarters in the small city of Corning in the hills of western New York state. Unlike a host of glass makers whose names survive only on antique bottles, Corning eagerly embraced new technology. The company built a healthy business making glass bulbs for Thomas Edison's electric lamps. In the late 1920s, Corning made glass tubes for the pioneers of electronic television.[3] Instead of retreating in the face of the Depression, Corning hired scientists in the 1930s to develop new kinds of glass. Fused silica was among the results.[4]

Among the new hires was a young organic chemist named Frank Hyde, whose job was to explore ways to apply organic chemistry to glass making. Plastics were new, and their potential competition with glass worried Corning. The company also was casting a giant blank for the 200-inch telescope on Mount Palomar, using Pyrex glass because its size and shape change little with temperature. Pure silica would have been better, because it expands less, but no one knew how to make it in quantity. Hyde came up with an idea. He had earlier discovered that silicon tetrachloride, a colorless liquid, boiled at 57.6°C (136°F), not far above room temperature. He filled a wash bottle with the liquid and squirted it into the flame of an oxy-hydrogen torch. The hot water in the flame reacted with the silicon tetrachloride, yielding a fine dust of the purest silica anyone had ever made. From that experiment, Corning developed a process called flame hydrolysis that is still used today because it yields materials of exceptional purity.[5]

Removing the last traces of impurities is very difficult with standard chemistry. Precipitate something from solution, and a little bit stays dissolved in the liquid, while little traces of the liquid remain in the solid. "Pure" chemicals may contain parts per thousand or parts per million of other materials. Flame hydrolysis works differently because it separates materials by differences in their vapor pressure. Pick the right materials and the right temperatures, and vapor pressures can differ far more than solubility. You can't separate a dash of salt from a glass of water by mixing chemicals, but the salt will stay behind when the water evaporates.

The first step in making fused silica is to dissolve ordinary sand and convert its minerals into chloride compounds. Then you separate the chlorides and heat them slightly. Silicon tetrachloride evaporates readily because of its low boiling point, but the chlorides of troublesome impurities have much higher boiling points and stay behind. For example, molecules of iron chloride are a thousand trillion (10^{15}) times less likely to evaporate than silicon tetrachloride.[6] Inject the silicon tetrachloride into a flame where hydrogen burns in oxygen to form water, and the silicon tetrachloride splits the water molecules. Hydrogen chloride remains in the air (but has to be removed as a pollutant). A thick white "soot" settles out, silica containing just parts per billion of impurities.

In Hyde's time, the main attraction of fused silica was its low thermal expansion. In 1939 another Corning scientist found that adding titania (titanium dioxide) to fused silica could reduce its thermal expansion almost to zero near room temperature. In the early 1950s, Corning started making large pieces of ULE (for ultra low expansion) glass and built a small but healthy business selling it for demanding applications including telescope mirrors and spy satellites.[7] (It also is used in Corningware ceramics.)

Research remained important at Corning, which in 1963 moved its scientists into a sparkling new glass-walled complex designed by the same architect as the sprawling Bell Labs building in Holmdel. Sitting on a hillside outside of the town of Corning, it overlooks the Chemung River valley, where the company began.

A Worry Catalog

Bob Maurer was a logical choice to seek the clearest of glasses. Born July 20, 1924, in St. Louis, he grew up in Arkansas and earned a doctorate in low-temperature physics from MIT before joining Corning in 1952. By the mid-1960s, he had thoroughly settled in rural western New York. He had spent a decade on glass research and managed a small glass development group. Maurer had tried to make lasers by doping glass with the rare earth europium. In the course of that work, he had met Eli Snitzer, who had taught him to view optical fibers as waveguides.

In his systematic way, Maurer collected information. Optical fibers were not new to Corning; a plant in the valley had begun making fiber-optic faceplates in 1963.[8] Chuck Lucy, who was in charge of that product line, became the corporate sponsor of Maurer's research.[9] Maurer talked with the Corning fiber engineers, although he knew communication fibers would require a different technology. He asked Stew Miller at Bell Labs about communications.[10] He quizzed Shaver about his visit to Britain, and when others followed Shaver, he talked with them.

He also began investigating on his own. He had two men in his group, Jack Stroud and Guy Stong, study the loss in bundled fibers. They concluded it came mostly from defects formed during fiber fabrication. To some extent that was good news; if imperfections caused the loss, better fiber fabrication should reduce it.

Fused silica was a logical starting point. Maurer had recognized its low scattering back in 1956, and the lab had plenty of samples of the two types Corning manufactured—pure fused silica and titanium-doped ULE glass. Ideally, he would have preferred to use the pure material for the core, where most light traveled, to limit impurity absorption. However, like most impurities titanium increases the refractive index of silica, so the titanium-doped silica had to be the core, with pure fused silica as the cladding.

"The choice of silica had a lot of disadvantages to it, which concerned me considerably," recalls Maurer. He had a long catalog of worries. No one knew the lower limits of glass absorption, or what glasses would be clearest. It wasn't certain if anything could meet Kao's goal of 20 decibel per kilometer fibers. "Everything absorbs light; it's just a question of what that level of absorption is," he explains.

Another concern was the high temperature needed to draw silica fibers. Oxygen can escape from very hot glass, leaving defects called "color centers" because they absorb light. "Putting the dopant in the core is a lousy idea in that it gives you great opportunity to generate these color centers," says Maurer.[11]

He knew it was a gamble and thought the odds were against success. However, Armistead thought the payoff was worth the risk, and Maurer agreed. It wasn't a big investment, or even a full-time job for anyone. They wanted to see if anything in Corning's considerable bag of glass tricks could beat the odds.

Maurer put optical fibers on a list of potential projects for graduate students working during the summer of 1967. His goal was not hugely ambitious, just to make single-mode fibers and measure their attenuation. It caught the eye of Cliff Fonstad, an MIT student interested in both electronics and materials. Fonstad started with Snitzer's papers on single-mode fibers, but realized he didn't have enough time to make the best possible fibers. He took the simplest approach he could think of, threading an unclad fiber from Corning's fiber-bundle plant through a glass capillary tube from the laboratory stockroom, melting them together, and stretching the preform into a fiber. It was essentially the rod-in-tube approach Larry Curtiss had used a decade earlier, but by using a small fiber instead of a thick rod, Fonstad could produce the thin core needed for a single-mode fiber. He didn't have a furnace hot enough to melt silica, so he used ordinary glass. He didn't worry about preparing the surfaces.

The fibers he drew were single mode, but they weren't very good. Flaws scattered the red light from a helium-neon laser out of the fiber, and the loss was high.[12] Fonstad went back to MIT to study other things, but his results encouraged Maurer to think that better materials and better preparation would yield better results. He convinced Armistead to invest a bit more.

Maurer also decided to try fused silica for the first time. He had found a unique resource. Another Corning scientist, Frank Zimar, had a furnace that could heat glass above 2000°C (3600°F), which Maurer's group could use to melt fused silica and draw it into fibers. At the time, no one else working on fiber optics had a furnace that could match it.[13] Initially, Maurer stayed with the rod-in-tube approach, machining a rod of titanium-doped silica and slipping it into a hole drilled through a cylinder of pure fused silica. His group succeeded in drawing fibers, but loss was about 10 decibels per meter. Maurer was not discouraged; everyone at Corning had expected fused silica to be difficult. It was time to expand the project by adding one of Corning's glass experts and hiring a young physicist to work on fibers full-time.

The Strategist and the Young Scientists

As a research manager, much of Maurer's job was to plot strategy. He had already focused on fused silica to exploit Corning's unique expertise in the material. He weighed that strength against the risks that silica might be intractable, that its loss might be irreducible, or that there might be no way to make it into a viable light-guiding fiber. As a good strategist, he marshaled human assets as well as physical ones.

Compared to Kao and Roberts, Maurer was a master of glass, but other Corning scientists knew more. In late 1967, he asked the glass chemistry department for help. They pointed him to Peter Schultz, hired just months earlier after earning a doctorate in ceramic engineering from Rutgers University. Ironically, Schultz had earned his degree by developing a glass doped with iron and lithium for computer memories that was absolutely black.

Tall, thin, and self-directed, Schultz was barely 25, a kid from working-class New Jersey whose fascination with space drew him into science. He grew interested in aerospace ceramics as a Rutgers undergraduate and stayed there for graduate school, where he fell under the spell of Norbert Kriedl, a world-class glass scientist who had headed research and development at Bausch and Lomb until he turned 60. When Schultz graduated, Kriedl told him Corning was the best place in the country to work on glass science. Corning put him to work on fused silica when he joined the glass chemistry research group in July 1967.[15] Just a few months later, Maurer asked him to make the clearest glass ever known.

With money for a full-time physicist, Maurer went recruiting and found Donald Keck. Born in Lansing on January 2, 1941, Keck went straight through Michigan State University, receiving a doctorate in 1967. Jobs were plentiful for physics graduates as NASA pushed for the moon and the Pentagon pushed new weapons for Vietnam; Keck had a choice of offers. A compact man full of energy and enthusiasm, he had grown intrigued by the propagation of electromagnetic waves. Maurer hooked him with the bait of building a low-loss optical waveguide. In January 1968, just after his 27th birthday, Keck began working full-time on the fiber project.[16] He soon got help from a young Canadian, Felix Kapron, who had joined Maurer's applied physics department the previous year after receiving his doctorate from the

University of Waterloo. Kapron had been studying laser glass, but Maurer asked him to calculate the ideal dimensions for core and cladding of single-mode fiber.[17]

Keck and Schultz started by drilling rods and tubes from chunks of fused silica, melting them together, and drawing them into fibers, hoping to improve the rod-in-tube process. However, collapsing the tube trapped bubbles, cracks, and abrasive grit remaining from mechanical polishing at the critical interface between core and cladding. Those irregularities scattered light. "Polishing is simply a process of going to finer and finer abrasives to make the cracks as fine as possible. If you get them small enough, you don't see them, but the light will," says Keck.[18]

Next they tried flame polishing. The rod was easy; they just ran it through a hot flame. Reaching inside the tube was much harder, because the central hole was only about a quarter-inch (six millimeters) across. Keck first tried the infrared beam from a carbon dioxide laser, which glass absorbs strongly. He bounced the beam off a mirror at one end of the tube and directed it at the inside wall with a second mirror that moved along the tube. The focused beam melted the surface, but the moving mirror made screw-thread patterns inside the tube. He eventually turned to sliding a small torch through the tube to flame polish the inside surface smoothly.

Meanwhile, Schultz and Keck decided to try a different approach to making single-mode fibers. They didn't need much material for the core, so it seemed simpler to deposit a layer of raw core glass inside a thick tube. Flame hydrolysis could deposit fluffy white fused-silica soot. Heating the tube should melt the soot to form a clear inner layer that would become the core when drawn into a fiber. That, they reasoned, should make a cleaner interface.

The trick was to deposit the glass soot uniformly. Schultz had set up his burners in a fume hood for ventilation and pollution control, so they had to custom-build a lathe to roll up to the torches. Keck mounted a stainless-steel tube inside a giant ball bearing with a two-inch (five-centimeter) central hole, and mounted a 1-inch (2.5-centimeter) fused silica tube inside it. He added a belt drive to turn the tubes and hauled the whole assembly in the elevator from his first-floor laboratory to Schultz's lab on the fifth floor.

Schultz carefully aligned a burner to point along the hole in the tube. They started the lathe rotating the tube slowly, then fired up the burner to spray titania-doped soot for the fiber core. The lathe spun perfectly, but all the soot collected at the end of the tube. The soot couldn't go down the ¼-inch (6-millimeter) hole because no air was flowing through it.

They needed to get air flowing down the tube to carry the soot and coat the inside uniformly. The frustrated pair looked around Schultz's lab, and their eyes fell on an old General Electric canister vacuum sitting in a corner, which Schultz used to clean up the mess inevitable in a glass lab. Inspiration struck. They shut down the lathe and burner, cleaned off the tube, and remounted it. This time they hooked the vacuum to the end of the tube to suck air and soot through the little hole. When they fired up the burner and removed a baffle, the vacuum sucked air down the tube, spreading soot along the inside surface.

The vacuum didn't survive the nasty mixture of chemicals in the burner exhaust, but it did achieve a breakthrough. The pair had learned how to deposit glass inside a tube. Making uniform deposits required air-flow equipment more precise than an old vacuum cleaner, but that was fine tuning. They had developed an important new way to make a fiber preform, called inside vapor deposition because it deposited the core glass on the inside of the tube that became the cladding. The next step was to heat the preform, melt the fluffy soot to coat the inside of the tube, and draw it out into a fiber.

Schultz did not have fiber-making equipment in his lab. Keck hauled the preform to Zimar's lab, where he mounted it vertically, with the lower end in the furnace. The heat first melted the soot, then softened the whole tube. Pulling on the bottom stretched the glass into a fiber, closing the central hole. After they drew a short length of fiber, Keck snipped off a bit and hustled back to his laboratory to measure its core diameter through a microscope. Then he went back to Zimar's lab and adjusted temperature and drawing speed to get closer to the desired diameter.

They carried out a series of experiments, making preforms and drawing fibers from them in various ways. They carefully measured fiber properties to see what happened as they changed things. Between experiments, Keck and Schultz analyzed their findings and devised the next round of trials. It was a pattern common to every lab trying to make low-loss fibers: design an experiment, perform it, measure the results, deduce what happened, then design a new experiment. They cycled through the process again and again. The results didn't always improve, but like every experimenter they sought to learn something each time. Maurer strategized; Zimar, Kapron, and others contributed ideas. Bit by bit, the Corning team worked out the details of making single-mode fibers.

Working Out the Process

Corning didn't reveal much to the outside world, although Roberts paid a visit and Maurer spoke with some British Post Office researchers. However, Maurer had progress to share whenever time came to report to Armistead. Lucy went on the road, hunting outside support for Corning's fiber research and help with other aspects of fiber communications. AT&T showed no interest, so he went overseas. He lured money from a few makers of copper cable—Siemens in Germany, Pirelli in Italy, and BICC in England. He worked out agreements with French government labs and with Furukawa Cable in Japan. It helped keep the project going.[19]

Through the hot, humid summer of 1969, Keck drew fibers in Zimar's laboratory. "It was hotter than Hades" with the furnace running, he recalls.[20] He wound fibers of various sizes onto 2-foot (60-centimeter) diameter cardboard drums used to ship raw materials to the glass lab, a step beyond oatmeal cartons. The drums spun about once every two seconds, pulling about 160 feet (50 meters) of fiber a minute. The experiments taught Corning important practical lessons. Heat and humidity made small fiber, 50 micrometers or 2 mils thick, stick to the tractor drives. Fiber 250 micrometers or 10 mils thick grew brittle and broke in a few weeks, with loose ends springing up on the reels. Corning settled on an intermediate diameter, 125 micrometers or 5 mils, which remains the standard size for telecommunications fiber.

Other challenges appeared. The transparency of titanium oxides depends on their chemical form. When one titanium atom combines with a pair of oxygen atoms to form titanium dioxide, TiO_2, the molecule absorbs almost no light. However, the high temperatures needed to melt silica drove off some oxygen atoms, leaving some pairs of titanium atoms to share three instead of four oxygen atoms. That compound absorbs light, increasing fiber absorption to a decibel per meter or worse—just as bad as the compound glasses Corning was trying to avoid. Keck and Schultz experimented and found they could replace the missing oxygen by heating the fiber to 800 to 1000° C in oxygen, returning titanium to its more transparent state. Unfortunately, the cycle of heating and cooling tended to make the randomly oriented atoms in glass align themselves into crystals at the fiber surface. The crystallization caused tiny cracks to grow at the surface, making the fibers very fragile.[21]

Keck and Schultz found a delicate but not perfect balance. By early 1970, they had found a titanium doping level that seemed right and had learned how to make good preforms. Progress was slow because it took them about three months to make a preform, measure its properties, draw it into fiber, analyze the results, and use the lessons from those experiments to make another preform.

That Eureka Moment

Experiments inevitably have frustrating moments. One came in early 1970, when a preform slipped against the furnace wall soon after fiber drawing started. The accident ruined the preform, but Keck decided to test the 20 meters (65 feet) of fiber he had pulled. After heat-treating the fiber, he aimed a helium-neon laser down it. At first he couldn't detect any loss, but more careful measurements showed attenuation probably was close to the magic figure of 20 decibels per kilometer.[22]

That was a big improvement. Keck had been estimating the loss of fibers only about three meters (10 feet) long by measuring changes in light transmission as he broke small pieces off the end. The light intensity had changed every time he broke a 10-centimeter (4-inch) segment off the old fiber, but with the new fiber the change was too small to measure accurately with the best instruments he had. With the preform ruined, he couldn't draw any more fiber from it. Nor did he have enough data to justify announcing spectacular results.

The only thing for Schultz and Keck to do was to make another preform using the same formula and hope the fiber was as good. They pulled a kilometer (3280 feet) of fiber without incident, although it broke in two when Keck rewound it onto a drum for heat treating. He concentrated on the smaller 200-meter (660-foot) segment.

The fiber came out of the heat-treating furnace on a Friday afternoon. It was early summer, and almost everyone else in the lab had gone home, but Keck was eager to see the results and worried that the fiber—fragile after heat treating—might not be in one piece when he returned on Monday. He set up a test jig that aimed a red helium-neon laser beam into the fiber to help him align it.

"I remember so vividly moving the fiber over, and when the laser spot hit the core, all of a sudden I got this flash of light. It was different, a spot that was different than the laser spot," recalls Keck. That puzzled him; he had never seen such a spot before, and he didn't immediately know what it was. Eventually he realized that the light had gone back and forth through all 200 meters of fiber. The distant end of the fiber reflected 4 percent of the light that hit it back toward Keck, and that light was passing back through the fiber to reach his eye. It was still bright after passing through 400 meters (1320 feet) of fiber. He had before him the clearest glass ever made. Delighted and excited, he went searching for someone to share his eureka moment, but there wasn't a soul left in the lab. He went home and enjoyed the afterglow of success all weekend.[23]

Careful measurements later showed the fiber had attenuation of 16 decibels per kilometer. This time they had plenty of it. The Corning team pinned the measurement down so tightly the error was only plus or minus one decibel per kilometer. It was an impressive achievement and a dramatic improvement over previous fibers. Don Keck, Bob Maurer, and Peter Schultz had hit the fiber-optic jackpot.

A Mutually Surprising Announcement

The breakthrough presented Corning with a strategic dilemma. Corning had built its success in the highly competitive glass industry on new technology, which for glass usually means new processes. While processes can be patented, it is hard to make the patents so tight that competitors cannot circumvent them, so the glass industry had a tradition of jealously guarding its techniques as trade secrets. This left excited Corning scientists wary about disclosing details, even to their development partners. Maurer was particularly cautious because he thought competitors were close on his heels. It didn't matter that the British, the Japanese, and Bell Labs were reporting mediocre results; Maurer assumed that they, like Corning, were keeping their successes quiet.[24]

In fact, everyone else in the low-loss fiber race were telephone companies, who lived in a different world than Corning. In 1970, phone companies didn't worry about competition; suppliers like STL had secure niches selling equipment to government monopolies. They had better patent coverage and a tradition of talking openly. For AT&T, it was more than a tradition; the company had agreed to share Bell Labs technology to keep its telephone monopoly in the United States, although it still took care to patent its inventions.

Corning managers assessed their position carefully. They thought they were ahead of other companies and wanted to stay there. Their fibers were exceptionally transparent, but far from practical. Brittleness was the most obvious problem, but a host of other issues remained. They wanted to announce Corning's success while giving potential competitors as few clues as possible about how to duplicate it. "We knew that it was going to be many years before this thing was going to be used, and wanted to give ourselves as much lead time as we possibly could," recalls Keck.[25]

The first order of business for Keck, Maurer, and Schultz was to sit down with the patent lawyers. Glass patents were not bulletproof, but Corning believed in patenting its inventions and guarding them jealously. In July Maurer shared the good news—but not the details of how the fiber was made—with Bell Labs, because the phone company was Corning's biggest potential customer.[26]

Meanwhile, Maurer wrote a paper to announce the breakthrough—but only very coyly. He concentrated on measurements of light scattering, not on total attenuation, which includes absorption as well as scattering. He reported scattering loss of about 7 decibels per kilometer in straight fibers, "very close to intrinsic material scattering loss." That was not surprising; Kao had blamed impurity absorption for most fiber loss. At first glance, Maurer's paper seemed to confirm this, because he noted that the scattering measurements were made on two 30-meter fiber segments with total loss of 60 to 70 decibels per kilometer. The paper went on to compare measured losses in bent fibers with a theory proposed by Henry Marcatili and Stew Miller of Bell Labs. The written version gave no details of fiber composition or fabrication, and hid the breakthrough in a single sentence:

> The lowest value of total attenuation observed in all waveguides constructed for this work was approximately 20 decibels per kilometer.[27]

Maurer listed Kapron and Keck as co-authors because they did the calculations and measurements he described.[28] Not wanting to reveal anything about composition, he listed the glass specialists—Schultz and Zimar—only in an acknowledgment. Maurer decided to announce the results at a conference held in London by the Institution of Electrical Engineers at the end of September.[29] In August, he sent a slightly different paper to *Applied Physics Letters*, which appeared after the conference.[30]

The conference on "trunk telecommunications by guided waves" was an ironic venue to report a fiber breakthrough. It had been organized by the developers of millimeter waveguides, confident that their technology was at last ready to provide the "trunks" linking the world's great cities. The grand old man of the millimeter waveguide, Harold Barlow, opened the meeting. It was an honor for a man whose vision seemed about to be realized. He rose to the occasion, tracing progress from the discovery of radio waves and citing major contributions by Stew Miller and Toni Karbowiak. He admitted that a decade earlier developers had been "rather too sanguine about the prospects of practical application." Now he was convinced its time had finally arrived. He joked about being accused of having a "pipe dream" and of almost being driven "round the bend" by the bending losses in millimeter waveguides. Then he looked to the future:

> With most of these difficulties behind us, I feel quite sure that the journalists from the other side of the Atlantic are quite realistic in describing the application as "pipes of progress." It is true that fully engineered prototypes have still to be established, but so much work has already been done in a practical environment that confidence in the ultimate success of the project runs high.
>
> Dielectric surface waveguides, including optical fibers, for telecommunications are perhaps a little more remote at present, but their turn will come, and even in the near future they could find valuable applications over the shorter distances.[31]

As group manager, Maurer made the trip to London and sat through the handful of talks on optical fibers buried among the many on millimeter waveguides. Dick Dyott came to describe his work at the Post Office. Charles Kao came as well, but having made fiber optics a respectable cause, he was about to turn to a more personal mission. His children were eight and six, and he wanted them to learn what it meant to be Chinese. He was taking leave from STL to teach at the Chinese University of Hong Kong and turning his fiber-optic group over to Murray Ramsay.

Most of the crowd was not tuned in to fiber optics or to Bob Maurer. He is a modest, matter-of-fact speaker, not a master showman, and he was talking about a technology they thought was at least a generation away. He buried the breakthrough datum—measured attenuation of 16 decibels per kilometer—in a talk describing the details of scattering. It was an eye-opener only to those who already had their eyes open.

"A few people in that crowd were really turned on," Maurer recalls.[32] Among them was Stew Miller. Not completely sure of the initial measurements, Maurer only claimed to have reached 20 decibels per kilometer when he spoke to Miller in July. By the end of September, he was convinced the actual loss was 16 decibels per kilometer. Miller didn't realize Maurer was talking about the same fiber; he thought that Corning had made further improvements since July.[33] The misunderstanding helped light a fire under Miller, who feared Bell was falling further behind.

Engineers from a British electronics firm hustled Maurer off to lunch to ask questions. Engineers from the Post Office and STL both asked Maurer to bring samples for them to test in their own labs. Others seemed less excited, not realizing the importance of the breakthrough, or knowing how easily mistakes can be made in optical measurement.

The few other fiber papers at the meeting surprised Maurer. Thinking he was in a tight race, he hadn't turned around to study the field behind him. After he crossed the finish line, he was amazed to see everyone else remained far, far behind.

Barlow opened a general discussion at the end of the meeting by asking what technology would come next, the metal waveguide or the optical fiber. One Post Office engineer said the millimeter waveguide was the obvious choice because it was ready for installation. Heady with bravado after hearing Maurer's report, Dyott brashly retorted, "I'm quite happy for you to lay the waveguides, and we will come along later and fill them with optical fibers."[34]

Guarding Snips of Fiber

The Post Office invitation was too good an opportunity for Maurer to resist. It came from a driving force in the field and a major potential customer. Dollis Hill had developed equipment to measure fiber properties, which in the early days was a tricky task, and independent confirmation of Corning's results would be an important plus.

The basic concept of measuring loss in an optical fiber is simple: Compare how much light enters the fiber with how much emerges from the far end. Divide to get the fraction of light emerging, take the logarithm, multiply by ten, and you get fiber loss in decibels. Light output was easy to measure, but not how much light entered the fiber. Especially in the early days, most light aimed at the fiber would not wind up in the tiny core. Early fiber developers found the best way to account for light that missed the fiber core was to work backward. They first measured how much light emerged from the end of a long piece of fiber. Then they cut the fiber close to the light source, leaving only a segment too short to have much loss of its own. The amount of light exiting the short fiber should be essentially what entered the other end. Dividing the output of the long fiber by that of the short fiber gave the loss.

The cut-back method yields good results, but it consumes fiber. Every new measurement requires chopping another segment from one end of the fiber, so you have to have enough fiber to spare. You can't repeat measurements. More worrisome for Corning, which wanted to keep its fiber composition secret, it breaks off little pieces of fiber, which someone might analyze to determine its content. It was bad enough that Corning's fibers were very brittle; cut-back measurements could sprinkle fiber segments around the lab for anyone to pick up.

Nonetheless, the Post Office measurements were crucial to establish Corning's credibility. Corning carefully wound a long length of precious brittle glass thread on a paint can, then packed it in a custom-built container that looked like a hat box and bought a seat for it on the plane. Maurer sat beside the box, and he or Chuck Lucy, who went with him to discuss business prospects for fiber, kept a hand on it at all times.[35] It was an odd-looking piece of luggage, and London was going through a bomb scare when they arrived. A bobby inspecting incoming luggage looked long and hard at the strange package. Worried about ruining the fragile fiber, Maurer took off the lid and showed the bobby the optical fiber wound on the can.

The bobby looked and replied, "Oh yes, they use them in medicine."

Amazed that a policeman would recognize such novel technology, Maurer explained he was going to the British Post Office to measure it. That satisfied the bobby, who sighed and said, "You Yanks will probably be selling them to us one of these days," before moving on to other luggage.[36]

Maurer and Lucy arrived at Dollis Hill late in the morning of Monday, November 9, 1970. After the Post Office team installed the fiber in their apparatus, Dyott suggested they break for lunch and do the measurements after returning. Maurer insisted he had to stay with the fiber. Dyott displayed a large key for the 1920s vintage lock and offered to lock the lab, but Maurer refused to budge, even when Dyott said he could hold the key. Not about to be denied their lunch, the British left Maurer behind, returning with a cup of coffee and a cheese sandwich for him.

They spent the afternoon making measurements with a helium-neon laser. Several tests confirmed that attenuation was "an astonishing 15 decibels per kilometer,"[37] as Maurer had claimed. Each time he carefully cleaved a piece of fiber off the reel, Maurer retrieved it and taped the piece onto the side of his drum. Then a short piece broke with an audible ping and fell to the floor. The horrified Maurer searched for it on a floor sprinkled with pieces of British fiber left from earlier tests. Corning's fiber was larger, and heat treating gave it a distinctive curve. Nonetheless, he couldn't find the missing piece and had to leave without it after the day's measurements were done.

As soon as Maurer was gone, Dyott stopped everything to sweep up the fiber fragments on the floor. The Post Office crew sifted carefully through the debris and, Dyott recalls, "in no time we found the piece of larger-diameter slightly curved fiber, about 1½ to 2 inches long. I handed it to George Newns to see what he could make of it."[38]

Dyott and Newns knew the secret of Corning's fiber lay in its composition, so Newns immediately took the little piece to British Titan Products, which had a neutron activation system that could sensitively analyze composition of even tiny samples. Within hours, British Titan reported back that it was the purest form of silica they had ever encountered. Dyott knew the fiber had to contain something else, either to dope the core to increase its refractive index, or to reduce the index of the cladding. He asked what else was in the glass, but the neutron activation specialists said they had found only pure silica. Dyott insisted there had to be something else. Reluctantly, British Titan admitted there was one possibility. The company specialized in titanium products, so traces of the metal were everywhere, and that background contamination could conceal small quantities of titanium from their instruments. That hinted at the identify of the magic ingredient, but a definitive answer was out of reach. The test had consumed the tiny piece of fiber.[39]

Maurer also lost a fiber fragment at STL, where engineers used a different trick to deduce composition of the Corning fiber. To make sure the measurements were accurate, light captured in the cladding had to be removed from the fiber. This can be done easily by immersing the fiber in a liquid with higher refractive index than the cladding, so light in the cladding escapes into the liquid. STL kept changing liquids during the measurements, each time giving Maurer one with lower refractive index. They figured that when they got to a liquid with an index lower than the cladding glass, the light would not escape into the liquid, and the measurement wouldn't work. However, all the liquids worked, indicating the cladding had to be the lowest-index glass, fused silica. Murray Ramsay recounts the trick with pride, but admits "it didn't help us, because we still didn't know how to dope the bloody core."[40]

A Crucial Turning Point

The low-loss Corning fiber was a breakthrough that marked a turning point for fiber-optic communications, although even the technical press almost completely ignored it at the time.[41] It made Charles Kao a visionary; no one else had been willing to bet his future on the clarity of glass fibers. It thrust Corning into the telecommunications business. It made Roberts a prophet as well, and gave fiber a big boost around the world.

Yet in reaching that crucial milestone, Corning did not demonstrate a practical technology. The heat treatment that made titanium-doped fibers clear also made them brittle. If the lab got too hot or cold over the weekend, Monday morning found little brittle ends of broken fibers sticking up on the reel. Light sources remained a problem (detailed in chapter 12). Much work remained to make viable communication systems.

Corning concentrated on fibers. Keck and Schultz found heat-induced crystallization caused the brittleness. They tried etching fibers with hydrofluoric acid to remove surface contamination, but the process was awkward and cumbersome.[42] Hydrofluoric acid is nasty stuff, and thin fibers could easily be etched away to nothing. However, their results persuaded Corning management to move fibers from research into a larger-scale "development" program.

Competitors stepped up their efforts as well. The British Post Office and STL steadily improved double-crucible fibers. Other labs around the world started looking seriously at fibers for the first time. The biggest change came at Bell Labs, where Dave Pearson's little group at Murray Hill had been the only ones working on fibers. Top management was not happy to see another company stealing the thunder of what was supposed to be America's biggest and best corporate research operation. "We could feel the pressure all the way from the vice president," recalls Pearson, who was surprised by Corning's announcement. Management asked, "If Corning can do it, why can't you?"[43] Pearson stepped up his efforts on multicomponent glasses and added to his team at Murray Hill.

Seeking a Better Magic Ingredient

As others tried to identify the mystery ingredient in Corning's first low-loss fiber, Schultz sought better dopants. Corning reduced loss of titania-doped fibers to about 10 decibels per kilometer, but they remained extremely fragile. Schultz wanted something that increased the refractive index of fused silica, without absorbing light or requiring the troublesome heat treatment. He also wanted a material that made a better glass.

The chemistry of glass is tricky, in an entirely different league from things that fizzle in high-school test tubes. Glasses are unusual solids that form only when certain liquids are cooled in the proper way. As most liquids freeze, atoms link up in ordered crystalline lattices. In a glass, the atoms retain the random arrangement of a liquid, but are frozen in place.[44] Only a few materials form glasses; silica is among the best. Titanium dioxide, a compound called "titania," is not a glass-former. Silica can form a glass if it contains a dash of titania, but mix in too much titania, cool the liquid too slowly, or do any of a thousand other little things wrong, and the solid crystallizes instead.

Germanium was a logical alternative. Just a row below silicon on the periodic table, it has similar chemistry and is also a glass-former. It absorbed little light and did not require the troublesome heat treatment. Schultz also found problems: Germanium oxide (or "germania") evaporated more easily than silica, which complicated glass formation. Nonetheless, in early 1972 Schultz adjusted the flame-hydrolysis process and started doping preform cores with germanium instead of titanium.

By then, Corning was scaling up fiber development, adding people and investing money, but still cautious because the future was far from assured. The changes meant Keck and Schultz no longer had to draw every fiber themselves. Technicians and other scientists helped them mount the germanium-doped preforms in the furnace and draw fiber from them. Keck took samples to his lab and measured them. Even the first of the germanium-doped fibers looked encouraging.

It wasn't long before Keck had results good enough to invite Maurer's boss to take a look one Monday morning. However, when he set up the test, not a speck of light went through the fiber. Keck sputtered an apology and set to work finding the problem. He discovered the problem arose from Corning's habit of shutting down air conditioning over the weekend. Humidity built up and the cardboard reel holding the fiber expanded, stressing the fiber such that it wrinkled, forming tiny "microbends" where light leaked out. Microbend loss had shut down light transmission totally. Keck rewound the fiber and called the man back down to show that he'd solved the problem.[45] With a little experimentation, he and Schultz reduced loss to a mere four decibels per kilometer. It was June, 1972, not quite two years after they had broken the 20 decibel per kilometer barrier.

"This was to my mind the breakthrough for fiber optics, because it was now a practical fiber," says Schultz. "It didn't break, you didn't have to go through all this other stuff. You could make fibers right off the draw with these nice low losses."[46] Bob Maurer's strategy had paid off. Half a dozen years earlier, fiber optic communications had been merely the dream of a Chinese immigrant to England. Charles Kao had spread his dream around the world. Now scientists working in a western New York town with a population of only 13,000 people had made it a reality.

At the moment of triumph, the deluge struck. Hurricane Agnes swept across Pennsylvania and dumped heavy rains on western New York, causing disastrous floods. The hillside research lab was safe, but Corning and other towns in the Chemung River valley were not. Corning scientists and managers dropped everything to spend weeks helping the community clean up. Yet even in the midst of the disaster word of the new fibers percolated upward. Bill Armistead spotted Schultz while the two were helping flood victims in Corning and congratulated him. "Pete, that's fantastic. You really did it!"[47]

When the mud and the water were gone, the Corning fiber team went back to their lab. They carefully measured fiber loss from the edge of the ultraviolet into the infrared. It was near four decibels per kilometer in two critical parts of the infrared where lasers emit light: 800 to 850 nanometers, and near 1050 nanometers. They identified traces of water as responsible for most of the remaining absorption. In December 1972, Keck, Maurer, and Schultz wrote a paper reporting prospects for low-loss fibers, in which they predicted, "Total attenuation of about 2 decibels per kilometer in the region beyond 800 nanometers thus appears possible."[48] That meant fibers made of doped silica could carry light ten times farther than Charles Kao had thought would be necessary.

Optical fibers had crossed the threshold for communications. Much work remained to be done. Fibers had to be fine tuned to meet system requirements. Telephone companies worried how they could transfer light into fiber cores. Light sources remained an issue, although progress was being made on semiconductor lasers. Fiber strength had to be improved. Coatings were needed to prevent damage during handling. Cable structures and manufacturing technology had to be developed.

Nonetheless, fibers were looking better as field trials made the millimeter waveguide look worse. With the right supporting technology, it looked like Dick Dyott might get his chance to thread fiber-optic cables through buried millimeter waveguides.

12

Recipes for Grains of Salt

The Semiconductor Laser

(1962 to 1977)

> Jack Tillman said "Did you know the semiconductor laser has just been invented? . . . I think we should get some and see if they're any use at all for communications." So I went over [in 1963 to the Joint Services Electronics Research Laboratory in Baldock, north of London] for a month, and brought some of these things back with me. They only operated at liquid nitrogen temperature, only operated in pulses of about 10 microseconds, and it took 100 amperes in each pulse to drive the bloody things. We concluded that they weren't very promising.
>
> —David Newman, British Post Office Research Laboratories[1]

As Corning all too keenly recognized, clear optical fibers were not the only building block vital for fiber-optic communications. A second crucial ingredient was a matching light source. From Bell Labs to the British Post Office, developers of optical communication systems thought the ideal light source would be a semiconductor laser. By a curious coincidence, the spring and summer of 1970 also marked symbolic breakthroughs on the semiconductor laser frontier, although as with fibers, practical devices took longer.

The allures of semiconductor lasers were considerable. One was the sheer magic of solid-state devices. Semiconductor technology was hot in the 1960s;

transistor electronics were pushing vacuum tubes into the grave of obsolescence. Semiconductor lasers generated light easily, with the intensity controlled by the electrical currents passing through them. Tiny as grains of salt, they matched the size of optical fibers, but the potential for compact, solid-state laser transmitters attracted even the developers of hollow optical waveguides.

Unfortunately, early semiconductor lasers were as useless for practical communications as the optical fibers of the early 1960s. The lasers operated only at the cryogenic temperature of liquid nitrogen, −321°F or −196°C. They burned out quickly and unpredictably. Worse yet, the warmer the laser got, the more current you needed to make it emit light, and the higher the current, the faster the laser burned out. It didn't look good, and after an early burst of energy, progress stalled in the mid-1960s. Developers needed better recipes for their grains of salt.

A Curious Class of Materials

Both the allure and the problems of semiconductor lasers came from the nature of semiconductors. From an electrical standpoint, materials fall into three classes—conductors, semiconductors, and insulators. The essential difference among them is the amount of energy needed to free an electron from the bonds that link atoms in the material. In conductors, notably metals, very little energy is needed, so electrons flow freely through copper wires. In glass and other insulators, the electrons are bound so tightly between atoms that essentially none of them have enough energy to escape.

Semiconductors fall between the two extremes, because a few electrons do escape from atomic bonds to conduct electricity in the crystal. The bonded electrons occupy a "valence band," where they form bonds between atoms in the crystal. Electrons that get enough energy to escape those atomic bonds fall into another energy state called the "conduction band." No energy states exist between the valence and conduction bands, so electrons have to get enough energy to jump this "band gap" before they can carry a current.

In a pure semiconductor, only a few electrons have enough energy to reach the conduction band. Those that escape the valence band leave behind vacancies called "holes," which effectively have a positive charge equal to the negative charge of the electron. Other valence-band electrons can move to fill the hole, leaving another hole behind, so essentially the holes move as well as the electrons in the conduction band. For practical purposes, you can think of holes as positive charge carriers and electrons as negative charge carriers, although it really isn't that simple.

One way that engineers can adjust the number of electrons and holes in a semiconductor is to add impurities to replace some atoms in the crystal lattice. If the impurity has one more outer electron than the atom it replaces, that extra electron is free to roam through the crystal, forming an *n*-type (for negative carrier) semiconductor. An impurity with one less outer electron creates a hole, forming a semiconductor of the *p*-type (for positive carrier, because the absence of an electron leaves a positive charge on the atomic nucleus). Adding such dopants releases more carriers in the crystal, making better electronic devices. Put layers of *n*-and *p*-type semiconductor together, and you have a simple two-terminal device called a diode.[2]

Things get interesting when you apply a voltage across a diode. A positive electrical voltage attracts the negative electrons, while a negative voltage attracts the positive holes. Apply a positive voltage to the n-type material and a negative voltage to the p-type material, and the electrons and holes move to the terminals and stay there, so no current flows through the junction between the two materials. However, a current will flow if you switch the voltages, because electrons in the n material move toward the positive terminal on the p material, and holes move in the opposite direction. Thus, a semiconductor diode conducts electricity in only one direction[3] (figure 12-1).

The real action happens at the junction between n and p material, where electrons from the *n* material combine with holes from the *p* material. As the electron drops into the hole, it releases its extra energy by a process called recombination. In silicon, the energy typically is released as heat. However, in gallium arsenide, indium phosphide, and certain other semiconductor crystals, some energy is released as light. This is the basis of a light-emitting diode or LED. Initially, the process was quite inefficient. The first LEDs turned only about 0.01 percent of the input electrical energy into visible or infrared light.[4]

Birth of the Semiconductor Laser

When the laser hit the headlines, a few physicists thought of making lasers from light-emitting semiconductors. However, nobody took the idea too seriously until 1962, when Robert Rediker, Tom Quist, and Robert Keyes changed the way they were processing gallium arsenide at the MIT Lincoln Laboratory. LEDs made from the new material generated light surprisingly efficiently. When Keyes announced their results at a July 1962 meeting in New Hampshire, an astounded member of the audience stood up to say that Keyes's statement violated the second law of thermodynamics—a cardinal principle of modern physics which holds that the degree of disorder or "entropy" always increases. Keyes returned to the microphone and deadpanned, "I'm sorry."[5]

Bob Hall, a semiconductor expert from the General Electric Research Laboratory, immediately figured out what the MIT group had done. After everyone stopped laughing, he stood up and explained why the results were possible. He also realized that high efficiency opened the door to making a semiconductor laser. He worked through the numbers on the train ride home,

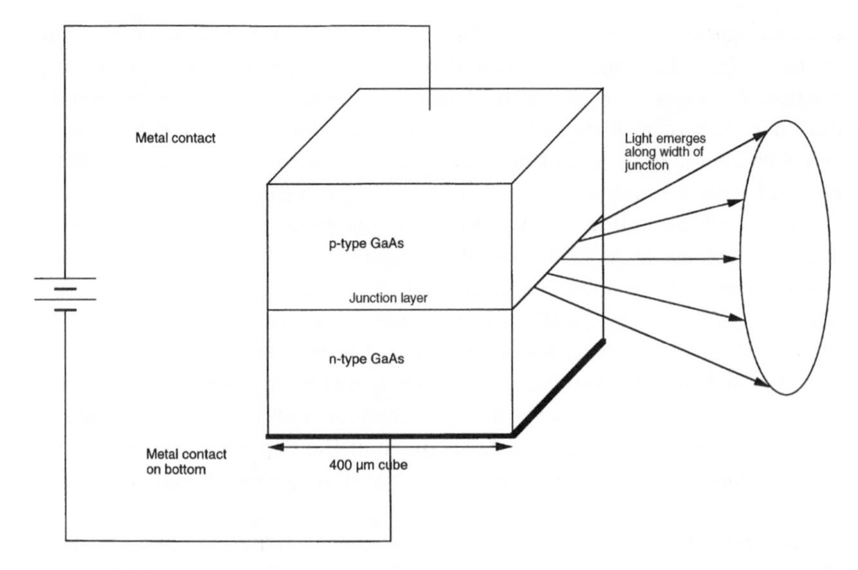

Figure 12-1: The first semiconductor laser was a simple cube of gallium arsenide (GaAs), half p-type and half n-type.

and back in Schenectady he rounded up a crew of other scientists to try some experiments.[6] In short order, they made gallium arsenide chips with edges polished to act as the mirrors needed to generate a laser beam. When they fired powerful electrical pulses through the chips, they emitted light at a narrow range of wavelengths, a sign of laser action. The whole process took just two and a half months.[7] It was a scientific tour de force.

Hall was good, but three other labs were hot on his heels. A team from IBM came in a close second with a slightly different variation, although no one from the lab had attended the New Hampshire meeting.[8] Nick Holonyak, Jr., returned from the New Hampshire meeting to try making lasers from gallium arsenide phosphide at GE's Syracuse lab but didn't succeed until he tried Hall's approach of polishing the crystals.[9] The Lincoln Lab group, distracted by other projects, wound up a close but frustrating fourth.[10]

That remarkably close finish—achieved in the shadow of the Cuban missile crisis—seemed to herald good things for semiconductor lasers, but there the technology stalled. The developers of communication systems wanted lasers that generated steady beams and operated for long periods at room temperature. The lasers made by Hall and the others only fired intermittent pulses and didn't last long even when cooled to liquid nitrogen temperature, −321°F or −196°C. Progress from there was painfully slow. By the end of 1964, the best semiconductor laser could fire a single pulse at room temperature when 25 amperes—more current than a standard household refrigerator draws—flowed through an area of 0.02 square millimeter for 50 billionths of a second. Then it had to cool before firing again.[11]

A Tough Array of Problems

Endless problems frustrated semiconductor laser developers. The devices didn't last long. Some died suddenly; others degraded gradually, emitting less and less light. The beams were messy, not pencil thin like those of gas lasers but spreading rapidly and unevenly, a fuzzy blur instead of the pinpoint of light wanted for communications. No one was sure what caused the problems. Was the design inadequate? Were requirements for crystal quality impossibly high? Or was the whole problem simply insoluble, dooming semiconductor lasers to the oblivion of interesting but impractical devices?

Many researchers interested in civilian communications bailed out, but not the military. Compact semiconductor diode lasers looked promising for use in portable systems with purposes ranging from battlefield communications to measuring the distance of an air-to-air missile from its target. Pulsed lasers could do many of those jobs, although room-temperature operation was vital. Besides, the Pentagon was flush with money.

As often happens, ideas that would help solve those problems had already been suggested but had not been tested. The first diode lasers were made entirely of one material, gallium arsenide, with different dopants. In 1963, Herbert Kroemer, an engineer at the Varian Central Research Laboratory in Palo Alto, suggested adding layers of different composition.[12] So did one of the Soviet Union's top semiconductor researchers,[13] Zhores Alferov of the Ioffe Physico-Technical Institute in Leningrad (now St. Petersburg), but the Soviet government classified his patent disclosure.[14] Their basic idea was the same, to trap electrons at the junction so they could recombine more efficiently with holes. They hoped this would make a more efficient laser, able to operate at warmer temperatures. They realized they could do this by adjusting the semiconductor composition, which affects the band-gap energy electrons need to be in the conduction band. Substituting aluminum for some gallium atoms in gallium arsenide increases the energy requirement, so electrons in gallium arsenide lack enough energy to move into gallium aluminum arsenide. Place a layer of gallium aluminum arsenide next to a *p-n* junction in gallium arsenide, forming what specialists call a "heterojunction," and you've trapped electrons on the gallium arsenide side of the junction, increasing the odds they will recombine with holes and emit light.

Unfortunately, no one knew how to make heterojunctions in 1963. Changing semiconductor composition also affects the spacing between atoms in the crystalline lattice, and mismatches cause fatal flaws in devices. The trick was to find a combination of materials where the lattice difference was small. Gallium arsenide was the logical place to start. The question was what to add.

Alferov's group considered two possibilities: replacing some gallium with aluminum, or replacing some arsenic with phosphorus. Adding aluminum hardly changes the lattice constant, but aluminum arsenide decomposes in moist air, so the Russians doubted gallium aluminum arsenide would be stable. Instead, they added phosphorus, but that changed the lattice constant so much that the lasers worked only at liquid-nitrogen temperature.[15] The Russians were getting discouraged when they looked at some gallium aluminum arsenide crystals one of them had stashed in a desk drawer a couple of years earlier. They were delighted to find nothing had happened to the crystals, showing gallium aluminum arsenide was stable.

Once they realized that, the Russians soon grew heterojunctions using a technique called liquid-phase epitaxy that Herb Nelson developed in 1963 at RCA Laboratories.[16] It crystallizes gallium arsenide compounds from a mixture of molten gallium placed on a semiconductor substrate. This avoids growth problems arising from differences in melting points of the elements. Although the Russians made the first heterojunctions,[17] the Iron Curtain and the need to translate their work into English kept word from reaching America until a team at IBM had reported the same thing.[18]

Heterojunctions revived diode laser development. They were hard to grow, but they reduced the threshold current needed to drive the laser—and its rapid increase with temperature. RCA became a hotbed of activity, funded by military contracts. Henry Kressel and Nelson began making lasers with a single heterojunction. They gradually perfected the process, reducing threshold currents so their lasers could fire repeated short, high-power pulses at room temperature,[19] meeting Army requirements. In April 1969, RCA announced plans to manufacture single-heterojunction lasers; two months later, a spin-off—Laser Diode Laboratories Inc.—announced they would, too.[20] The Pentagon was happy.

The communications industry was not. They wanted semiconductor lasers that generated a steady beam they could modulate with a signal. The best hope for that seemed to be a double-heterojunction laser, where a pair of heterojunctions sandwiched the active layer. Developers hoped that by trapping electrons better this would improve efficiency and reduce threshold current. Kressel and Nelson were working on the idea, but military contracts had taken top priority.

Bell Labs Places Its Bet

Military contracts were not a distraction at Bell Labs, where John K. Galt, director of solid-state electronics research at Murray Hill, was convinced that optical communications needed room-temperature diode lasers. In July 1966, he asked two Bell Labs researchers to figure out why diode lasers had such high thresholds at room temperature. Chemist Mort Panish, 37, and Japanese physicist Izuo Hayashi, 45, who had joined the Bell Labs group a month before, accepted the challenge. Galt didn't expect them to solve the problem, but he wanted Bell Labs to learn enough to compete.[21]

The two men had complementary skills, although Hayashi's awkward English sometimes slowed communication.[22] Panish had studied the chemistry, physics, and luminescence of gallium arsenide. Hayashi wanted to move into semiconductors from high-energy physics. In the careful, cautious way of Bell Labs, they spent months studying compound semiconductors and their luminescence. Not until the summer of 1967 did they start growing lasers, at first entirely of gallium arsenide. Their early heterojunction experiments in November surprised Hayashi by producing strong luminescence, suggesting the possibility of lower laser threshold currents.[23] At the end of the month, the two went to a semiconductor laser conference in Las Vegas, where inspiration struck Panish as he listened to an IBM talk and realized the importance of lattice matching. He walked out of the lecture room and sketched plans for a series of laser structures.[24]

Back at Bell Labs, they started growing single-heterojunction lasers and quickly saw a drop in threshold current. It was a step in the right direction, but they spent most of 1968 slowly improving their techniques and devices. They found that threshold current in single-heterojunction lasers increased by a factor of 11 between liquid nitrogen temperature and room temperature.[25] That was a big improvement over the thousandfold increase in lasers without heterojunctions. They duly reported their results,[26] but only after Kressel and Nelson.[27]

Double-heterojunction lasers were the next logical step, but they were difficult and required more complex processing. Panish and Hayashi tested their first double-heterostructure laser on January 29, 1969. Its threshold current was much higher than their best single-heterostructure lasers. But they were encouraged to find the threshold increased less with temperature than for a single-heterojunction laser. They felt they were on the right trail.

The Race to Room Temperature

They didn't know Alferov was pursuing the same goal. Little word of his work seeped through the Iron Curtain until his first visit to America in August 1969. Alferov told a research conference that he had made double-heterojunction lasers that had a low threshold current at room temperature but could not operate continuously.[28] The report "was like an unexpected bomb explosion for my American colleagues," Alferov recalls.[29] Afterward, he visited Bell Labs, where Hayashi and Panish grilled him for more details. His answers were sobering. "We had not realized that the competition was so close and redoubled our efforts," Hayashi wrote.[30]

In December, Panish began testing a new type of liquid-phase epitaxy, which deposited layers from a "boat" with three liquid-filled slots that slid over the wafer. It took time to perfect layer deposition. Hayashi tested ways to remove the excess heat that raised threshold current and could destroy the lasers. As Panish made new wafers, Hayashi tested their laser properties. By January, they had lasers with room-temperature threshold currents only half what Alferov had claimed in the summer. As the wafers got better, they got the lasers to operate at higher and higher temperatures. Panish kept pushing

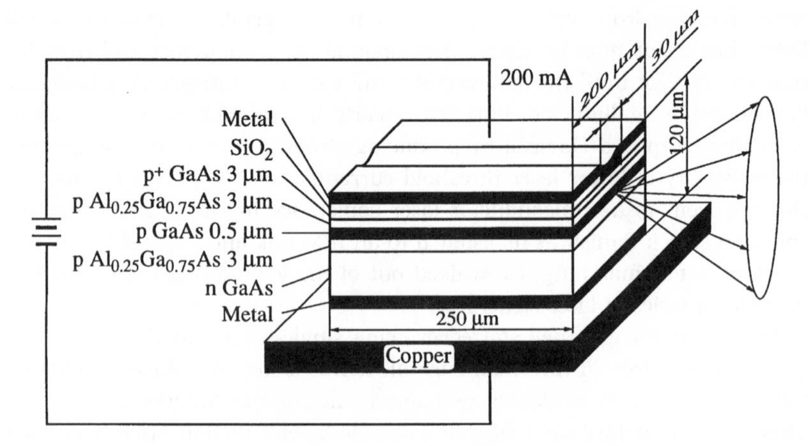

Figure 12-2: The first room-temperature laser made by Alferov's group included several thin layers and a narrow stripe, to better control the flow of current and light. (Courtesy of Zhores I. Alferov.)

after they passed the freezing point of water, making a better-looking wafer. In May, Hayashi got a few diodes from that wafer to emit steadily near room temperature.

Typical of hard-working Japanese scientists, Hayashi was not about to be interrupted by the Memorial Day weekend. On Monday morning, June 1, he tried another diode—and it emitted a steady or "continuous-wave" (CW) beam at 24°C, just above the nominal "room temperature" of 23°C (73°F). The excited Hayashi scribbled a note which Panish found on his door the next morning: "C.W. definite!!! at 24°C 10:30 A.M. June 1, 1970."[31] Word spread quickly through Bell Labs, and a top manager arrived at Hayashi's lab. "In violation of all AT&T rules, he brought a couple of bottles of champagne. I never saw it before or since," says Panish.[32] After growing more than a thousand wafers and testing countless diodes, Bell Labs thought they had scored an important first. Management geared up its public relations machinery to announce the news at an opportune moment.

A Narrow Stripe in Leningrad

Nobody at Bell knew the Russians had broken out the vodka a bit earlier. Alferov's large group made rapid progress after he returned to Leningrad. They couldn't match the equipment at the sprawling Murray Hill complex, but they were one conceptual step ahead. Their design concentrated laser action in a narrow stripe running the length of the semiconductor crystal (figure 12-2).

The double heterojunction laser worked well because it concentrated electrons in the junction layer, making them more likely to release their energy as light which increased laser efficiency and reduced threshold current. In the Bell Labs lasers, this region was the plane of the junction across the entire chip, 0.37 millimeter long and 0.08 millimeter wide.[33] Alferov concentrated current flow further by covering parts of the laser chip with silica, an insulator, so electrons had to flow through a 0.03-millimeter stripe in the 0.20 millimeter wide chip.[34] This extra confinement increased the likelihood the electrons would recombine with holes—improving laser operation. Later diode laser developers have followed the same philosophy, refining structures to better confine current. Performance also improves when the layers confine light by total internal reflection, trapping it in the active stripe like a cladding traps light in an optical fiber.

Ironically, the narrow-stripe laser originated at Bell Labs in 1966. Jack Dyment hoped that limiting current flow to a narrow stripe would reduce the current needed to drive the laser. He didn't make much progress toward that goal; the real problem was the homojunction structure of his laser. But the stripe geometry eased another problem of early semiconductor lasers—poor beam quality. While gas lasers generated steady pencil-thin beams, the uneven pulsed light from diode lasers spread out so fast that calling it a beam was barely justified. When he added the stripe, Dyment says, "We could start to see repeatable mode patterns coming out of these devices, the first time anyone could see anything repeatable happening in diode lasers."[35] Mode control was an important advance, and stripe-geometry lasers helped focus light into single-mode fibers.[36]

Panish and Hayashi knew about stripe-geometry lasers, but they didn't bother with the extra step of adding a stripe to their laser. Alferov's group did, covering their wafer with a thin insulating layer of silicon dioxide, in which they etched a 30-micrometer stripe for current to flow through. That extra concentration of current flow did the job, and the Russians had the first semiconductor laser to emit steadily at room temperature.[37]

It probably didn't last very long. The informal rules of the laser-making game specify only that a laser has to emit a steady beam long enough to measure its properties. It might last for only seconds or minutes. Hayashi was a stickler and insisted that his operate repeatedly; it emitted for a couple of hours altogether. The Russians left the precise lifetime discretely unmentioned.

The Russians also were vague on other details and announced their success only in a Russian-language journal. By the time word trickled through the Iron Curtain, Bell Labs had already claimed a first. The lack of details and the penchant of Soviet officials for claiming their scientists had invented almost everything left many Americans skeptical. You have to go back and compare submission dates to discover that Alferov submitted his report on May 6, a month before Hayashi submitted his. Neutral observers consider the Russian claim credible.[38]

Both groups realized they had something to learn from each other. In October, Hayashi and Panish visited Alferov in Leningrad. In November, Alferov came to America[39] to spend six months at the University of Illinois with

Nick Holonyak, Jr., a diode laser pioneer who had learned Russian from his coal-miner father.[40] It was a rare window of scientific cooperation during the Cold War. Alferov stopped in Philadelphia in April 1971 to receive an award from the Franklin Institute for his laser work. When he returned home, Soviet officials restricted his travel to the West for the next three years.[41] With limited resources, the Russians had gone about as far as they could.

"A Pocket Laser"

Bell Labs had more ambitious plans when heralding its new laser at an August 31 press conference. Bell management probably didn't know about the Russian claims, but they must have heard about Corning's low-loss fiber. Maurer already had told Stew Miller, and Bell executives understood public relations. Announcing the laser breakthrough would turn the press spotlight to Bell achievements and away from its fiber-optic shortfall. The *New York Times* duly reported "a low-cost, pocket-size, reliable and versatile infrared laser—the first such device that may be practical for use in communications systems." Rudy Kompfner told the *Times*: "This is the laser we've been waiting for, although it will be a few more years before we can actually use it in a communications system." An unidentified Bell Labs spokesman predicted lasers would play a major role in communications, "when picture phones become common, when high-speed computer conversations are more widespread, and when communication needs in general expand beyond current carrying capacity."[42]

Interestingly, Kompfner was more cautious with the technical press. He told *Laser Focus* that he did not expect the new diode lasers to be used in practical communications for many years, probably well into the 1980s. Some system specialists at Holmdel were less enthusiastic in private, saying diode lasers might be interesting in 20 years.[43] In fact, as Kompfner well knew, the room-temperature diode laser was a symbolic breakthrough, not a practical device. Panish and Hayashi did well to make one laser that lasted a couple of hours; AT&T wanted light sources that lasted for many years routinely, and those clearly were far away.

Competitors were close behind. Kressel soon had room-temperature lasers at RCA. So did Standard Telecommunication Labs.[44] In Japan, the Nippon Electric Co. duplicated the Bell Labs feat in October, making lasers with slightly lower threshold currents.[45] NEC gained another key player a year later, when Hayashi returned to Japan, lured by a tempting job offer and worried his daughters were becoming too American.[46]

The Reliability Problem

The key issue with lasers was reliability. Developers measured lifetimes in seconds, minutes or—at best—hours. Shifting to a stripe geometry did not solve the problem. Having done their job of showing the feasibility of room-temperature lasers, Panish and Hayashi were happy to give device specialists the responsibility for developing practical lasers.[47] The greatest strength of a place like Bell Labs is not individual superstars, but its tremendous depth of talent to apply the perspiration vital to practical inventions. Management handed the job to Barney DeLoach, a hero for inventing a vital high-frequency semiconductor device that helped restart the millimeter waveguide program.

At first, DeLoach's group measured laser lifetimes in one-minute intervals. "It was a rare device that lasted that long," recalls Robert Hartman. "It would emit light constantly for a couple of seconds, then boom, it's off. The better ones would last 30 to 40 seconds, and die in a second or two."[48] He spotted two distinct failure modes: sudden burnout and slower dimming to darkness.[49] They soon pinned down the cause of sudden failures. The laser beam emerged from a spot only about half a micrometer high and 20 micrometers wide on the edge of the chip. The total power wasn't high, but concentrating it on such a tiny spot damaged the edge or "facet" where the beam emerged. Applying special coatings and refining the laser structure reduced the damage.

Gradual degradation was a more stubborn problem until Hartman noticed it resembled changes in mechanical springs and wondered if it might arise from strain in the crystalline lattice. Studies with polarized light showed that the more strain in the crystal, the faster the laser died. Then he set up a microscope to watch laser emission through holes in its electrical contacts. The laser stripe started as a bright zone, then dark lines grew across the bright region, choking out laser emission. Growth of these dark lines was an uncanny sight for a solid-state physicist trained to think of atoms as locked immobile in a crystalline lattice. "Matter was moving around at room temperature, as you watched. Nobody ever heard of diffusion rates like that in a solid," recalls DeLoach.[50]

Detailed studies showed that the problem was flaws, usually in the gallium arsenide substrate on which the whole laser structure was deposited. The intense laser light triggered their formation and spread. Solving it required years of meticulous development of better technology to grow laser crystals.[51] The British Post Office, STL, RCA, and Nippon Telegraph and Telephone had their own programs. Life testing became an art, with racks of lasers, each mounted with its own drive electronics and a heat sink to keep its temperature constant. Sensors monitored drive current and output power, with feedback circuits keeping one or the other constant. When output power dipped too low or drive current rose too high, the testers declared the laser dead and plugged another one into the socket.

Thousands and thousands of lasers lived brightly and briefly in labs around the world. By the end of 1972, DeLoach had 25 to 30 people "going like gangbusters" on reliability. Lifetimes crept up from minutes to hours and then days. Then top management slammed on the brakes. DeLoach summarizes their attitude: "We've already got air, we've already got copper. Who needs a new medium?"[52]

Downs and Ups at Bell Labs

The early 1970s brought tough times to AT&T. Long-distance traffic slumped with the economy, and the first signs of competition appeared. It was time to cut back on unpromising research, so one grim day in January 1973, management told DeLoach to shut down diode laser development. They shifted his whole group to work on silicon integrated electronic circuits. The next morning DeLoach returned, pleading to stay with the laser project if anything was to be kept alive.[53] He got his wish, although he was one of three people left.[54]

The cutbacks were not irrational. Other gallium arsenide devices had earned a reputation for poor reliability, and diode lasers were doing nothing to change that.[55] Ironically, progress was close at hand. By mid-1973, Bell Labs had operated a double-heterojunction laser continuously for more than a thousand hours at 30°C, the temperature of a warm room. Its output power dropped a mere 10 percent over that 42-day period, although threshold current increased and mode structure changed. No fundamental limits were in sight.[56]

Longer lifetimes posed new problems in testing. At the British Post Office, says Alan Steventon, "We got to the stage where the predicted life of lasers was increasing at the same rate as time was passing."[57] The longer the lasers lasted, the longer experiments took to yield results. One thousand hours is 42 days; 10,000 hours is over a year. System developers wanted lasers to last up to a million hours (100 years), but nobody could wait that long. Test specialists turned up the heat, figuring that higher temperatures would accelerate aging.

As the research engineers debated the merits of accelerated aging, a little New Jersey company decided the time was ripe to start selling double-heterojunction lasers. Commercial solid-state and gas lasers had reached the market within a couple of years after the first laboratory successes. In June 1975—five years after Hayashi taped his note to Panish's door—Laser Diode Laboratories sent out its first new product announcements. The RCA spin-off was too small to wait for long-term life tests; it had to cash in on its investment early. The press release conceded that life tests had only reached 1000 hours, but boldly predicted that the lasers would last 10,000 hours. Nobody else was brash enough to offer room-temperature diodes, but Laser Diode bet that system developers would pay $250 to $350 for diode lasers emitting steady powers of 5 to 10 milliwatts.[58]

System designers wanted to verify much longer lifetimes, so laser developers turned up the heat. They put fresh new lasers into ovens at 50 to 90°C (122 to 194°F) and watched how long they lasted. Then they multiplied that number by a suitable factor to extrapolate room-temperature life. It was a simple way to test laser reliability without waiting forever for the results. Researchers debated the proper scaling factors for a few years, but it was never a crucial issue.

Bell Labs kept making better lasers, and the group grew once again. By the spring of 1976, they estimated average room-temperature lifetimes of 5 to 10 years for a batch of 90 diode lasers—after screening out 14 lasers that died within the first 10 hours.[59] By mid-1977, they had lasers expected to last a million hours at nominal room temperature.[60] Bell Labs put out a press release,[61] and the technical press took it as a good sign.[62]

The race was not quite over. Other labs doubted the Bell Labs calculations, but in the end they proved reasonably accurate. The technology still had to be transferred to manufacturing, and transmitters still required temperature stabilization, but developers could say they had tamed the gallium aluminum arsenide laser. They had little time to sigh in relief, because the fiber-optic technology was racing ahead at breakneck speed and developers were starting to talk about other kinds of semiconductor lasers.

13

A Demonstration for the Queen

(1970–1975)

The glass-fibre guide . . . will be very cheap, small, and flexible. With the increasing confidence that the attenuation can be reduced to an acceptable figure it is certainly the most promising form of light guide for long-distance optical communication links.

—Martin Chown, Standard Telecommunication Labs, July 1970[1]

The low loss which is attractive for long-range applications can only be achieved if the light propagates partly or completely in vacuum or gas.

—Detlef Gloge, Bell Labs, October 1970[2]

We have seen that low-loss glass fibers can be made; we must make them in usable form. We must have semiconductor lasers as inexpensive commercial devices rather than as laboratory miracles.

—John R. Pierce, Bell Labs, address at Centenary of Institution of Electrical Engineers, May 17, 1971[3]

The low-loss optical fiber and the room-temperature semiconductor laser made 1970 a very good year for fiber-optic communications. Although both developments would prove crucial, at the time it was painfully obvious that neither was ready for practical use. The fibers were far too brittle; the lasers were too shortlived.

To the American communications industry, fiber optics was at best a vague hope for some distant future. It had no obvious connection with the near-term trends that *Electronics* magazine predicted at the start of the year:

> Face-to-face communication will be ushered in with the introduction of Picturephone service in 1970, while information exchange will be greatly expanded through the use of pulse-code modulation, millimeter waves, and satellites.[4]

The magazine forecast echoed the plans of AT&T, secure as the country's virtual telephone monopoly. Bell Labs saw optical fibers only as a long shot, a possible niche technology to run several miles between some local telephone switching offices, or more remotely a long-term alternative to gas-lens optical waveguides after Picturephone circuits clogged millimeter waveguides. If top AT&T management knew fiber optics existed, they considered it just another far-out Bell Labs program.

The British were the wild-eyed optimists. Having just heard about the Bell Labs room-temperature laser, Martin Chown waxed optimistic in a July 16, 1970, supplement to the British weekly *New Scientist*:

> By combining recently announced improvements in technology, it should be possible to build a system to carry over 3000 [voice] channels per fiber, with repeater spacing of two kilometers and no cooling of the lasers. With future improvements in components, the channel capacity is expected to be around 10,000.[5]

In 1971, the Institution of Electrical Engineers celebrated its one hundredth anniversary[6] by inviting British companies to show their latest and greatest technologies to the assembled dignitaries. Jock Marsh, managing director of Standard Telecommunication Labs, picked fiber optics. Murray Ramsay, who had taken Charles Kao's place, was in charge, and on May 19 he demonstrated fiber-optic video communications to Queen Elizabeth II.[7] The fiber worked flawlessly.

The IEE Centenary

Looking back, Ramsay calls demonstrating fiber-optic transmission at the IEE Centenary "a hell of a gamble."[8] Marsh was willing to bet on it after hearing of the room-temperature laser and the low-loss fiber, but Ramsay worked with it every day and knew its practical limits.

Ramsay had inherited some problems along with the hot new technology. Kao was as enthusiastic in managing his five-man group as he was evangelical to outsiders, so he hadn't stopped spending when he went beyond the budget. When Ramsay asked about one project that had consumed £8000, Kao shrugged and said, "We'll have to sort that out later."[9] Fortunately,

Marsh saw the promise in fiber optics and put promising ideas before rigid rules. The word went up to Harold Geneen at the very top, who asked STL to demonstrate fiber optics for top ITT managers in Brussels. The technology was on a roll.

A successful television experiment helped convince Marsh to push ahead with the IEE demonstration. BBC engineers were testing digital transmission of color television signals. They started with a length of millimeter waveguide, supposedly the digital transmission medium of the future, which the Post Office had installed at its new research lab at Martlesham Heath. They needed three weeks to get their television signals through the waveguide,[10] and even then it never worked for more than a few minutes at a time because the signal hopped between modes. Then the BBC engineers asked Chown about trying fiber optics.

They set aside another three weeks to test a fiber system that Chown and another STL engineer brought to their lab. The two arrived at 9 A.M. but had to go out and buy a vacuum flask to hold liquid nitrogen to cool their laser before they could set up their system. By 10:45, a small reel of fiber was carrying the digitized video signal and Chown was marveling at the crystal-clear color version of "Playschool," a program his children watched in muddy black and white at home.[11] Beating the millimeter waveguide hands down was a big morale boost. Ramsay says, "That convinced us that fiber was a transmission medium that had a future."[12]

Chown carefully took apart the BBC demonstration system and carefully put the pieces back together in exactly the same way at the exhibit hall. Once again it worked. As the royal party approached, security pushed Chown off among the unauthorized. Ramsay wondered where he had disappeared but soldiered on and proudly showed his new marvel to Queen Elizabeth.

The BBC had lost its color signal, so STL had hastily borrowed a color signal from the rival Independent Television (ITV). Otherwise, the display worked perfectly. The Queen dutifully watched for five minutes before going on to the next exhibit. As the crowds milled about, Chown found himself alone to demonstrate the system to her husband Prince Philip. The prince stayed for nearly half an hour, prompted by Lord Louis Mountbatten to ask probing questions which amazed Chown with their acuity.[13] Trained as an engineer, Mountbatten already knew about fiber optics. Two years earlier, he had told a laser meeting in Southampton about prospects for "fiber optics connecting every British home with huge information centers."[14] STL was showing he was right.

The audience with the Queen was a symbolic milestone, but Lord Mountbatten's interest had far more practical import. He was a patron of engineering, a top-level advisor with serious clout behind the scenes in the British government.[15] Nobody could pretend that all the problems were solved when the laser sat in a flask of liquid nitrogen, but the demonstration alerted the British establishment to the potential of fiber-optic communications.

The Importance of Being Digital

The spread of digital pulse-code modulation in the telephone industry brought honors to Alec Reeves from his fellow engineers, recognizing the technology he invented a generation earlier.[16] Invited to give the 1969 John Logie Baird Memorial Lecture at the University of Strathclyde, the aging visionary talked of future fiber-optic communication.[16] In the fall of 1969, he told South African engineers that within a half century, "The main overland and submarine information highways will consist of wide-band optical quadratic-law [graded-index] fibers with laser-type repeaters spaced just over two kilometers apart."[17]

ITT gave Reeves an award and a $5000 honorarium when he retired at 68 at the end of 1970. Intending to remain active in engineering as well as dabble in the paranormal, he formed Reeves Telecommunication Laboratories Ltd. in London with Charles Eaglesfield and another engineer. They landed a Post Office research contract, but years of heavy smoking caught up with Reeves. He fell ill with lung cancer and died October 13, 1971, in London, having done his part to launch the new technology.[18]

By then, F. F. Roberts had become the central figure in British fiber-optic research. Dedicated, energetic, and efficient, the crusty career technocrat focused British fiber-optic development on meeting the practical needs of the Post Office's telephone division. He concentrated on testing the technology, looking for potential fatal flaws as well as pushing the state of the art.[19] Top management had already committed to digital transmission, made practical by the advent of transistors and the integrated circuit,[20] so he concentrated on digital systems.

Other system trials also yielded encouraging results. The University of Southampton sent live BBC color-television signals through 1.25 kilometers of experimental fiber and later repeated the experiments with French and German television.[21] By late 1972, Chown could switch a narrow-stripe double-heterojunction on and off up to a billion times per second. That meant the laser could transmit a billion bits per second.[22]

The experiments showed fiber-optic systems handled digital signals well. In fact, semiconductor lasers worked much better for digital signal than for analog ones. The light output of early semiconductor lasers did not increase evenly with their drive current, so they could distort analog signals, like a bad audio amplifier or speaker jumbles sound. Digital signals are immune to such distortion, because systems need not detect the precise signal level, but only if the signal is on or off. That doesn't require high-fidelity reproduction—a doorbell buzzer, a damaged speaker, or a semiconductor laser will suffice.

A Well-Planned Future Begins to Go Awry

On the other side of the Atlantic, AT&T had begun going digital in 1962. The first step was the "T1 carrier," running between urban switching centers.

A pair of wires carried 24 digitized voice channels, combined into a single stream of 1.544 million bits per second, with a repeater every 6000 feet (1.8 kilometers) to boost signal strength. By 1970, a whole hierarchy of faster digital systems were on the drawing boards. Each step up involved combining or multiplexing slower signals, to be separated at the other end. The digital pipes interfaced with digital switches, special-purpose digital computers hard-wired to route telephone signals to their proper destinations. The second step in the hierarchy, called T2, merged four T1 lines on wires, a total of 6.3 million bits per second. Combining seven T2s gave a T3 carrier at 45 million bits per second, a speed Bell had not decided how to transmit. Six T3s combined to make a T4 signal, 270 million bits per second or 4032 voice circuits, routed over coaxial cables or microwaves.[23] Millimeter waveguides were to be the next step, with each pipeline carrying sixty T4 signals, 240,000 voice circuits or about 17 billion bits per second.[24]

Those plans assumed that video communications would follow speech as surely as television had followed radio. Bell Labs had sought that goal since the 1920s, when it competed with C. Francis Jenkins to devise mechanical television transmitters.[25] Top AT&T management had planned the Picturephone revolution to start in 1970, with service in Manhattan and Pittsburgh. The whole Bell System was designed to handle the new technology once you pushed the # button on your Touch-Tone phone to signal a video call.[26] Each Picturephone circuit needed the equivalent of a hundred voice lines, so the network would need plenty of new capacity, assuring a place for millimeter waveguides.

That well-planned future started to unravel before regular Picturephone service began. New York state regulators blocked the new service in Manhattan until the phone company improved its regular service. AT&T inaugurated Picturephone service June 30, 1970, with only 38 sets at eight companies in Pittsburgh, but that did not deter the optimists. The front page of the *New York Times* proclaimed, "A major stride in the development of communications was taken today."[27] AT&T predicted 100,000 Picturephones would be in use by 1975.

Potential customers didn't agree. A year later, only 33 Picturephones were operating in Pittsburgh, and only 12 could call beyond their buildings. "The thing hasn't really grown the way we thought it would," admitted a phone-company marketer, although higher officials maintained that its time would come. The *Times*—perhaps a bit embarrassed itself—buried its follow-up story on page 26.[28]

Still sure that the time would come for the millimeter waveguide, Bell planned a full-scale field trial in 1974, the first outside the lab.[29] To avoid troublesome mode problems, the waveguides were buried in trenches as straight as arrows. The tubes were filled with pure dry nitrogen, because oxygen in the air absorbs millimeter waves. Nobody pretended it was going to be cheap or easy, but with government regulations assuring a return on its investment, AT&T was ready to spend untold billions.

The telephone monopoly was a giant battleship, steering full steam ahead into its carefully planned future. From the captain's tower of top management, Picturephone seemed merely a little behind schedule. Other unpleasant realities were easy to mistake for patches of choppy water. Long-distance traffic slumped along with the economy. Angry consumer advocates and upstart MCI challenged Bell's telephone practices and long-distance monopoly. Looking back, we can see the well-planned future starting to unravel, but those on board the great ship didn't spot the icebergs on the horizon.

Nor did Bell Labs worry, at least initially. AT&T's well-oiled invention machine had earned a Nobel Prize for the transistor and scored a technical triumph by demonstrating the first communications satellite. After years of development, the labs had turned Picturephone over to manufacturing and the millimeter waveguide to final field trials. Millions of dollars invested in gas lenses and hollow optical waveguides had yielded only modest success, but no one expected overnight miracles. Management was convinced they had the best and brightest scientists on the job, and no one could do it better.

In short, Bell Labs was not merely fat and happy, but growing complacent and overconfident. The "not invented here" syndrome was rampant; many managers didn't want to hear ideas that came from outside what they considered the world's premier industrial laboratory. That made fiber optics a hard sell for Charles Kao and F. F. Roberts. They succeeded by convincing Bell that fiber might fill the vital but unglamorous role of linking telephone switching centers a few miles apart. Small and flexible, optical fibers could thread easily through the underground ducts that carry phone lines in urban and suburban areas, an impossible task for bend-sensitive millimeter waveguides and hollow optical waveguides.

Yet Bell still saw fiber playing a secondary role to the more glamorous high-capacity systems running between cities. Bell was sure glass could never be as transparent as air. Detlef Gloge calculated hollow optical waveguides would have loss below one decibel per kilometer, so they could carry light over 20 times farther than fibers with loss of 20 decibels per kilometer.[30] Fibers might go a mile or two, but waveguides would run from city to city, carrying signals at rates far higher than optical fibers. "The scale on which an optical intercity system will become useful is in the range above 500,000 two-way voice channels, or equivalently over 5,000 two-way video telephone channels," Miller wrote in 1970. That meant carrying 60 billion bits a second, although eventual capacities might be a hundred times larger.[31] He expected individual fiber links to merely carry "a single video telephone signal, or a hundred or so voice channels" within urban or suburban areas.

A Low-Profile Program

Bell stepped up the pace of its fiber-optics program after learning of the Corning breakthrough, but it did not change its direction. As Kao had originally proposed, Bell concentrated on interoffice transmission. Initially it was a quiet program, virtually invisible to the outside world. The general press ignored it, and even the technical press took time to catch on.[32] In early 1971, John Kessler at *Electronics* magazine picked up rumblings of the fiber-optic revolution, but only after a few months did he realize its importance and write a cover article.[33] Even advocates were cautious. "Corning has several fibers about 200 meters long which exhibit losses of 20 dB" per kilometer, said Chuck Lucy, "but this is a long way from a system."[34] Technology trends were unclear; Corning and others were working with the small-core single-mode fibers that Kao had proposed, but some developers were considering larger-core fibers that could collect light more easily. Bell had not changed its plans for millimeter waveguides, which Kompfner still predicted would fill the communication needs of major population centers through the 1980s.

Behind the scenes, the pressure was on at Bell Labs,[35] which still had not matched Corning's low-loss fibers. The stress highlighted tensions between fiefdoms in the AT&T research bureaucracy. Miller wanted to draw fibers at Crawford Hill, reasoning that as waveguides they fell into his communications research domain. Murray Hill disagreed, believing that glass fell into its domain of materials research. Eventually, upper management assigned glass chemistry and fiber-making physics to Murray Hill but let Crawford Hill keep experiments with novel fiber structures.[36]

Groping for something they could demonstrate, Crawford Hill turned to "liquid-core" fibers, fine silica tubes filled with a dry-cleaning solvent, perchloroethylene, C_2Cl_4. The liquid core guides light because its refractive index is higher than pure silica, and the solvent was clearer than anything else Bell could make. Crawford Hill measured loss of 13 decibels per kilometer, letting them briefly claim the record for lowest reported fiber loss.[37] An Australian group came up with the same idea independently and soon measured loss of 6.5 decibels per kilometer.[38] The University of Southampton later did slightly better.[39] Nobody today admits seriously considering using liquid-core fibers in communication systems; their problems were legion. However, they were useful for laboratory tests of fiber transmission.[40]

Miller also seized on an idea of Peter Kaiser's, placing a flat planar waveguide of pure silica inside a hollow silica tube and stretching it out into a hollow fiber-like structure. Kaiser made samples,[41] but as with Karbowiak's thin-film waveguide, they worked much better in theory than in reality.

Murray Hill assigned more people to fiber development, but they lost time shuttling back and forth to measure their fibers in Kaiser's lab in Crawford Hill.[42,43] This fast became a nuisance, so management asked a Murray Hill optics specialist to design a new fiber measurement lab and funded it with amazing speed.[44]

At Crawford Hill, Miller shifted more people from hollow optical waveguides to fibers. Their first task was to take a long, careful look at fiber technology. No one knew if fibers would hold up over time.[45] Experimenting with plastic coatings, Kaiser found they protected pure silica fibers from the environment, and they remain standard on modern communication fibers.[46]

A Change of Mode

As Miller's group looked closely at fiber-optic systems, they began worrying about getting light into the fibers. Kao's proposal for transmitting light in a single mode was theoretically elegant, but the light-carrying core of a single-mode fiber was microscopic. Corning's fibers had cores only three micrometers across—0.003 millimeter. If they were much larger, they would carry light in more than one mode. That was far larger than unclad single-mode fibers would have been, but it was uncomfortably small for telecommunications engineers. There was no obvious way to increase core size much; the upper limit for single-mode transmission depends on the light wavelength and the difference in refractive index between core and cladding. Moving to the 850-nanometer wavelength of gallium arsenide diode lasers would only increase core diameter to four micrometers; the index difference had already been cut to one percent.

Bell Labs took a hard look at the trade-offs and decided single-mode transmission had to go.

A crucial issue was aligning the tiny fiber cores with each other or with light sources, and in 1970 that seemed impossible. Their positions has to be adjusted to within a micrometer—a task extremely difficult in a well-equipped optical laboratory, and totally inconceivable for a technician in a manhole or on a telephone pole. The light-emitting zones of semiconductor lasers were several times wider than single-mode cores, and Miller wasn't optimistic about their prospects.[47] Bell considered light-emitting diodes (LEDs) to be much better light sources. Their output was feebler than semiconductor lasers, and they emitted from an area much larger, but they could operate for years. In fact, Crawford Hill had already made a special LED with a hole etched into its top so that a fiber could be inserted to collect light efficiently.[48] The things had to be made by hand, but they survived almost indefinitely and delivered a usable amount of light to a fiber.

Another concern was that single-mode fiber seemed hard to make. Corning was keeping mum while its patent was pending, and obvious alternative techniques—the rod-in-tube and double-crucible processes—were poorly suited for drawing single-mode fibers.

What made large-core fibers appear more practical was the invention of the graded-index fiber at Tohoku University. By nearly equalizing the speeds of the many modes in the fiber, the refractive index gradient avoided the biggest problem of large-core fibers—large pulse spreading that severely limited signal speed. Alec Reeves spotted the idea and called graded-index fiber "the most promising practical wide band optical waveguide that can be foreseen now."[49] The Japanese were making graded-index fibers, although they weren't as clear as Corning's, and Miller had also thought about graded-index waveguides, which may have made him more receptive.

Graded-index fibers could not completely eliminate pulse spreading, but Miller wasn't worried. He still expected millimeter waveguides or hollow optical waveguides to carry long-haul, high-speed traffic. He wanted fibers to link switching centers in urban and suburban areas. That meant they had to carry a hundred to a thousand voice lines, 10 to 100 million bits per second, over 2 to 20 kilometers (1.2 to 12 miles). LEDs could handle the lower end of that range, but faster semiconductor lasers would be needed for the high end. On the technological horizon, Miller envisioned multimode graded-index fibers carrying 300 million bits per second between repeaters six to eight kilometers (four to five miles) apart—four to five times farther than coaxial cables.[50] The ever-optimistic Reeves thought the ultimate speed limit might be 10 billion bits per second over a few kilometers.[51]

Bell turned its attention to improving graded-index fibers. Two top theorists, Gloge and Henry Marcatili, refined the design by calculating a new refractive-index gradient that in theory increased transmission capacity a thousand times above large-core fibers with a sharp "step" boundary between core and cladding. In practice, the improvement was much less, ranging from 20 (in production) to 75 (in the laboratory).[52] However, that seemed good enough because the theorists expected other pulse-spreading effects to limit transmission capacity, even for single-mode fibers.

The most important of these is the same phenomenon that makes a prism spread white light into a spectrum—material dispersion. The refractive index of glass varies slightly with wavelength, so some colors travel faster than others. Semiconductor lasers are not purely monochromatic, and although the effect is small, it builds up with distance. Material dispersion limited the transmission capacity of single-mode fibers at the 800 to 850 nanometer wavelength of gallium arsenide semiconductor lasers to only two to three times that of graded-index fibers. That wasn't enough of an advantage to offset the other problems of single-mode fibers.[53] Indeed, for short links between switching offices, even LED sources looked viable, although their broader range of wavelengths made material dispersion 10 to 50 times worse than for semiconductor lasers.

Most of the rest of the world came to the same conclusion. Some staunch single-mode advocates remained at the Post Office, but to others light-coupling problems made single-mode seem "an unattainable dream."[54] STL also was slow to move away from single-mode fibers, wary of multimode transmission problems in millimeter waveguides and seeking the highest possible capacity for its parent company's submarine cable business.[55]

The Corning Connection

Corning knew glass and it had a healthy head start on making low-loss fibers, but it didn't know much about communications. Chuck Lucy could see enticing possibilities. "It really seemed like it had the possibility for infinite bandwidth and zero loss," he recalls. "If you had that, there had to be a market somewhere."[56] However, merely supplanting millimeter waveguides in high-end, long-distance systems would amount to what another glass company derisively called "just a few buckets of sand."[57] Corning wanted to crack the much bigger market for shorter cables running between telephone switching centers. Filling the need for shorter, moderate-speed cables was AT&T's goal in shifting to larger-core graded-index fibers, so Corning followed, trying to reduce fiber costs to expand the market. The fiber developers didn't mind the change; Bob Maurer found large-core fibers were simpler to make than single-mode ones.[58]

The partnership between Corning and AT&T was an uneasy blend of competition and cooperation. About the time Corning developed its 20 decibel per kilometer fiber, the company cross-licensed patents on semiconductors and optoelectronics with AT&T. The deal was a flat-out trade—both companies could use the other's patents without paying royalties. Corning licensed the electronics patents for Signetics, a subsidiary that manufactured integrated circuits. The managers who arranged the deal thought they came out ahead, because Bell insisted fiber communications was decades away, but it proved to be a fiasco when Signetics floundered and AT&T decided to make its own fibers instead of buying them from Corning.[59]

The two cooperated in developing fiber measurement techniques, crucial to establishing the new technology. Measurement specialists Paul Lazay from Murray Hill and Marshall Hudson of Corning drove the back roads between western New York and northern New Jersey with reels of fiber in their back seats, comparing results from their labs. Sometimes the results were unsettling, as when Hudson started to see different results on Bell fiber from one day to the next. An uneasy Lazay worried that the fibers were degrading, but several round trips showed the changes were in the measurements. They puzzled over the discrepancy until one day they compared fiber spools in Hudson's lab. When Lazay asked about the thermal expansion of Corning's Styrofoam spools, they were surprised to find it was much larger than Bell's aluminum spools. There lay the key to the mystery. Hudson's lab wasn't air-conditioned, and the minicomputer he used to automate his measurements slowly warmed the room during the day, heating the reels and the fibers. The measurement results had changed along with the temperatures.[60]

On the other hand, the competition in fiber development was intense. Corning couldn't block Bell from using its patented technology, and Bell didn't have to pay royalties. Yet Corning didn't have to share information beyond what the patents contained, and that only when the patents were published. When Corning reduced loss to four decibels per kilometer in mid-1972, they shared the attenuation measurements and revealed the fiber core was a large 91 micrometers, inside a cladding with an outer diameter of 125 micrometers.[61] But Corning kept its fabrication process and composition the deepest and darkest of secrets, trying to keep its technical lead in a market it expected to take years to develop.

Materials were the most sensitive topic. Corning didn't say a word about titanium until after it changed its core dopant to germanium.[62] Peter Schultz playfully called the new magic ingredient "fairy dust" at a few technical meetings—until George Sigel of the Naval Research Laboratory deduced it was germanium and spilled the beans at a meeting of the American Ceramic Society.[63]

Corning also developed a new process for the new fiber, depositing glass soot on the *outside* of a ceramic or graphite rod, instead of on the inside of a pure silica tube. They could change the glass composition gradually as they deposited soot, so the process was easy to adapt for graded-index fibers. Heating melted the soot, which solidified to form a glass tube covering the rod. Cooling shrunk the rod more than the tube, so it slipped out easily, and could be reused. Then heating collapsed the tube so it could be drawn into a fiber. The process promised better quality at lower cost, and Corning still uses it.

With strong support from top management, Corning added more people to the fiber program. Looking for a theoretician, Maurer hired Robert Olshansky,[64] who wanted to get into industry after two discouraging research fellowships in particle physics. He says, "It was just a fantastic time to arrive on the scene."[65] He started with no fiber-optic background but soon eased the troublesome problem of losses from small bends in the fiber. He found 85-micrometer cores were too close to the fiber diameter of 125 micrometers but that shrinking core diameter to 50 or 60 micrometers would cut the losses by a factor of 10. Then Olshansky refined the recipe for the ideal gradation of refractive index in a multimode fiber.[66]

The Conversion of Bell Labs

No single event made Bell Labs collectively stand and shout "Hallelujah, I believe in fiber optics." Kao's measurements of fused silica transparency had opened eyes; Corning's 20-decibel fiber opened more. The four-decibel fiber was a shock. After being stalled at 60 decibels per kilometer in late 1971,[67] Bell had briefly claimed a low-loss record with liquid-core fibers. Bell had some other good results but nothing consistent that came close to Corning's new fiber,[68] and little insight into what Corning was doing. AT&T management was not happy.

From 1970 through 1972, Stew Miller shifted the balance of his guided-wave group increasingly toward fiber optics. The people didn't change, but their research did. His three department managers, Jack Cook, Tingye Li, and Henry Marcatili, moved into fiber system development.[69] They organized an interlaboratory committee on fiber systems with department heads from the parent AT&T, device groups at Murray Hill, the engineering research center at Princeton, fiber and cable manufacturing in Georgia, and the digital transmission lab in Holmdel.[70]

Yet as late as the summer of 1972, Rudy Kompfner had not abandoned hope for hollow optical waveguides that used pairs of mirrors to relay a laser beam. He admitted their mechanical stability "has been a subject of concern for some time" but held out hope for corrections that would allow use of their potentially vast transmission capacity. However, optical fibers were looking better, and he conceded that they "may satisfy a demand for much smaller capacities much sooner."[71] By the time Kompfner retired at 64 the following summer, hollow optical waveguides were essentially dead.[72] Research management changed as John Pierce also retired to join the faculty at Caltech.

A New Process Explodes at Bell Labs

Bell finally started to make progress on fiber fabrication in the winter of 1972, after France published Corning's application for a fiber patent.[73] Bill French, who had worked with Dave Pearson at Murray Hill, saw a translation and wondered if a similar process, chemical vapor deposition, would work for fibers. He suggested the idea to another Murray Hill scientist, John MacChesney, who had used chemical vapor deposition to make other types of silica.

To see if it would work, they flowed silicon tetrachloride, titanium tetrachloride, and oxygen through a hot glass tube. The gases reacted, depositing glassy soot inside the tube, which they then melted and drew into a fiber. "The experiment worked the first time, largely for the wrong reasons, but that didn't matter," MacChesney recalls.[74] Further experiments reduced fiber loss.[75]

Another Bell glass specialist, Ray Jaeger, realized that a dash of boron would reduce silica's refractive index. That boron-doped glass could serve as a low-index cladding on a core of nearly pure silica, avoiding the need to add dopants to the light-carrying core. MacChesney used that trick to make fibers with loss of only 5.5 decibels per kilometer, a breakthrough for Bell.[76,77] By 1974 they reduced loss to four decibels per kilometer at 900 nanometers, and to just over two decibels at 1060 nanometers.[78]

Refining the process took time and countless experiments. MacChesney veritably mass-produced preforms, which he sent to another lab that drew a few hundred meters of fiber for measurements. Sometimes he let deposition run unattended overnight; it speeded research but was risky with volatile chemicals. One Saturday night an oxygen line broke, fueling a fire so hot it melted window glass and sealed the lab door shut. Firemen needed much time and even more water to quench the inferno, which devastated MacChesney's lab and flooded one Paul Lazay ran downstairs.[79]

It was the second lab MacChesney melted, but the fire didn't diminish his stature at Bell. It was the sort of occupational hazard that comes with high-temperature chemistry; Southampton also burned down a lab.[80] The occasional fire was a small price to pay for a process that worked, and MacChesney's did. It became AT&T's standard way to make fiber.

Others developed their own vapor deposition processes. In England, Southampton doped fiber cores with phosphorus to make fibers with minimum loss of 2.7 decibels per kilometer at 830 nanometers.[81] There was room for some variation, although Corning soon received fundamental patents that would become gatekeepers for entry into the American optical fiber industry. The British Post Office plugged along with double-crucible fibers but never quite caught up.[82]

Seeking Supporting Technologies

With fibers improving and LEDs available for short-distance transmission, developers started looking at the imposing host of supporting technologies needed to make practical communication systems. Their work could be a book in itself. Fresh glass fibers were surprisingly strong, but they needed plastic coatings to protect their vulnerable surfaces. The fibers, in turn, had to be encased in a cable that could be hung from poles, threaded through buried ducts, buried directly in the dirt, or run through walls. AT&T knew how to cable wires, coating them with insulating plastic and wrapping the coated wires with strong plastics or metallic armor. Corning didn't have any cabling expertise, and the few cables it tried making weren't very good. Chuck Lucy trotted the globe hunting cable manufacturers to form partnerships, and made deals with BICC in England, Siemens in West Germany, Pirelli in Italy, and government labs in France and Japan. Mostly, Corning supplied fiber and let others do the cabling.[83]

Connecting fibers end to end was another major worry. Nature smiled on one approach—melting the ends together to form a permanent splice. Surface tension of the glass tended to align the tips. Typical of much fiber development, different groups each took small steps. Bell Labs did it first for multicomponent glasses,[84] then the Post Office added the delicate alignment for single-mode fiber,[85] and Corning adapted the process for silica by zapping the fiber tips with an electric arc.[86] It worked well as long as fiber cores were precisely centered, and there, too, nature smiled on careful manufacturers.

Temporary junctions were a tougher problem. Electrical connectors, like the jack on your telephone, need only touch each other. Fibers have to be aligned such that the light goes straight from one core into the other, and without special tools that's as hard as lining up a couple of hairs. Bell Labs put a small army to work at the task; the Naval Research Laboratory and the Defense Advanced Research Projects Agency set defense contractors to work on military connectors. Some results were awful, but a few ideas worked. Engineers devised special tools and alignment techniques, and their performance steadily improved.

Field Trials for Fiber and Waveguide

By the fall of 1973, Bell Labs was sold on fiber-optic communications and Stew Miller had put light pipes on the shelf. Jack Cook laid out plans for digital fiber links carrying 24 telephone conversations in *Scientific American*.[87] Developers demonstrated digital repeaters operating at 6.3, 45, and 274 million bits per second. The two slower ones used LEDs; the fastest one a room-temperature diode laser.[88] The next step was to take the technology out of the lab for a field trial like the one of the millimeter waveguide nearing its start in New Jersey.

AT&T, being a giant corporation, had a committee set the parameters. Some choices were easy. The main goal was to test digital transmission in an urban environment. Fiber-optic cables packed a lot of capacity in a small diameter, so they could alleviate crowding in underground ducts—saving the phone company the untold millions needed to dig up the streets and lay new ducts. Fibers also could stretch much farther between repeaters than the 6000 feet (1.8 kilometers) possible with wires.

Committees need something to debate, and this one concentrated on transmission speed. Some people wanted fibers to replace the 1.5 million bit per second T1 lines that AT&T had been installing since 1962; that way they could interface directly with existing systems. Another suggestion was to operate at the 6 million bit per second T2 speed, where electronics were readily available. However, executive director Warren Danielson pushed for the 45 million bit per second T3 rate, which had never been used for lack of a cable to carry it. He argued the slower speeds didn't improve on wires and that the higher speed would share the high cost of the laser and fiber over more channels. The others agreed.[89]

Ira Jacobs's group at Holmdel spent 1975 assembling equipment for tests at a suburban Atlanta AT&T plant. They packed Corning and AT&T fibers into plastic-coated ribbons of 12 fibers, and stacked a dozen ribbons together to make a pencil-thin 144-fiber core for the cable.[90] They threaded 650 meters (2100 feet) of that cable through ducts buried under a parking lot, then spliced fiber ends together to test transmission over distances up to 10.9 kilometers (6.8 miles).[91] They put a separate fiber cable through environmental torture tests in a duct where they could adjust the temperature and humidity. On January 13, 1976, they turned on the Atlanta fiber system and spent the next several months waiting for something bad to happen.

The communications world watched. The British Post Office and others had their own tests in the works, but no one had the resources to match Bell Labs. Bell might not lead in inspiration, but it excelled in perspiration. Its careful and cautious engineers would poke and probe the new technology and spot any problems.

They did find one problem. A distressingly large fraction of the lasers died during the test. That was not unexpected, nor was it fatal. Laser lifetimes were improving steadily, and Bell was hedging its bets by developing LED-based transmitters. Otherwise, "things went quite smoothly," Jacobs recalls. Fiber loss was smaller than expected, as was the variability among components.[92] Bell had hoped to get 100 good fibers, but 138 of the 144 were good. Average loss was six decibels per kilometer, below the planned eight decibels.[93] All systems were go for the next step, installing a fiber-optic link to carry live telephone traffic in the AT&T network.

The End of the Millimeter Waveguide

The wheels of progress ground more slowly for the millimeter waveguide. It took years to prepare the test in Netcong, New Jersey, along a route planned for a waveguide between New York and Philadelphia. Bell buried 8.7 miles (14 kilometers) of waveguide four feet underground inside a 6-inch (15-centimeter) pipe covered with yellow plastic to protect against corrosion. Then they filled it with nitrogen gas. Completed behind original schedule in February 1975, the installation cost $100,000 a mile.[94]

Careful tests followed. Officially, Bell insisted the waveguide worked as advertised. "The verdict was that the TE_{01} [waveguide] system was a success," says a corporate history. "It had the predicted low losses, only 5 percent above" theoretical predictions.[95] However, achieving that low loss was horrendously difficult because minute deviations shifted energy to undesired modes. Settling of the soil around the buried pipeline caused transmission problems. The desired mode faded away, and new modes interfered with the signal.

AT&T had wanted the new technology to provide tremendous capacity over long distances. That would have justified the tremendous cost for a regulated monopoly. By the time the test was done, the need for that extra capacity had evaporated with the economic slump and the failure of Picturephone. New technology packed more signals onto microwave links.[96] The "success" of the millimeter waveguide became academic; AT&T donated its surplus waveguide to construction of the Very Large Array radio telescope in Socorro, New Mexico.[97]

The story was the same around the world. The British Post Office buried 14.2 kilometers (8.8 miles) of millimeter waveguide five feet under a main road near Martlesham Heath, disrupting traffic for three months. To keep the waveguide straight even when temperature changed, they set up equipment to pull constantly on both ends.[98] Like Bell, the Post Office considered the tests starting in October 1975 to be successful.[99] However, at the last minute the Post Office canceled plans to run a millimeter waveguide from London to Reading. Long-term monitoring of the Martlesham system showed the loss was increasing as the soil settled.[100] The end was abrupt; research director John Bray called the 60 or 70 people working on millimeter waveguides together one Friday and said that on Monday they would start working on fiber optics.[101]

The millimeter waveguide was a dead duck, although the organizations that had poured vast sums into the technology took care not to publish an obituary. Like vanquished generals, they declared victory and retreated. Dick Dyott never got to carry out his brash promise to run optical fibers through millimeter waveguides because none were ever installed in the global telephone network.[102] The technology sputtered to a stop barely a half-dozen years after Harold Barlow had proclaimed that "confidence in the ultimate success . . . runs high."[103] What Dyott had intended as a bold statement turned out to be too cautious. Fiber optics blew the millimeter waveguide away before it could carry a single phone call.

Three Generations in Five Years
(1975–1983)

Optical communications, for all its glamour, has already passed most of the hard tests of practical application. We know how to draw fibers of the unprecedented degree of transparency required. We know how to combine those fibers into cable. We know how to splice the cable. We have designed and built devices to generate and modulate optical signals. And we have designed and built repeaters to regenerate these signals. In short, we have designed and built a completely integrated optical transmission system.

—John D. deButts, chairman, AT&T, October 17, 1975[1]

It is now clear that optical fiber systems will win in the competition for performance and economy. . . . The future will always be constrained by the physical properties of materials and by the technological skills developed to overcome or make use of them. A step-wise move toward wavelengths of about 1.2 micrometers can be foreseen, limited largely by the development of efficient and reliable enough optical sending and receiving elements. The splicing problem may remain a significant barrier to the wide use of monomode fiber systems, but other barriers may be set by the terminal costs and overall network reliability costs associated with any systems of very high transmission capacity. But none of these barriers will be insuperable.

—F. F. Roberts, 1977[2]

As Bell Labs carefully prepared its Atlanta field trial in the summer of 1975, lightning struck a two-way radio antenna used by the Dorset police in southern England. The current surge destroyed the communication system at the department's eastern division control room in Bournemouth.[3] To keep in touch with roaming officers, the police had to park a patrol car atop a nearby hill to relay signals to other cars.

That was no way to run a police force, and the chief constable did not want it to happen again. He sought help from Geoffrey Philips at the Home Office Police Scientific Development Branch in London. Philips had been Don Williams's boss at the Ministry of Defense, so he knew about fiber optics.[4] The military was interested in fibers because they did not carry electric currents, and Philips knew that meant they would not carry power spikes from lightning strikes, so he suggested the chief constable contact Standard Telephones and Cables.

Standard Telephones and Cables had sold fiber equipment for experiments, but no one had asked them for a commercial system before. The chief constable didn't care that the technology was new. What he wanted was a cable that would keep another lightning strike from frying the replacement electronics.[5] He didn't haggle with STC about specifications; he trusted their assurances it would work. STC didn't disappoint him; in mid-September, within just a few weeks of the order, the fiber cable was on line and the Dorset police resumed normal operations.[6]

The timing gave Charlie Sandbank a perfect opportunity to report a triumph at the First European Conference on Optical Fiber Communications, September 16–18, 1975, in London. The first fiber-optic system put to work in the real world was up and running.[7]

A Premature Consensus

The Dorset system was good news, but it wasn't good enough to convince the traditionally conservative telephone industry of the mid-1970s. Phone companies wanted standardized equipment, tested for reliability and compatibility with existing systems. AT&T set the standards for North America; the International Telecommunications Union set them for most of the rest of the world. To Bell Labs, the Atlanta trials were the first step toward a new standard, fiber-optic transmission at the previously unused 45 million bit per second T3 rate. Bell had taken risks and made trade-offs to reach the consensus design. Graded-index fiber seemed far more practical than single-mode, which had essentially the same loss at 800 to 900 nanometers and offered only a modest increase in transmission speed at those wavelengths. The toughest choice was between gallium arsenide lasers and LEDs; lasers were brighter, but in 1975 they didn't last long. Bell hedged its bets and worked on both.

The consensus reflected a widespread belief that the technology was near a plateau. The best laboratory fibers were approaching the fundamental limit on transparency at 850 nanometers, two decibels per kilometer.[8] Developers added in other losses expected in production fibers, connectors, and splices and figured signals couldn't go much beyond 10 kilometers (6 miles) without a repeater. Graded-index fibers could easily carry tens of millions of bits per second over that distance, and perhaps a few hundred million. After 15 years of development, gallium arsenide lasers were approaching desirable lifetimes.

A few labs pushed a bit further. The Post Office tested gallium arsenide lasers and graded-index fibers to transmit similar distances at the European standard speed of 140 million bits per second.[9] A team at Fujitsu sent 400 million bits per second through 4 kilometers (2.5 miles) of graded-index fiber and held out hope for even faster transmission.[10] However, the industry generally followed the Bell Labs consensus.

The little village of fiber-optic developers had become a fast-growing town. When Charles Kao returned from Hong Kong in 1974, he found himself an elder statesman in his early forties. He was honored for recognizing fiber's potential, although he had underestimated the transparency of silica and the industry had turned away from his suggestion of single-mode fiber. ITT gave him a high profile, sending him to Roanoke, Virginia, to launch a new American fiber-optics group.

A few others held to the original vision of single-mode fibers. Dick Dyott felt alone, "bleating in the wilderness" about their virtues before he left the Post Office in 1975.[11] Developers of submarine cables were interested because they had to stretch transmission speeds and distances as far as possible, but they were not ready to invest much. F. F. Roberts and Stew Miller thought single-mode fibers were long-term possibilities but gave them low priority for use on land. The biggest single-mode project was a Navy effort to develop acoustic sensors to track the movement of Soviet nuclear submarines.

The Fiber-Optic Gold Rush

At the end of May 1976, Bell Labs reported that the Atlanta trial had worked better than expected. Deep collective sighs of relief were breathed. Much more detailed analysis remained,[12] but the basic message was loud and clear: Fiber works. That was more than sufficient to launch a fiber-optic gold rush.

Hot new technologies lure would-be entrepreneurs like rumors of a gold strike draw miners. Corning fueled the fires when it attacked corporate financial problems by companywide layoffs that hit the fiber group. Corning's fiber sales had been stubbornly low, but two victims of the layoff still saw a bright future for fiber optics. Rich Cerny had been marketing fibers after earning an MBA in the ample spare time afforded an Air Force captain tending missile silos in North Dakota. Eric Randall had worked in fiber development after earning a doctorate in glass science.

Boyishly bright and friendly, Cerny was a born salesman. Randall had the technology down cold. Young and energetic, they went looking for cash to get into the fiber business. American Optical turned them down flat, writing, "There's no market for communications fiber optics."[13] They got a better reception from Jim Godbey, an ambitious former Air Force officer and president of the Valtec Corp., a small company with stock traded over the counter. Several years earlier, Godbey and two other veterans of Mosaic Fabrications had started a company called Electro-Fiberoptics in an old pickle factory in Worcester, Massachusetts, to make fiber-optic bundles for lamps and signs. They changed the name to Valtec after a 1972 merger with Valpey, an optics company headed by Ted Valpey, who contributed some family money and became chairman while Godbey ran day-to-day operations.[14]

Cerny and Randall did not have to start cold. Paul Dobson, a Valtec engineer, had already talked Godbey into buying a high-temperature furnace to draw silica fibers. Unable to make complex preforms, Dobson pulled fibers from pure silica rods and clad them with a plastic with lower refractive index. It was a simple way to make fibers, and other companies also were experimenting with it, but Dobson had a knack for making things work that outweighed his limited formal training and tiny budget. When Cerny and Randall arrived, Dobson was producing the best plastic-clad silica fibers available,[15] with loss of only 3.5 decibels per kilometer at 850 nanometers.[16] However, the process could not make graded-index or single-mode fibers, so its potential was limited.

Godbey saw that Cerny's salesmanship and Randall's expertise in glass science complemented Dobson's practical skills. He hired them and set the three up in a separate subsidiary called Valtec Communication Fiber Optics. They built a glass lathe and started making graded-index fibers using the Bell Labs vapor deposition process. After a local cable manufacturer failed disastrously in its efforts to make fiber cable, they decided to build their own cable plant.[17]

You couldn't buy ready-made equipment to cable fibers, so Dobson built his own on a low budget. Lacking much machinery, he had to assemble the cable in stages. First he packaged the fibers into loose tubes. Then he assembled several tubes into a complete cable. Winding the fiber-containing tubes around a central steel strength member required keeping the spools holding the tubes aligned horizontally while winding them around the strength member. Big companies custom-built elaborate heavy-duty machines with all sorts of wheels and gears. Dobson built Valtec's from plywood and 2 × 4s.[18] It did the job, and the design had an added bonus. Separating the tubes at the ends turned out to make cable installation easier.

The Rise of Irving Kahn

Fiber optics also caught the eye of Irving B. Kahn, a colorful promoter and nephew of composer Irving Berlin. Kahn had made a fortune from the TelePrompTer Corporation, which he launched in 1950 to sell TelePrompTers to cue forgetful actors with their lines. He later made TelePrompTer an early giant in cable television but courted local officials a bit too hard. At the height of his success in 1971, he was convicted of perjury and bribing officials of Johnstown, Pennsylvania. He resigned from TelePrompTer and spent 20 months in jail.[19] Forced to sell his stock, he emerged from prison with cash and new ideas, including using fiber optics to carry signals for cable television.

Short, stocky, and physically powerful, Kahn had a blend of charm, clout, and arrogance that meant power in the entertainment industry. He found a couple of former Bell Labs scientists and helped them set up a small fiber-making company called Fiber Communications Inc. His contacts helped the little company land a contract to supply TelePrompTer with one of the first fiber systems used in cable television. He prophesied that fiber would soon replace coaxial cables for cable television. Many cable companies already were working on fiber. Most of the rest scoffed at him, but Kahn found open ears at Insilco, a large Connecticut company that owned Times Wire & Cable Co. Kahn persuaded Insilco to merge its subsidiary with Fiber Communications. Insilco contributed a complete coaxial cable plant that supported an ongoing business. Fiber Communications contributed little more than a draw tower and Kahn's virtuoso salesmanship.[20] Veterans of the fiber industry still marvel at the results. When the deal was done in December 1976, Kahn's group owned 49 percent of the new Times Fiber Communications Inc.[21,22]

From Atlanta to Chicago

The behemoth AT&T paid little heed to would-be competitors; the telephone monopoly had long made most of the equipment it used. The next step in the Bell System plan was to run fiber cables through urban underground ducts and test them with live telephone traffic. Telephone engineers worried that a cable full of glass threads would shatter like a plate-glass window or degrade rapidly when exposed to the harsh conditions in manholes, sometimes filled with water and inhabited by wildlife from rats to alligators. AT&T ran 1.6 miles (2.6 kilometers) of fiber cable through buried ducts linking three buildings in the downtown loop district of Chicago. It was typical of the sites where AT&T planned to use fibers first: urban centers where underground ducts were crammed with copper cables. A single fiber cable no larger than a garden hose could hold 144 fibers, and four fiber cables could fit into a duct built to house one fire-hose sized copper cable. Yet each fiber cable had many times more capacity than the bulkier copper cable. Replacing the old copper cables with new fiber cables promised to multiply duct capacity, without digging up a single downtown street.

AT&T refined its Atlanta system hardware, using fibers that were a little clearer, lasers that were a little better, and connectors considerably improved.[23] Engineers set up test equipment at both ends and carefully measured each of the 24 fibers after technicians threaded the thin cable through the twists and turns of the buried ducts. Not a single fiber broke. On April 1, 1977, the system carried its first test signals, as Bell prepared to send regular digitized phone service—including Picturephone and video signals—through the cable starting on May 11.[24] Proud of getting the testbed up and running less than six months after approval, AT&T officials geared up for a major press announcement of the feat. Then they opened their papers to discover that the nation's second-largest phone company, General Telephone and Electronics, had beat them to it.

GTE Steals a March in Long Beach

While AT&T coordinated its Chicago demonstration within its corporate bureaucracy, the much smaller GTE Laboratories quietly laid its own plans just off Route 128 west of Boston. They picked a 10-kilometer (6-mile) route served by General Telephone of California between Long Beach and Artesia and bought six-fiber cable from the General Cable Corp.

John Fulenwider, a top GTE research engineer, went along to test fiber splicing in manholes and get the system running. The California phone company sent a couple of Hollywood cameramen to film him. After water was pumped from the manhole, the soft-spoken Fulenwider descended with his splicing tools and microscope, dressed in a white lab coat. The cameramen followed with their klieg lights and cameras, joking uneasily about their fate if an earthquake hit while they were underground. They had never been down a manhole before, and in the tight quarters one of them leaned a lamp housing on a four-inch cable. He didn't know that telephone companies pump air through cables to keep water from seeping in and damaging the wires. Nor did he know that his lamp housing grew hot enough to melt the plastic jacket on the cable. The first thing he noticed was a loud "whoosh" as the pressurized air escaped from the cable. Convinced the "big one" had struck, the two dropped everything and scrambled out of the manhole.[25]

That didn't stop GTE from getting its system up and running on April 22.[26] It was not as ambitious as Bell's; it used LED transmitters and carried only 6.3 million bits per second. But it was a first.

AT&T was not amused; Chicago was supposed to be the first. Press officers scrambled to inform reporters that the Chicago system had carried its first signals on April 1, but they had to admit that regular telephone traffic did not start until six weeks later. GTE had won a race that Bell didn't realize it was running.

British Field Trials

F. F. Roberts turned 60 in the spring of 1977, and Post Office rules forced him to retire. As he turned management of the fiber-optics group over to his quick-witted and scholarly lieutenant John Midwinter, he could take pride in a job well done. Fiber optics were on a roll.

In January 1977, the Post Office pulled a graded-index fiber cable through old, crowded ducts near Martlesham Heath, picked to highlight the contrast with the millimeter waveguide.[27] By April, test signals at the European standard level of 8.4 million bits per second were going through 13 kilometers (8 miles) of fiber.[28] When that worked well, Post Office engineers tried the next level, 140 million bits per second, and were amazed how easily that worked over eight kilometers (five miles).[29] Bell hadn't tried such high speeds in the field; pulse spreading was a problem in graded-index fibers at the next level in the North American transmission hierarchy, 274 million bits per second. On June 16, the Post Office started routing live telephone traffic through the system.[30]

Standard Telecommunication Labs ran its own field trial between Hitchin and Stevenage, splicing the cable in the field and mounting two repeaters in standard Post Office roadside housings along the 9-kilometer (5.6-mile) route. The system worked so well at 140 million bits per second that they pulled out one repeater and sent the signals through 6 kilometers (3.7 miles) of fiber.[31] Others followed, the French National Center for Telecommunications Research,[32] Siemens in West Berlin, and Pirelli in Turin.[33] Everything worked better than expected. Problems were trivial compared to millimeter waveguides, which officially had been merely delayed until demand justified them. The results assured that millimeter waveguides would not rise from the dead, and forced top management to decide what to do with fiber.[34]

Organizers of the third European Conference on Optical Communications invited Roberts to open the September 1977 meeting with a talk on the present and future of optical communications. It was an honor for the godfather of European fiber optics. He prepared his talk in the weeks after he retired, noting that fiber was successful in cost and performance. Looking at new research, he predicted that a new generation of systems would operate at wavelengths around 1.2 micrometers, and doubted that barriers to practical single-mode systems would prove insurmountable.

He did not have a chance to see how prophetic his predictions would prove, or even to hear his compatriots' reactions. In early summer Roberts suffered a fatal heart attack while in the shower.[35]

A Second Window Opens

The first decade of fiber optic communications built up a critical mass of researchers pushing the limits of the technology. As they approached the limits of what was possible at 850 nanometers, they began looking for alternatives.

Two effects restrict light transmission through glass—absorption and scattering. Absorption depends on glass composition, but scattering is a fundamental limit that depends on wavelength. It decreases quite rapidly as the wavelength increases.[36] The best fibers were approaching the scattering limit at 850 nanometers, but longer wavelengths promised much lower scattering. Double the wavelength and the scattering limit should drop by a factor of 16.

Early measurements at longer wavelengths were discouraging; higher absorption offset lower scattering loss. Total attenuation—loss plus scattering—hit a low point near 850 nanometers, increased at longer wavelengths, then dipped briefly around 1000 nanometers (1 micrometer) before rising again. Another dip loomed beyond 1400 nanometers (1.4 micrometers), but it didn't look promising.[37] In 1974, Bill French and John MacChesney pushed loss down to 1.1 decibels per kilometer at 1020 nanometers,[38] close to the theoretical minimum, but beyond the range of practical diode lasers.

Researchers traced the residual absorption to bonds between hydrogen and oxygen atoms from the traces of water remaining in the fused silica. The University of Southampton reduced water levels to 0.5 part per million,[39] but the formidable effort of removing the rest hardly seemed worthwhile without good diode lasers beyond one micrometer. By 1975, few researchers bothered measuring transmission beyond 1.1 micrometers.[40]

Masahara Horiguchi, who had been working on fiber communications since 1971 at Nippon Telegraph and Telephone's Ibaraki Electrical Communication Lab, was not ready to stop. What lured him further were suggestions that pulse spreading caused by material dispersion should drop nearly to zero—allowing much higher transmission speed—between 1.2 and 1.3 micrometers. As far back as 1970, Dick Dyott's computer calculations had shown material dispersion should drop to zero at 1.23 micrometers.[41] Shortly afterward, Felix Kapron at Corning did a more detailed analysis predicting a minimum near 1.3 micrometers, with the exact zero point depending on the fiber structure.[42] Nobody paid much attention at the time because they lacked suitable lasers, but in 1975 Dave Payne and Alec Gambling at Southampton took another look and found material dispersion should be zero at 1.27 micrometers.[43] Horiguchi saw their report and decided to look at longer wavelengths.

In the fall of 1975, Horiguchi developed an automated system to measure fiber loss between 0.4 and 2.5 micrometers, a much wider range of wavelengths than anyone else. Knowing that water caused most absorption between 1.1 and 1.8 micrometers, he realized that fibers containing less than 100 parts per billion of water should be very clear. He teamed with Hiroshi Osanai of the Fujikura Cable Works, who in early February 1976 succeeded in making low-water fibers. "On rare occasions, a beautiful loss spectrum was observed in the 0.6 to 1.1 micrometer spectral range," recalls Horiguchi.[44] An annoying water absorption peak at 0.945 micrometers was gone.

On March 27 Horiguchi measured a low-water fiber at longer wavelengths and found it had loss of 0.47 decibel per kilometer at 1.2 micrometers, with all but 0.1 decibel due to scattering. That was their best fiber so far, and it was no fluke. Total attenuation was below 1 decibel from 0.95 to 1.37 micrometers. They looked very very closely at loss over a range of wavelengths and concluded that their clearest fiber contained only 80 parts per billion of water, one-sixth as much as the previous record low.[45] They had hit the jackpot, a new window for fiber optic communications.

Crucially, their new window offered both low loss *and* low material dispersion. Not only could signals go farther at the longer wavelength, but also they could carry much more information because shorter pulses could be packed more closely together. Opening the new window was encouraging news, and even while Bell Labs was designing its Chicago system, researchers on the cutting edge began considering the new window at 1.3 micrometers.

A New Laser Family

The most obvious problem in moving to a new wavelength was the need for a new laser source. That seemed a formidable obstacle because after a dozen years of development, gallium arsenide lasers still didn't meet telephone industry reliability requirements. However, as Horiguchi and Osanai opened the second window in fibers, a Chinese-born scientist at the MIT Lincoln Laboratory invented a new family of long-wavelength lasers.

J. Jim Hsieh was in grade school when his father, an air force officer, fled the communist revolution and took his family to Taiwan. After a year in the Taiwanese Navy, the tall young Hsieh came to America, where he earned a doctorate studying the then-exotic semiconductor gallium nitride.[46] He started working on gallium arsenide when he arrived at Lincoln Lab in 1971, but his interest soon wandered. The Air Force wanted 1.06 micrometer diode lasers for space communications, and Hsieh thought such wavelengths might be useful in fiber optics. After standard materials showed little promise, about 1973 he turned to an unconventional approach, making semiconductor lasers from a mixture of four elements, gallium, indium, arsenic, and phosphorus.

The conventional wisdom held that as a bad idea. Two-element semiconductors such as gallium arsenide must be mixed in the right proportions to make working devices. Adding a third element to the blend caused more problems because they had to be grown on substrates of simpler compounds with almost identical atomic spacing. That worked for gallium aluminum arsenide on gallium arsenide, but not for other semiconductor mixtures. Most specialists thought there was no hope of balancing four elements in what they called a "quaternary" compound.

Hsieh saw an opportunity instead of a problem. A two-element compound has a fixed lattice spacing and gap between valence and conduction bands—which determines the laser wavelength. Adding a third element generally changes both wavelength and lattice spacing, but leaves only one degree of freedom to adjust both parameters—like one knob both tuning a radio frequency and changing its volume. Mix four elements together in the right proportions and you get two degrees of freedom, so you can adjust both wavelength and lattice spacing. Hsieh realized that mixing gallium, indium, arsenic, and phosphorus would let him both select the laser wavelength and match the lattice spacing of the thin quaternary layer to that of a substrate of indium phosphide.[47]

Like Panish and Hayashi, Hsieh had to expend considerable perspiration to take advantage of his inspiration. Few others had worked on indium phosphide, so he had to painstakingly test mixtures and growth techniques. He and a technician set up four furnaces and grew samples, collecting one data point per day per furnace, compiling the data he needed to make recipes for his lasers.[48]

With data in hand, Hsieh grew simple double-heterojunction lasers with a blend of indium gallium arsenide phosphide (InGaAsP) mixed to emit at 1.1 micrometers. The first ones emitted only pulses at room temperature, but as Horiguchi and Osanai measured their record low loss, Hsieh saw a steady 1.1-micrometer beam from a InGaAsP laser at room temperature.[49] When Hsieh tried making longer-wavelength lasers to match the new window, he found they were easier to produce. By the end of 1976, he had made room-temperature InGaAsP lasers emitting steadily at 1.21 and 1.25 micrometers and generating pulses at 1.28 micrometers.[50]

Other developers expected lifetime problems like those that continued to plague gallium arsenide. But Hsieh was delighted to find that his quaternary lasers "looked great" in early life tests. From the very start, they were much more durable than gallium arsenide, although initially no one knew why. Long wavelengths began looking good.

A Big Push in Japan

Osanai and Horiguchi were on a roll in the summer of 1976. Hydrogen-oxygen bonds absorbed very strongly at 1.39 micrometers, even in dry fibers. But removing boron from the core glass opened a third window beyond 1.5 micrometers. The window opened further when they used cores of fused silica doped with germanium. Their best fiber had loss of only 0.46 decibel per kilometer at 1.51 micrometers.[51] They also dried the glass further, reducing water to 30 parts per billion, although their clearest fiber at the time contained 150 parts per billion.

Opening the new window changed the ground rules for fiber optics. Systems operating at 1.3 micrometers had a whole different set of operating characteristics than 850-nanometer systems. Japanese engineers were quick to see that lower loss and pulse spreading could be the basis of a second generation of fiber technology. A team at Nippon Telegraph and Telephone calculated that 1.3-micrometer LED transmitters could send several tens of millions of bits per second through 20 kilometers (12 miles) of graded-index fiber. Brighter laser signals could carry hundreds of millions of bits per second through 30 to 50 kilometers (19 to 31 miles) of graded-index fibers. They even looked farther into the future and envisioned a third generation of systems that would use single-mode fibers to carry up to two billion bits per second over similar distances.

The NTT engineers saw more than just record-setting numbers. They looked beyond the short interoffice systems that were the center of American efforts to systems carrying high-speed signals long distances between major urban centers. Millimeter waveguides had been assigned that role, but they were dead. Fiber optics could do the job with repeaters at regular intervals, but repeaters were expensive and potentially troublesome. Short-wavelength systems would require a repeater about every ten kilometers, and many would have to go into the unfriendly environments of manholes. Stretch repeater spacing to 30 kilometers and they could go inside buildings, offering important practical advantages.[52]

The Japanese also spotted Hsieh's long-wavelength lasers much faster than American developers. They invited Hsieh to Japan, and when he arrived they grilled him about his process. Within a year, they had copied it.[53] It was part of a sea change in the attitudes of Japanese engineers, who for decades after World War II had suffered a cultural inferiority complex, automatically following the lead of overseas groups. By the mid-1970s, they were growing more confident in their own technology and willing to compete head on with Americans. Having captured much of the consumer electronics market, they decided to go after communications and placed the first bets on a second generation of fiber-optic systems.

Mixed News from Field Trials

AT&T steered a more cautious course, even as its engineers returned from Chicago to report a boring litany of successes. Automatic protection switches rerouted signals only four times in the first six months—once because a laser died—but no one using the system noticed because the signals were switched in milliseconds.[54] After a year, Ira Jacobs told an eager audience of fiber specialists: "The system has performed excellently—to date there have been no service outages owing to the lightwave components, and the error-performance surpasses the most stringent Bell System objectives."[55] The biggest problem remaining was gallium-arsenide laser lifetimes. While Bell Labs reached the million-hour target during the Chicago field trial, it was only for lasers carefully made in the laboratory. Mass-produced chips did not last as long, and even the best lasers needed thermoelectric coolers to keep them from heating to the point of self-destruction. Nonetheless, Jacobs announced AT&T would start installing systems operating at 45 million bits per second as regular parts of its network starting in 1980.

Looking farther into the future, Bell Labs set new fiber goals. One target was submarine cables using single-mode fibers and gallium arsenide lasers; another was long-distance systems operating in the 1.1 to 1.3 micrometer window.

In Britain, the field trials at Martlesham Heath and Hitchin-Stevenage were similarly encouraging. The Post Office decided to lay eight-fiber cables along routes totaling about 500 kilometers (300 miles), including the London-Reading route where the millimeter waveguide had almost gone.[56] Like Bell Labs, they began looking to future single-mode and long-wavelength systems.

Some annoying glitches did nag at engineers evaluating the field trials. It proved impossible to predict reliably how much light would be transferred from one fiber to another at Hitchin-Stevenage, even when their specifications seemed identical. Something was inducing noise at the junctions. Initially baffled, Richard Epworth sat down to analyze the problem at Standard Telecommunication Labs. Common sense said that "bad" lasers were the most likely cause, and switching lasers usually cured the problem. However, the lasers that seemed best in laboratory measurements fared the worst in the system.

Epworth finally realized that the problem lay in the coherence of laser light. Coherent light waves can interfere with each other if they travel slightly different paths, producing light and dark zones. Illuminate a small area with coherent light and you see shifting grainy patterns called "speckle." It's an effect well known to laser specialists, which makes laser-illuminated holograms look grainy. Epworth realized the same effect could occur when many modes interfered with each other at the end of a multimode fiber. The more coherent the light, the more pronounced the speckle pattern and the more intense the "modal noise" caused by minute shifts in laser wavelength or fiber position. From a theoretical standpoint, it resembled the modal problems that plagued the millimeter waveguide.[57]

Modal noise hit the STL test hard because it used more coherent lasers and fibers with 30-micrometer cores instead of the 50 or 62.5 micrometer cores used in American trials. The coherence increased speckle intensity, and the small cores meant that loss of only a few speckles could cause noise. When Epworth first described the problem in 1978, he recommended shifting to less coherent light sources,[58] or using fibers that carried more modes. However, it soon became clear that only single-mode fibers could eliminate modal noise.[59] It was a message many people did not want to hear.

Back to Single-Mode

Single-mode fibers had a bad reputation because light coupling was difficult. They also didn't offer much advantage over graded-index fibers at 850 nanometers, where high material dispersion largely offset any attractions of single-mode transmission.

Opening the 1.3-micrometer window made single-mode fibers look much better. The lower loss—about 0.5 decibel per kilometer—meant that signals could travel tens of kilometers. The low material dispersion promised capacity many times higher than at 850 nanometers. Moreover, core size increases with wavelength, to nine micrometers at the longer wavelength, compared to a mere four micrometers at 850 nanometers. That eased alignment tolerances, which had been improving with splice and connector technology.

The single-mode revival spread rapidly. Having finished its trials of graded-index fiber, the British Post Office turned to single-mode.[60] Corning, committed to a strategy of staying at the forefront of the new technology, shifted Bob Olshansky to single-mode, and he discovered it was easier to design and make than graded-index fiber.[61]

The Japanese stepped up single-mode research after they opened the 1.3-micrometer window. By late 1977, NTT was making low-loss single-mode fibers.[62] The Ibaraki lab pushed to remove the last traces of water, paying off at the end of 1978 with single-mode fiber showing a dip at 1.55 micrometers where loss was lower than anything anyone had ever seen before. They had made the clearest glass in the world, with attenuation only 0.2 decibels per kilometer, just a little higher than the theoretical lower limit on scattering. NTT knew they couldn't do much better, and called it "ultimate low-loss" fiber.[63]

The lower the loss, the more enticing single-mode fibers became. Pulse spreading increases with distance; it's a hundred times larger over 100 kilometers of fiber than over one kilometer. Good graded-index fibers could carry a hundred million bits per second for 10 kilometers, but only 20 million bits over 50 kilometers—and at 1.3 micrometers, 50 kilometers (30 miles) became a reasonable transmission distance. Single-mode fibers could easily carry a billion bits 50 kilometers, leaving graded-index fibers in the dust.

In America, single-mode fibers caught the eye of Will Hicks, recovering from a descent into alcoholism that followed his sale of Mosaic Fabrications. Never satisfied with other people's explanations, he calculated the properties of single-mode fiber for himself and found its transmission capacity went far beyond the billion bits a second that impressed others. He stubbornly ignored people who insisted single-mode wouldn't work and started evolving his own vision of future fiber-optic systems.

Hicks knew from his early experiments with fiber bundles that light could leak between fiber cores. Electromagnetic theory explained the process, and Hicks realized it could be applied to switching light into and out of fibers, something important for practical communications. He also realized that one fiber could simultaneously carry signals at many wavelengths, an idea called wavelength-division multiplexing. Others could see the possibility in theory. Glass transmits the whole visible spectrum as well as some infrared light, while the air simultaneously carries radio and television signals at many different frequencies. A single fiber could carry many different wavelengths, but getting many separate signals into the same fiber and separating them at the other end were extremely challenging problems. At best, most specialists thought, each fiber could carry only a few signals. Only Hicks seriously thought it might be possible for one fiber to carry tens, hundreds, or even more separate wavelengths. Taking the idea a step farther, he envisioned that a single optical amplifier, using the laser principle, might simultaneously boost the strengths of the whole range of wavelengths passing through the fiber, amplifying all the signals at once.

In his mind, Hicks put it all together, looking into the future and seeing an all-optical network, carrying light signals all the way to homes without bothering to change them back to electric currents. Fibers and optics would go beyond being mere pipes to become part of the switches that routed calls through the network. He talked some about his ideas, earning a reputation as a wild-eyed visionary, but all that he wrote was patent applications. Then in his early fifties, he was broke, but he talked Chuck Lucy out of a quarter-million dollars from Corning. It wasn't much to build a new technology, but Hicks had cut his entrepreneurial teeth on a shoestring budget. He set up shop in the loft of an old building in central Massachusetts, and with a couple of helpers settled down to inventing a new fiber-optic network technology. He whimsically called his company 1984 Inc., because it would have so much transmission capacity that it could keep track of everybody all the time.[64]

Fiber development accelerated with the opening of the new windows. Each issue of *Electronics Letters*—the British journal that published the hottest developments—announced new wonders. Corning and Bell Labs pushed loss in their best laboratory fibers below 0.2 decibel per kilometer at 1.55 micrometers, to about 0.16 decibel.

Semiconductor lasers advanced apace with fibers. Much to the surprise of the establishment, long-wavelength lasers proved much more reliable than gallium arsenide. As the trend became obvious, Martlesham Heath shifted all its research from gallium arsenide to InGaAsP in a week. Initially frustrated because they had been making progress in gallium arsenide, developers changed their mind when they saw that the longer wavelengths did not trigger the growth of fatal flaws. "The wind was behind us," says Alan Steventon.[65]

Valtec Rides a Dangerous Growth Wave

The revolution in telephone operations was more modest. Industry was trying to do the logical, cautious thing, grow by making systems based on technology that already worked, not the latest idea from the labs. That meant staying with gallium-arsenide lasers and graded-index fibers.

There was no denying fiber optics after its successes in Chicago, Long Beach, and Martlesham Heath. Valtec landed a contract to install a 4.2-kilometer (2.6-mile) system for Central Telephone in Las Vegas, which wanted to show its fiber prowess when a big trade show came to town in December 1977.[66] Marshall Hudson, freshly hired away from Corning, taught the workmen how to splice fibers and promised everyone who beat his target a six-pack of beer. The whole crew won. To show the system was built for the real world, Cerny handed out photos of a Centel workman guiding the cable into underground ducts—sitting in a grimy manhole, his hands covered with greasy black cable lubricant.[67] The brash little company became the first independent contractor to deliver a working fiber-optic system to a telephone company.

The young fiber industry was growing explosively. Big cable and telecommunications companies in America, Europe, and Japan jumped into the field, worried that they might miss a revolutionary new technology. For a while, the flood of companies threatened to splinter the market so much that no one made any money. In one year, sales of fiber-optic connectors rose an impressive 50 percent, but the number of manufacturers tripled, meaning that average sales per company dropped.[68]

Valtec stood at the peak of the wave, struggling to keep its feet. The company was in an enviable yet dangerous position—the purest stock-market play in fiber-optic communications. Fiber was only a small part of the billion-dollar Corning Glass Works; it was a minuscule fraction of corporate giants like AT&T or GTE. Moreover, Godbey shrewdly kept the company growing by such moves as buying Laser Diode Labs. Sales doubled to about $30 million in 1977, and Valtec made a $1.3 million profit.[69] However, neither stock market hype nor Valtec's manic energy could generate the tons of money it needed to invest in fiber development. The company needed a partner with deep pockets, and the logical place to look was in the metal-cable business.

The first candidate was Canada Wire and Cable, which agreed to buy a small interest in Valtec. However, a close look at Valtec's business and finances from the inside convinced Canada Wire it could do better itself, so it founded its own fiber division called Canstar.[70]

Next, Godbey talked to Comm/Scope, a North Carolina company that made coaxial cable for cable television. The company was profitable, privately held, and about the same size as Valtec. In size and scope, it looked like a merger of equals. Valtec brought fibers to the deal; Comm/Scope brought cabling. By spring 1977, they were jointly developing fiber-optic cables. Later in the year, they hammered out terms of a merger, in which the publicly held Valtec technically acquired Comm/Scope in exchange for 38 percent of the stock in the merged company.[71] Yet when the dust settled in mid-1978, Frank Drendel, the head of Comm/Scope, was in control as vice chairman and chief executive officer, with Ted Valpey as chairman. After spending a decade building the company, Jim Godbey had lost control. The stress hit him hard, and while attending a cable-television trade show in California he suddenly fell ill. The next day he was dead of a heart attack at 43, the victim of an unrecognized heart defect.[72]

The merger and a supplementary stock offering gave the fiber program badly needed money. Valtec hired more people to help Cerny, Dobson, Hudson, and Randall with the Las Vegas telephone system. However, morale started slipping after Godbey's death. Randall left for another company in 1979, after Godbey died.[73] At the start of 1980, Ted Valpey swapped some

$3 million in cash and stock for divisions that made quartz crystals and decorative fiber-optic lamps and resigned as chairman.[74]

Firmly in control as the new chairman, Drendel focused Valtec on communications and tried to sell the company's other operations. Entranced by the promise of fiber optics, the stock market ran Valtec shares up from $16 at the end of March to $26.25 in mid-June. However, the company looked better from outside than in; it was again running out of cash. In mid-June, Valtec announced plans to merge with M/A Com Inc., a communications company with about twice its sales. That pushed both company's stocks up further, and when the deal was closed in September 1980, Valtec owners received a staggering $224 million in M/A Com stock.[75]

Despite its tremendous stock market valuation, the merged company remained short of cash. Valtec had become a money pit; a massive investment was needed to realize the bright promise of fiber, and M/A Com didn't have the resources to do it alone. In February 1981, M/A Com found a well-heeled partner in the Dutch giant Philips N. V., which sent money and experts to help the beleaguered Americans.

A Bet on Long-Wavelength Lasers

The entrepreneurial bug bit Jim Hsieh in the midst of the fiber boom. Lincoln Lab's role was to develop new technologies until they were ready for industry; the time came to transfer long-wavelength lasers in 1978. Hsieh didn't want to abandon them. After seven years at the lab, he was growing restless. After agonizing for months, he called Ken Nill, who had left Lincoln a few years earlier to start a laser instrument company. The call was a pleasant surprise for Nill, who had sold his interest in the instrument company and was getting restless himself. Nill knew about Hsieh's work on long-wavelength lasers and thought it might be a good starting point for a new company. Other companies would have to grow their own expertise in long-wavelength lasers; Hsieh would have a running start at his own company,

Their first task was raising money to start manufacturing lasers at the company they called Lasertron. Hsieh already had talked with one interested but unusual party. The Ministry of Post and Telecommunications of the People's Republic of China wanted the new technology to modernize its antiquated phone system. China had changed after the death of Chairman Mao, but the communist government was an unlikely investor in an American start-up.

Venture capitalists were more logical funding sources. Fiber-optic technology was hot, Hsieh had a good technology, and Nill had an entrepreneurial track record. But when the two sat down and weighed the offers, the Chinese came out ahead. Nill had picked up a sharp business sense at his first company; he didn't want to give up too much equity. The venture capitalists wanted a business plan with nice-looking numbers and a large chunk of the company for themselves. The Chinese wanted good people.

The Chinese also offered more than money. Their engineers knew how to set up factories, package lasers, and design transmitters. Those skills nicely complemented Hsieh's expertise in making lasers and Nill's background in making and selling scientific instruments. They formed the company in 1980 and signed up with the Chinese, who promised to send engineers when Lasertron was ready to get its production lines running.[76]

Counting Installations

Through about 1980, you could easily count the number of major fiber systems installed by telephone and cable television companies. The cable-TV industry took a while to decide what they wanted, but the telephone industry quickly settled on systems like Chicago. AT&T's standard design, mirrored through the American telephone industry, was for graded-index fibers and gallium arsenide lasers to carry 45 million bits per second distances of several miles between switching offices. The application was exactly what Charles Kao had proposed, although the technology had changed considerably.

Initially, every new installation was an adventure. Engineers worried if their brand-new technology could survive in the outside world. GTE installed its first permanent fiber system in Indiana after finding it could save $1.5 million in construction costs by squeezing two fiber cables through a single duct in downtown Fort Wayne.[77] Outside of town, a bulldozer plowed a 4-foot (1.2-meter) trench and laid fiber cable directly into the fresh cut. Allen Kasiewicz, General Cable's technical rep, watched uneasily but was relieved when the first segment tested okay. On the next leg, the bulldozer bogged down in a muddy stream. The installers summoned help and chained what Kasiewicz still calls "the biggest 'dozer I've ever seen" to the one stuck in the mud. The behemoth rumbled onward, pulling the smaller one which still plowed the cable into the ground. Much to Kasiewicz's amazement, the system worked.[78] The cable engineers had done their job well.

Fibers were still a novelty at the time of the 1980 Winter Olympics in Lake Placid, New York. Bell Labs talked about sending digital video over fibers months in advance, but New York Telephone stalled and loaded its poles with temporary coaxial cables. At the last minute ABC Television decided that wasn't enough. With the world watching—and millions of dollars of advertising revenue at stake—the network wanted backup video feeds from the Olympic arena to its local control center. The poles couldn't hold any more heavy metal cables, but they could hold a light fiber cable. It was too late to install digital electronics, so Ira Jacobs's group at Bell Labs hastily hooked up standard video. The backup fiber system worked much better than the metal cables and quickly became the primary video feed. For the first time, the world saw fiber optics working—transmitting signals so clearly there was no sign the glass was there.

The public took little notice, but engineers saw and believed. Lake Placid was a torture test for coaxial cables. The sun baked them in the day and bitter cold froze them at night. Normally, engineers had to adjust the electronics to compensate for effects of the huge temperature swings on coax. They didn't need those adjustments with fiber.[79] It was the sort of thing engineers noticed. Fiber was ready.

So too, at last, was the ponderous machinery of industrial production. AT&T and independent suppliers had supplied one-off systems; now they were making standard hardware using gallium arsenide lasers and graded-index fibers. They knew better technology was in the works in the labs, but they thought years would be needed to make it as reliable and economical.

Second-Generation Technology

Even as production lines were starting to roll for first-generation systems, a second generation was sprouting vigorously from "hero experiments" that sought to show how far and fast fiber signals could go. In August 1978, NTT sent 32 million bits per second through a record 53 kilometers of graded-index fiber at 1.3 micrometers.[80] In a matter of months, they raised the data rate to 100 million bits per second.[81] You couldn't do that at gallium arsenide wavelengths—Fibers absorbed too much light and pulse spreading limited transmission speeds.

Steady improvements followed in laboratory system tests. Commercial companies began to take notice. Corning, determined to fight for the market by pushing the technology, jumped on the long-wavelength bandwagon with a dual-window fiber usable at either wavelength.[82] Others offered long-wavelength fibers, and long-wavelength lasers were on the market before Jim Hsieh had production up and running at Lasertron. Even AT&T worked on long-wavelength systems, although Bell Labs, wary of laser lifetime problems, concentrated on 1.3-micrometer LEDs. The LEDs weren't as powerful as lasers, but they did not require cooling, and with the lower loss at 1.3 micrometers, their signals could span ten kilometers (six miles) of graded-index fiber at the same 45 million bit per second rate as short-wavelength lasers.[83]

System developers grew bolder with second-generation systems, partly because the changes were relatively minor. Long-wavelength graded-index fibers fit into the same cables and used the same connectors and splices as first-generation systems. Many transmitters and receivers could be converted to the long wavelengths simply by plugging in new light sources and detectors; the new components cost more but paid handsome benefits in longer repeater spacing.

Rural phone companies were the first to take a serious interest, because they have to span longer distances than the few miles between urban or suburban switching offices. Valtec charged boldly into the field, installing a pair of long-wavelength systems in rural Virginia that ran 18.7 and 23.5 kilometers without repeaters, using its own special graded-index fiber optimized for 1.3 micrometers. Those were record repeater spacings for any working system at the start of 1982. Rich Cerny had left to start his own company, but the infusion of cash from Philips was reviving Valtec, helping it build a solid technical reputation.[84]

Phone companies began shifting their interest to the longer wavelength to run long distances. The change came as Saskatchewan Telecommunications was building a 3400-kilometer (2100-mile) loop to carry a dozen digital video signals to the largest towns in the prairie province. Project manager Graham Bradley bet boldly on fiber and contracted for the best available fiber systems. Construction started in 1980 with a first-generation system but soon shifted to 1.3 micrometers over longer distances.[85]

The longer wavelength roughly doubled transmission speed and distance through graded-index fibers. Second-generation systems spread fastest where those capabilities were critical, as in rural Canada. First-generation systems remained in the pipeline, with many potential customers wary about shifting to new systems that used costly new lasers with little track record. The consensus of telephone companies remained solidly in favor of graded-index fibers for the same reasons Stew Miller had advocated them years before—their large cores collected light easily.

A Fateful Trial at British Telecom

Doubts about graded-index systems were growing in the labs. They "are exceedingly complex," John Midwinter complained as he reviewed the state of the art circa 1980.[86] The Post Office was splitting off its telephone division as British Telecom, and Midwinter was tired of battling mode monsters. He wondered how far 140 million bits per second could go through single-mode fiber at 1.3 micrometers, so his group set up a laboratory test to find out.

The answer was 49 kilometers (30 miles). "That sent shock waves through British Telecom," recalls Midwinter. There were buildings about every 30 kilometers (19 miles) along the company's long-haul phone lines, originally built to provide electrical power to repeaters on coaxial cables. With single-mode fibers, repeaters could be put in those buildings and kept out of dingy manholes, making the system simpler and more reliable. The demonstration "suddenly made people realize that graded index was a dead duck," recalls Midwinter.[87] That was not what people scaling up production of graded-index fiber wanted to hear.

Midwinter's boss, Sidney O'Hara, wanted to go single-mode immediately in 1981. Despite his frustration with graded-index fibers, Midwinter urged caution because a few problems remained with single-mode fibers.[88] His talented team tackled the problems head on and made rapid progress. They collaborated with Standard Telecommunications Labs on a single-mode trial between Martlesham and Ipswich in 1982. The results were impressive. Single-mode fiber carried 565 million bits per second—then the highest speed used in European telecommunications—62 kilometers without repeaters at 1.3 micrometers. Their signal went even farther, 91 kilometers at 1.55 micrometers, although it was limited to 140 million bits per second because pulse spreading was higher than at 1.3 micrometers.[89] The tests convinced British Telecom to shift entirely to single-mode, and it began installing single-mode systems in 1983.

Nippon Telegraph and Telephone and its Japanese suppliers, who opened the long-wavelength window, also moved quickly to single-mode. However, Bell Labs still saw submarine cables as the only place for single-mode. Bell's position was that single-mode was at least a decade away on land. Those who disagreed were frowned upon. After telling a group of AT&T manufacturing engineers about the virtues of single-mode fibers, Paul Lazay was taken aside by a manager who informed him that his "comments on the superiority of single-mode technology were probably a career-limiting presentation."[90]

Midwinter briefed Crawford Hill on British single-mode programs and left "with the overwhelming impression that they were very interested but frankly they were very convinced that this was all very blue-sky stuff, and the real world was graded-index and that's what they were going to engineer the hell out of."[91] AT&T's top priority was building a massive high-capacity digital system to run along the nation's busiest communications route, the Northeast Corridor from Boston to Washington. The phone company was committed to using graded-index fibers.

The Northeast Corridor

The Boston-to-Washington route was no ordinary link in the phone network. Its heavy traffic came from heads of government, industry, and finance. It was big, it was visible, and the existing system was antiquated and overloaded.

Microwave relays still carried most long-distance traffic in 1980, but AT&T wanted to replace the old analog technology. Digital microwaves could do the job in the wide open spaces but not along the Northeast Corridor. Traffic was heavy, cities were close together, and the microwave spectrum was crowded. Once the route had been planned for millimeter waveguides, which could have provided the tremendous capacity needed. Now AT&T chose it as the launching pad for optical fibers in the long-distance telephone network.

AT&T already had a cable right of way along the route, which had been used for an old coaxial cable system. Huts had been built for coax repeaters every seven kilometers (four miles), closely matching the repeater spacing needed for graded-index fiber carrying 850-nanometer signals. Each fiber could carry only 45 million bits per second—672 phone calls—but Bell had designed its cable to hold up to a dozen 12-fiber ribbons. By filling the cable with fibers and using the technology already tested in Atlanta and Chicago, AT&T thought it could achieve both high capacity and robust reliability. The company did not want captains of industry, senators, and congressmen waiting for phone lines.

In January 1980, AT&T asked the Federal Communications Commission to approve construction of the $79 million system.[92] The designers played it safe. In its first stage, the system would send one 45 million bit channel per fiber at 825 nanometers. Later, they proposed to add two more channels per fiber at different wavelengths, 875 nanometers and 1.3 micrometers. (Signals went only one way through each fiber.)

The idea, dating back to the light-pipe schemes of the 1960s, was that each wavelength could carry an independent signal through the fiber, with optics at the far end directing them to different receivers. Will Hicks was playing with the same concept of wavelength-division multiplexing in his loft but using single-mode fiber. He thought Bell's choice of graded-index fiber was stupid and said as much in a letter to the president of AT&T in which he predicted it would be the last graded-index installation in America. AT&T never responded.[93] Corning also suggested using higher-capacity single-mode fiber in a complaint filed with the Federal Communications Commission.[94]

However, the central issue for Corning was not the choice of technology but the decision of who should build the system. AT&T wanted the job to go to its Western Electric subsidiary (now the separate Lucent Technologies). Corning argued that competitive bidding was vital to a healthy American industry. Not surprisingly, Corning's main concern was its own health—it had to sell fiber to make money, and AT&T was its biggest potential customer. If the phone company made its own fiber, Corning would be shut out of the market. The FCC, starting to like the idea of competition, bought Corning's argument for open bids. It allowed AT&T to build part of the system from New York to Washington but insisted on open bidding for the rest.[95]

The Boston–New York segment attracted bids from Europe and Japan as well as America. AT&T decided to stay with three wavelengths through graded-index fibers. A major reason was that Bell did not trust the longevity of 1.3-micrometer lasers. Its design called for lasers at 825 and 875 nanometers and an LED at the longer wavelength, with the same repeater spacing on all three channels.[96] Intense political maneuvering followed the bidding, and in the end the foreign-owned winners were disqualified in favor of Western Electric.[97]

Fiber technology moved steadily forward as AT&T assembled the system. Transmission at 825 nanometers was doubled to 90 million bits per second before the hardware went in. Long-wavelength lasers made the grade, so AT&T used them to transmit 180 million bits per second and never bothered with 875 nanometers.[98] That brought total capacity to 270 million bits per second per fiber.

But the wheels of progress ground slowly at the FCC, the courts, and the giant telephone company. It took years to approve, build, and install the system. The world had changed by the time the whole Northeast Corridor was up and running in 1984. The Bell System had split into AT&T and seven regional operating companies, and competing long-distance companies had begun sending 400 million bits per second through single-mode fibers. The Northeast Corridor was almost instantly obsolete.

A Bold Bet on Single-Mode

In retrospect, the trend toward single-mode should have been obvious earlier. British Telecom and NTT reported encouraging progress on single-mode fibers at the annual Topical Meeting on Optical Fiber Communications in April 1982. AT&T's manufacturing division called single-mode fiber "very promising for intercity and ocean applications involving repeater spans of 20 to 30 kilometers."[99] Yet most American system makers and industry watchers thought years would pass before single-mode fibers became common.[100] They underestimated the momentum of single-mode technology and the daring of the upstart MCI Corp.

The company's charismatic leader, Bill McGowan, did not start MCI, but he led its rapid growth.[101] By 1982, he knew MCI needed a nationwide long-distance network to compete head-on with AT&T. McGowan didn't believe long-distance carriers should develop new technology; he wanted his engineers to pick the best equipment developed by outside companies. MCI engineers visited British Telecom, where they learned the attractions of single-mode fiber. They talked with Japanese companies, who were manufacturing laser transmitters and receivers operating at 400 million bits per second, faster than any AT&T product. They concluded the best approach was 1.3-micrometer transmission through single-mode fibers. Although loss was lower at 1.55 micrometers, lasers were hard to make and pulse dispersion limited the capacity of standard single-mode fibers. Those problems looked solvable, but MCI didn't have time to wait.

MCI boldly asked how much Corning would want for producing 100,000 kilometers of single-mode fiber. It was almost exactly what Corning wanted to hear. The company had poured millions of dollars into fiber optics, making dramatic technological advances, but sales had stayed minuscule. Corning had assigned a top manager, David Duke, to build fiber optics into a big business. Duke had pushed ahead with a full-scale fiber production plant in North Carolina, betting that large orders would follow. Unfortunately, he had listened to Bell and other American phone companies and geared up to make graded-index fiber. Single-mode would require changing equipment, but that didn't stop Duke. Corning developers had been working on single-mode fiber for years. He signed a contract with MCI and went back to order changes at the factory.[102]

MCI plunged in whole-heartedly, starting in December 1982 by leasing Amtrak right-of-way from New York to Washington. MCI brashly boasted single-mode technology would make its new system better than AT&T's Northeast Corridor system, which ran along the same route. The single-mode fibers carried data 50 percent faster at one wavelength than AT&T's graded-index carried at two. MCI's repeater spacing of 30 kilometers (19 miles) was over four times AT&T's.[103]

British Telecom had switched to single-mode earlier, but its needs were modest compared to MCI, which needed to build a whole new national long-distance system in a much larger country. AT&T Long Lines signed its first contracts for single-mode systems—from its manufacturing subsidiary—soon after MCI, but they spanned only about a fifth the distance.[104] The world's biggest phone company already had a massive long-distance network and was worried about interfacing with existing equipment. MCI could start from scratch with single-mode fibers.

The MCI contracts marked a turning point in America. Bill McGowan's gamble was right on the money; single-mode technology was ready. Graded-index fibers lingered for short systems, but almost overnight all the long-distance companies—including AT&T—switched to single-mode fiber for their nationwide backbone systems. The change came so fast that Northern Telecom finished building the loop around Saskatchewan with single-mode fibers carrying 565 million bits per second.[105]

The Corning Patent Wars

As the technology raced forward in the early 1980s, a long-stalled patent battle came to a head. In 1973, the US Patent Office began issuing a series of patents to Corning covering fused-silica processes and fiber dopants. As Corning strategists had intended, the patents essentially controlled access to the technology; to make communication fibers, you needed the Corning patents. Cross-licenses exempted Bell from paying royalties,[106] but Corning was not going to let anybody else get off that easily. Corning filed its first lawsuit in July 1976, charging ITT with infringing Corning patents in making fiber it sold to US military agencies.

International Telephone and Telegraph was a logical target because it looked like it could become a formidable competitor. Standard Telecommunication Labs was an important technology center; the conglomerate had sent Charles Kao to head development at its American fiber group. He brought with him a technique Phil Black had developed at STL to deposit glass soot inside a tube that could be collapsed into a fiber. Black's method closely resembled the process John MacChesney developed at Bell Labs, but his August 1973 application earned ITT a British patent, and the company applied for an American one.[107] Corning claimed the ITT process infringed its fundamental patents. It didn't matter if ITT had devised the process itself, or even if ITT held a valid patent—Corning insisted that ITT couldn't make fibers without using a patented Corning process.

The rest of the industry watched uneasily as lawyers for the two companies spent years maneuvering for position. They grilled engineers and executives on both sides. The case threatened to go on forever, but ITT caved in just before a trial was set to start in 1981.[108] In a consent decree, ITT admitted infringement, agreed to pay a lump sum penalty, and licensed the Corning patents.[109] The last-minute decision surprised Black and other observers, who thought ITT had a good case.[110] However, to win ITT lawyers would have had to break every applicable claim of every Corning patent involved in the case. If they failed, Corning could legally refuse to license its patents and shut down ITT's American fiber operations, as well as claiming financial damages. If Corning threatened to play that card if the suit came to trial, ITT may have swallowed hard and signed a license rather than risk being shut out of the American market.

With that suit settled, Corning started looking at other fiber companies. They could do nothing about AT&T, but they could go after other competitors. Valtec was at the top of the list, and Corning filed suit in July 1982.[111]

Valtec was growing, but it still was losing money—$1.3 million in the last three months of 1982—and its patent position was weak. M/A Com held on for a few months, but in early 1983 it bailed out, selling its interest to Philips.[112] The Dutch giant soon realized that it couldn't win against Corning's formidable patent position, and put Valtec up for sale. With Corning's lawyers closing in, Philips was desperate to close a deal before the courts shut down its fiber plant. They turned to ITT, now licensed for the Corning patents and needing production capacity to fill a big contract from AT&T.

With 14 days to go to the deadline, ITT sent three executives to Massachusetts to close the deal. The three worked day and night, amazed to find how much Valtec had achieved with minimal resources. They closed the deal on a Sunday morning, facing a midnight deadline, but that didn't end the crisis. Many top Valtec people didn't want to work for ITT. Jim Kanely, the cable industry veteran who had rebuilt Valtec, quit immediately as president. Others followed, including Dobson and Hudson.[113] Paul Lazay, who had left Bell Labs to try his hand at business, shuttled from Virginia to Massachusetts, trying to reassure employees and integrate operations.[114]

Corning turned its patent artillery elsewhere and soon forced small companies to surrender. The toughest battle was with the Japanese giant Sumitomo, which had built a fiber plant in North Carolina. Corning needed heavy-duty maneuvering with the courts and the International Trade Commission to extract hefty damages and force Sumitomo to idle its plant until the last of Corning's fundamental patents expired.[115]

A Technology Takes Off

The fiber-optics market took off in the mid-1980s. The deregulation of long-distance telephone service in America created a market for long-distance transmission. MCI, Sprint, and smaller carriers spread tendrils of fiber networks across the country, along railroad lines, gas pipelines, and other rights of way. AT&T was not far behind.

As the largest independent American maker of long-wavelength lasers, Lasertron was able to cash in on the growing market, although not without problems. The Reagan Administration had settled into Washington, and fresh Cold War chills blew through the Pentagon. Military planners counted on American technology giving them an edge on the battlefield, and they waxed paranoid about potential enemies gaining American technology. Their main worry was the Soviet Union, but the Chinese were communists, and communications was a vital technology. When Chinese engineers arrived in Boston to help Lasertron, government officials hit the ceiling and demanded the Chinese be sent home "tomorrow."[116] Hsieh and Nill argued, but in the end Lasertron had to perfect manufacturing technology by itself. By the mid-1980s, it reached 350 employees and $28 million in sales. The Chinese eventually sold their stock back to the company, earning a healthy profit but not access to the new technology.[117]

Short of money to fund further development, Will Hicks sold 1984 Inc. to Polaroid in 1982 for several million dollars.[118] Polaroid's legendary founder Edwin Land was gone, but parts of the company retained his innovative spirit and bet on Hicks's plans to send tens of thousands of high-definition television channels through one single-mode fiber. Hicks persuaded Polaroid to hire Eli Snitzer to run the group while he continued research, but he still didn't fit into a big company. Within a couple of years Hicks quit to return to running his own tiny ventures.[119] In time, Polaroid abandoned the program, although the technology was promising, because its tremendous capacity "was too far ahead of the commercial needs."[120]

The Right Technology at the Right Time

Looking back, single-mode fiber was the right technology at the right time for the telephone industry. The breakup of AT&T and the deregulation of long-distance service changed the rules, ending the era when stodgy corporate bureaucrats slowed technology to a glacial pace. The new carriers needed new national networks; AT&T needed new transmission lines to compete.

The burst of orders tipped the balance decisively toward single-mode. In five short years, fiber-optic communications had gone through three technological generations. The consensus view of 1978 was obsolete in 1983. Graded-index fiber left no room to grow;[121] single-mode left plenty of room for expansion. As F. F. Roberts had planned to say in the valedictory he did not live to deliver, the barriers to single-mode technology were conquered. The Northeast Corridor system that the cautious bureaucracy of AT&T had intended to blaze a path to the future instead became an evolutionary dead end. Like the millimeter waveguide, it was soon forgotten, an eight-track tape player in the attic of outdated technologies. The wild-eyed visionaries had won.

15

Submarine Cables

Covering the Ocean Floor with Glass (1970–1995)

> In the much more difficult case of the sub-ocean routes, it may take up to about 20 years to produce repeaters with long enough average lives to give complete cables 5000 km long that will operate without maintenance for an average period of 20 years—but it can be and certainly will be done.
>
> —Alec Reeves, 1969[1]

In the 1960s, it took a wild-eyed optimist like Alec Reeves to see a future for fiber optics in one of the toughest jobs for any cable—crossing the ocean depths to link continents. Yet by the mid-1970s, optical fibers were the last hope of salvaging an aging submarine cable industry besieged by rapid advances in satellite communications.

Submarine cable engineers had made considerable progress since TAT-1, the first telephone cable, crossed the Atlantic in 1956. They had multiplied capacity of coaxial cables a thousandfold, cramming 4000 voice circuits through the latest in the series. But there coaxial cables ran into a technological stone wall while satellites charged ahead. Only fiber optics offered any hope for a new generation of submarine cables.

The Submarine Cable Business

Ocean-spanning cables are unique parts of the global telecommunications network. The vast network on land is composed of many comparatively small elements. A call from Boston to Los Angeles hops across the continent from point to point: through Hartford, New York, Philadelphia, Pittsburgh, Columbus, Indianapolis, St. Louis, Kansas City, Cheyenne, Salt Lake City, and Las Vegas. This modular structure makes the network easier to build, operate, and maintain. If a town grows, a phone company can add cables from that town to the nearest node on the national network. If a careless contractor breaks a buried cable, the network switches signals around the break while technicians fix the damage. Total investment in the network is tremendous, but individual links are relatively inexpensive.

In contrast, submarine cables can span thousands of kilometers (or miles) of ocean and cost hundreds of millions of dollars; the biggest today have cost over a billion dollars.[2] Some run between two coastal landings, such as from New Jersey to France; others land at several points on the coasts of different countries. Virtually all the cable is submerged, and much lies in the deep sea, far beyond the reach of technicians in four-wheel-drive trucks. Repairs mean sending a cable ship to find the damage and haul the cable to the surface for repairs.

Repairs take so much time and money that undersea cable operators demand the utmost in reliability. Their specifications insist on no more than two failures in the cable's quarter-century lifetime that require hauling the cable to the surface. To meet those demands, cables must withstand pressures to 10,000 pounds per square inch, without corroding in salt water. Repeater housings must protect electronics from the same tremendous pressures, without letting a drop of water touch sensitive components.

The deep sea shields cable from most disturbances save massive underwater landslides or earthquakes. However, fishing trawlers and ship anchors threaten the ends in shallow water, which are armored with thick steel wires, then buried in a protective trench cut into the ocean floor. Fishing is banned from the zones around cable landings, and just to be sure, cable owners hire patrol boats to warn away violators.

An Old Tradition of Bold Ventures

Submarine cables began carrying electrical telegraph signals in the mid-nineteenth century. The first crossed the English Channel in 1850. It didn't work well because it was poorly insulated, and a fisherman soon cut it. But the next year a better cable followed, bringing news to England as fast as to the rest of Europe. American entrepreneur Cyrus Field soon decided to run a cable from Newfoundland to England. Some eminent scientists thought the idea was daft, but on the second attempt, Field laid a cable that in 1858 succeeded in relaying a handful of messages across the Atlantic before failing. The Civil War delayed the first permanent cable link across the Atlantic until 1866.[3]

A continuous cable running some 2000 miles (3200 kilometers) under the sea was an impressive achievement for Victorian-era technology. One reason for its success was the use of gutta percha, a hard plastic made from the sap of tropical trees that remained the standard insulator until the invention of polyethylene in the late 1930s. Another was that telegraph signals did not need amplification. Telephone signals did, so long-distance calling on land had to await the invention of the vacuum tube. Early tubes were not ready to go to the bottom of the sea; they drew too much power and could not withstand the pressure. Instead, transatlantic telephony began with wireless transmitters.

World War II highlighted the limits of radio telephones. Transmission was noisy, vulnerable to changes in the atmosphere, insecure, and limited to 15 or 16 channels.[4] The war also brought new durable, compact, and low-power vacuum tubes, which the British Post Office put into the first submerged repeater in the Irish Sea in 1943. AT&T used them in a 1950 cable between Key West and Havana that remained America's only direct link to Cuba until 1989.[5]

Satisfied that the shorter systems worked, the Post Office and AT&T agreed in 1953 to lay TAT-1, the first transatlantic telephone cable. The cautious phone companies didn't trust new-fangled transistors, so they equipped the pair of coaxial cables that ran 1,950 miles (3,100 kilometers) from Newfoundland to Scotland with 51 vacuum-tube repeaters each. On September 25, 1956—a little over a year before Sputnik—the system started carrying 36 telephone circuits, one cable carrying the eastbound voices, the other the westbound signals. It cost about a million dollars per two-way channel and was used until 1979.[6]

More cables followed, with fatter coaxial cables that carried more channels. Massive, costly cables crossed the Pacific, stopping at Hawaii on their way to Japan and Australia, tying the world together. Manufacturers replaced heavy steel armor with a thick plastic covering for deep sea areas where cable damage was unlikely, but were slow to adapt other new technology. They did not lay the first transistorized repeater until 1968.[7]

Submarine coax technology reached its high point in the sixth cable from the United States to Europe, TAT-6, which in 1976 added 4000 new voice circuits to the 1200 of the five previous TATs. At $179 million, TAT-6 cost more than double TAT-5, but the cost per circuit was $45,000—less than half that of TAT-5. The TAT consortium immediately started work on a duplicate, called TAT-7, to add 4000 more circuits in 1983.[8]

The End of the Line for Coaxial Cable

Engineers had developed many tricks to squeeze the most out of submarine cables, such as packing five conversations into a single transatlantic voice circuit. However, TAT-6 and-7 were at the practical limits of coax technology.

Increasing the number of voice circuits required transmitting at higher frequencies, but higher frequencies suffer more loss in coaxial cables. Offsetting the higher loss required adding more repeaters and reducing the distance between them, and that was bad news. Repeaters are expensive and their electronic innards are the parts of a submarine cable most likely to fail (excluding cable breaks). The more repeaters, the higher the cost and the more likely trouble. AT&T designed a new coaxial cable to carry 16,000 voice circuits, but it would have required twice as many repeaters as TAT-6—one every 4.6 kilometers (2.9 miles), or a thousand to cross the Atlantic.[9]

You also need thicker cables to carry higher frequencies, and their bulk posed another problem.[10] The TAT-6 cable was 2.08 inches (5.28 centimeters) thick, and engineers worried that a fatter cable would not fit on a single cable ship, increasing installation costs. They also worried that bulky, inflexible cables could be damaged more easily during installation or handling.

As coax looked worse and worse, satellites were looking better and better. However, cable operators couldn't simply switch to satellites; they had been frozen out of the satellite business in the 1960s. Nor was it easy for cable manufacturers to turn to making satellites. Moreover, while satellites benefited from the allure of space-age technology, they also suffered some technical limitations that affected how they relayed telephone signals.

One is transmission delay. Standard communication satellites circle the Earth exactly once a day, so they appear to park above a spot on the equator, where they remain continually in range of ground stations.[11] This requires them to orbit 22,000 miles (36,000 kilometers) above the surface. Radio signals travel at the speed of light, but at that distance radio waves take a quarter-second to make the round trip. The number sounds vanishingly small, but it's enough to throw off your verbal timing in a call with one satellite bounce. A second satellite bounce adds further delay, making conversation difficult.

Poor connections were a problem with the analog electronics used by satellites through the 1980s. Some circuits suffered annoying echoes; a few carried screeching feedback. Sometimes only one side of the conversation went through. If you made many international calls in the 1980s, you learned to recognize satellite circuits and were ready to hang up and try again if you needed a better circuit.

Satellite channels also lacked security. In the heyday of the Cold War, Soviet and American security agencies tried to eavesdrop on each other's satellite links. They didn't always get the information they wanted, but the threat made military agencies prefer cable links.

Unwilling to abandon transoceanic communications, AT&T, the British Post Office, and Nippon Telegraph and Telephone sought alternatives to coaxial cables for undersea transmission. With hollow millimeter waveguides out of the question under four miles of water, optical fibers were virtually the only possibility. Engineers did not have to start from scratch. They could adapt existing submarine cable structures to accommodate fibers instead of coaxial cables. Electro-optic repeaters could fit into the same pressure-resistant housings as coaxial repeaters. As fiber-optic communications evolved, the main issue became assessing different approaches.

The Allure of Single-Mode Fiber

When Alec Reeves suggested submarine fiber-optic cables in 1969, he expected repeaters to be only 2 to 3 kilometers (1.2 to 2 miles) apart, so a couple thousand would be needed to cross the Atlantic. He thought that high fiber capacity—allowing transmission at 10 billion bits per second, well over a million voice channels—would offset the problems raised by so many repeaters.[12] His estimates were based on fiber loss of 20 decibels per kilometer, and within a few years it was clear he had been far too pessimistic.

It also became clear that pulse dispersion could not be neglected in graded-index fibers. That was not entirely Reeves's fault for being overoptimistic. The effect is proportional to distance; go ten times farther and pulses spread out ten times farther. Dispersion that would be no problem over a couple of kilometers could limit transmission speeds in cables spanning ten times that distance.

Submarine cable developers also were much more concerned with reducing the number of repeaters than Reeves had been. That pushed them to seek the utmost in transmission capacity and repeater spacing, which led them to reconsider single-mode fiber. The usual objections to single-mode fiber centered on the demanding precision needed to align fibers with each other and with light sources. Attaining that precision was hardest in demountable connectors—but they weren't needed in submarine cables. Alignment was easier in the factory than in the field, and submarine cables were assembled with repeaters in place, then loaded into cable ships. Warnings that single-mode fibers were years from practical use didn't frighten submarine cable developers accustomed to spending many years perfecting new systems. The next generation of submarine cables was not scheduled until the late 1980s; without the best possible fibers, it might never come.

Nonetheless, as long as fiber systems had to operate at 850 nanometers, they did not offer dramatic advantages over coaxial cable. Loss of a few decibels per kilometer meant a repeater was needed roughly every 10 kilometers (6 miles), just a slight improvement over the coaxial cables used in TAT-6 and -7. Serious worries about laser lifetimes offset the attractions of using a smaller cable. Material dispersion was high at 850 nanometers, so single-mode fibers did not offer dramatically higher transmission capacities than graded-index.

The steady advance of fiber technology, the opening of the 1.3-micrometer window, and the spread of digital transmission tipped the scales decisively toward fiber for the next generation of submarine cables. Both loss and pulse dispersion were much lower at the long wavelength than at 850 nanometers. With the low loss at 1.3 micrometers, signals could travel 50 kilometers (30 miles) or more between repeaters. Graded-index fiber had lower dispersion at 1.3 micrometers than at 850 nanometers, but single-mode fiber had near zero dispersion, promising much higher transmission speeds. The choice was clear to AT&T, the Post Office, and Standard Telecommunication Labs in

1978, when they became the first developers to commit to using single-mode fiber for any real system.[13] They had ten years to go before the scheduled laying of the first big submarine fiber system, TAT-8 across the Atlantic.

Serious Single-Mode Development

Bell Labs cautiously waited a few months before pulling the plug on development of coaxial submarine cables. Some engineers doubted a fiber system would be ready by the TAT-8 target date, but fiber was the last chance for submarine cables. Coax had no prayer of keeping up with satellites.

Bell added fiber experts to the submarine cable group in Holmdel, which began working with Bill French and Paul Lazay, who were drawing single-mode fiber at Murray Hill. Peter Runge shifted from Stew Miller's group to take charge of developing submarine fiber cables. The task promised Runge personal as well as professional rewards; the transplanted German engineer had family on the other side of the Atlantic, familiar voices that passed beneath the sea.

Adapting the tried and true designs of submarine cables for fibers was a top priority. Cables are built outward from their centers. The starting point for the submarine fiber cable was a copper-clad steel "king wire," to provide essential strength. A soft plastic cushion covered it, with up to a dozen optical fibers embedded in it, gently wound round and round the king wire in a helix. Nylon covered the plastic, forming a core 2.6 millimeters (0.1 inch) thick. Then heavy steel strands were wound around the core, strengthening and shielding the fiber-optic heart of the cable. A welded copper tube covered the steel wires; its role was to carry electric current to repeaters across the ocean. A thick layer of solid white polyethylene covered the copper, making a 21-millimeter (0.827-inch) cable ready to lay on the sea floor like a thick, fat garden hose. The cablers wound extra heavy steel wires around lengths of cable to be laid in shallow water, where fishing trawlers might drag their lines across it. Tests of 100-meter (330-foot) samples began in September 1979,[14] in an artificial ocean Bell Labs had built at Holmdel to test the coaxial cables for TAT-6. Elaborate instruments controlled temperature, pressure, and cable tension.[15] The results looked good.

Standard Telephones and Cables took undersea cables even more seriously. They were an important part of the company's business, and by 1978 the submarine systems division was responsible for fiber manufacturing and the main supporter of fiber development at STL.[16] A major STL worry was that the high pressure at the bottom of the ocean would cause the fiber to wrinkle inside the cable, causing "microbends" that let light leak out. To assess the problem, they designed a cable with four graded-index and two single-mode fibers inside an inner aluminum tube, which was surrounded by steel strength wires, a concentric copper conductor, and an outer polyethylene layer, with 18 armored protective layers wound around the outside. They assembled 9.5 kilometers (5.9 miles) of cable, which the Post Office carefully laid in Loch

Fyne, Scotland in February 1980; later they added a 140 million bit per second repeater. Careful measurements showed the cable worked well in the water.[17] The single-mode tests added to the evidence that persuaded the Post Office to lay single-mode fibers on land as well as at sea.

A Bold Move at Bell Labs

By 1980 the course of the future was clear for AT&T Submarine Systems. Peter Runge told the world: "The next generation of coaxial cables will not be developed because of fibers." When TAT-7 was laid in 1983, it would be the last coaxial cable to cross the Atlantic. The next transatlantic cable, TAT-8, would contain two pairs of single-mode fibers each carrying 280 million bits per second, the combined equivalent of 35,000 phone calls—nearly nine times the capacity of TAT-7. Plans called for repeaters to be 30 to 35 kilometers (19 to 22 miles) apart. He said the cost per channel should be only 20 percent that of TAT-6 and only half as much as the partly developed coax system Bell had abandoned as impractical.[18]

Sitting in the audience as Runge spoke to a laser meeting, I recognized a milestone. AT&T had just announced the Northeast Corridor system, but that was only one link in a nationwide network. Submarine cables were the highest-performance telecommunication systems on the planet, and fiber optics would be the next generation, slated for operation in 1988. Fiber optics had gone beyond a few guys with a draw tower in an old pickle factory. The brash upstart had won a starring role.

Bell Labs tackled cable development itself. The submarine systems group asked the single-mode fiber group at Murray Hill to supply 20 fibers, each 20 kilometers (12 miles) long, for a cable to be tested in the North Atlantic. It became a crash priority for a growing team that turned their Murray Hill labs into a small-scale factory. The longest fibers they could draw were five kilometers. Splicing them together was a challenge, but the canny Paul Lazay realized it also was an opportunity. He could measure subtle differences among fibers and predict the results of connecting two slightly different types. By putting fibers together in the right order, he could fine tune their transmission. Working intensely, Murray Hill made 22 fibers, each 20 kilometers, for the submarine cable group.

The Holmdel group testing long-distance single-mode transmission cast eager eyes at those 400 kilometers (250 miles) of fiber. Lacking that much fiber, they had been bouncing signals back and forth through shorter fiber segments. Splicing those long fibers together promised more realistic results, but Lazay was not thrilled when Holmdel asked to borrow them. He was a researcher himself—he knew the other researchers would find excuses to play with the fibers, and he worried about possible damage. That wouldn't do; the fibers were committed to the submarine cable experiment. On the other hand, Lazay didn't want to be seen as obstructing other Bell research groups. He decided to let them borrow the fiber, but to protect it he spliced the 20 reels of fiber together and mounted them in a heavy locked wooden case. He left a few meters of fiber hanging out at each end so Holmdel could send light through the precious fiber, but the box kept them from messing around with it. Then he loaded the crate into a station wagon and drove—very carefully—to Holmdel, where he chained it to the massive optical table used for the experiments.[19]

When the Holmdel tests were done, Bell shipped the single-mode fiber to New Hampshire, where Simplex Wire and Cable put a dozen fibers into an 18-kilometer cable. In September 1982, an AT&T cable ship laid part of the cable and a prototype repeater 5.5 kilometers (3.4 miles) underwater on the Atlantic floor. It was the first fiber cable tested at that depth, and it worked perfectly in the water.[20] It continued to work when hauled back out; it remains stored in a Holmdel basement for occasional tests.[21]

A Nasty Scare

The North Atlantic tests showed fibers worked well for a short time underwater. But by mid-1982, long-term tests in Japan and Britain were showing some distressing trends.

Every profession has its own nightmares. Engineers dread flaws that remain hidden until the hardware is in place. Imagine a crack in a crucial beam supporting a bridge, which grows slowly as trucks rumble over it until the beam fails, plunging vehicles into the river. The lack of long-term data on optical fibers worried engineers, so they monitored early systems carefully. Land trials looked good, but undersea cables were a different environment.

Nippon Telegraph and Telephone spotted the problem in a couple of cables it laid for field trials in 1979. Initially transmission was excellent, but cautious NTT engineers checked back every several months to monitor aging. In June 1982, they discovered the loss was increasing at 1.3 micrometers. They looked for a bad component or splice, but their instruments told them the loss was increasing in the whole fiber.[22] STL saw the same pattern at Loch Fyne, although it was not as severe.

Uneasy rumors started spreading through the fiber-optic world. Something was wrong, and no one knew what. Anxiety churned in developers' stomachs. Uneasy engineers went back and measured their cables. Many found no change, but some noticed disturbing increases in loss. In America, the problem hit a cable that crossed Lake Washington in Seattle.[23] In Britain, a rise in fiber loss knocked out a link operated by the Central Electricity Generating Board. The problem "hit like a sledge hammer" in Britain, recalls John Midwinter.[24] The Japanese were even more anxious; they had seen a cable go totally dark.[25] Developers of submarine cables had put all their bets in the fiber-optic basket, and it looked like the bottom might drop out.

Baffled engineers turned to careful detective work to pin down the problem. Crucial clues came from a Japanese test of the long-held assumption that water did not affect fiber transmission. In early 1982, Naoshi Uesugi of NTT filled a cable with water and put an uncabled fiber in the same water to compare what happened to transmission. After eight months, loss of the cabled fiber increased dramatically at 1.08 to 1.24 micrometers and beyond 1.4 micrometers. Attenuation of the uncabled fiber had hardly changed.[26]

Those were the wavelengths where loss was mysteriously rising in other fibers. The shorter wavelengths were where hydrogen molecules absorbed light. The longer wavelengths matched the absorption of the tiny amounts of residual water that remained trapped in virtually all silica fibers. That pointed to hydrogen as the problem. Somehow the affected fibers were soaking it up. Yet the changes were inconsistent; they affected some fibers and some cable structures but not others.

Developers in Japan, America, and Britain set up task forces to attack the problem. Once they focused on hydrogen, they began to see patterns. The only fibers affected were those containing phosphorus, which softened the glass, making it easier to draw. British work showed that phosphorus was the bad actor; it reacted with hydrogen, forming chemical bonds that absorbed light. The problem didn't show up with a dash of phosphorus but became serious if concentration reached about one percent.

Cable structure was important because some types accumulated hydrogen. The high voltages carried by wires to power undersea repeaters caused metal in the cable to react with the traces of water that inevitably seeped into the cable.[27] Sometimes submarine cables collected so much hydrogen that it could be flared off when the cable was hauled to the surface and opened for repairs.[28] Moisture also could build up in cables on land. The extent of the problem depended on where the hydrogen collected in the cable; STL's original design had a problem because an internal barrier trapped hydrogen in the same region as the fiber.[29]

With the culprit identified, developers attacked the hydrogen problem on all fronts. Glass makers purged phosphorus from their preforms and their vocabulary.[30] Fiber drawers applied new coatings to keep hydrogen from seeping into fibers. Cable makers added hydrogen seals, eliminated plastics that might release hydrogen, and sealed their cables tighter to keep water out.[31] Engineers hooked measurement instruments to phosphorus-free fibers in hydrogen-free cables and watched carefully. They saw no changes. The great hydrogen scare was over. Hearts resumed beating naturally; sighs of relief echoed in Japan, Britain, France and America. Submarine fiber cables pushed onward.

A Go-ahead for Submarine Cables

Plans for the massive TAT-8 project unfolded with little heed of the hydrogen scare. The cable builders had no other options. In December 1982, an international consortium[32] requested bids for a fiber-optic cable running from Tuckerton, New Jersey, to a point off the European coast, where it split into separate cables to France and England. AT&T, Standard Telephones and Cables, and the French Submarcom consortium were the only bidders. AT&T tried to land the entire project, citing the hydrogen problem in the Loch Fyne cable.[33] However, the cozy little world of international telephony usually split contracts to assuage politicians and corporations. In the end, AT&T got the lion's share of the $335 million contract, from America to the branch point. STL got the British section, and Submarcom the French part.[34]

A transpacific consortium was just months behind TAT-8, with service to start in 1989. Cables had lagged far behind satellites across the wider Pacific. Submarine cables carried just under a thousand voice circuits from California to Japan in 1984; 1460 more went from Canada to Australia via Hawaii.[35] AT&T landed a contract to build the Hawaii-4 cable from California to Hawaii, and the part of Trans-Pacific Cable 3 from Hawaii to a branching point off Japan. The Japanese Ocean Cable consortium won the contract to build the rest of the system.

Meanwhile, work on smaller submarine cables went much faster. For each massive cable like TAT-8, there are dozens of short ones, typically linking offshore islands to the mainland or other islands, or linking major coastal cities. The high capacity and long repeater spacing possible with fibers also were important for many of these systems. British Telecom was the first to use a fiber cable for a short submarine link carrying regular traffic. In 1984, it laid an eight-kilometer (five-mile) fiber cable from the coastal city of Portsmouth to the Isle of Wight in the English Channel. STC supplied the hardware, which carried 140 million bits per second over each of four pairs of single-mode fibers. The whole system cost just $292,000 and contained no repeaters, but it was a first.[36]

More important was the first fiber-optic addition to the array of cables crossing the English Channel, UK-Belgium 5. Traffic was growing between Britain and Belgium, Germany, and the Netherlands; the carriers needed more capacity, and they wanted digital transmission. STC won the job by offering to lay a "fully engineered prototype" that would start with experimental service, then phase into regular use. A dozen cables already ran from Britain to Belgium and the Netherlands, carrying just over 23,000 voice circuits.[37] The fiber cable carried 11,500 voice circuits on three pairs of single-mode fibers operating at 280 million bits per second. The three repeaters in the 122-kilometer (76-mile) cable were the first on a submarine fiber cable to carry regular telephone traffic.[38]

Laying the main part of the cable took just five days in early 1986, although it had to be buried in the seabed to prevent damage from trawlers and ship anchors. A second cable ship laid the rest of the cable in shallower water, where a submersible trencher buried it deeper in the sea bed. Tests began as soon as engineers hooked up the cable; commercial service formally began October 30, 1986, with a two-way video conference between London and Ostend, Belgium.[39]

The Japanese bounced back from the hydrogen scare and kept pace. NTT designed a system to transmit 400 million bits per second between islands in the Japanese archipelago. After two years of shallow-water tests, in November

1984 they laid 28 kilometers (17 miles) of cable and two repeaters in the deepest water handy, the eight-kilometer-deep Ogasawara Trench. Careful tests showed little change, so reassured NTT engineers went ahead with plans for a submarine cable between Japan's two main islands.[40]

Shark Attack in the Canary Islands

With the lion's share of the two biggest submarine fiber cables, AT&T was not about to take chances on an untried technology. The Submarine Systems division carefully planned another test of TAT-8 technology in the Canary Islands off Africa, where the Spanish telephone authority wanted a link between Grand Canary and Tenerife islands. Canary Islanders got the latest submarine cable technology in return for putting up with exhaustive tests of the hardware, including trials of recovery of the cable from the ocean floor and shipboard repairs.

AT&T got more than it bargained for. Installation of the 119-kilometer (74-mile) cable went well, including a pair of repeaters 40 kilometers (25 miles) apart. Soon afterward, though, engineers noted a short-circuit in the cable, caused by ocean water, which conducts electricity, reaching the central conductor that carried electric power for the repeaters. The AT&T team worried how water was getting into the cable, but it wasn't an immediate catastrophe because the cable was designed to withstand a single short. Then a second short followed, knocking out repeater power and forcing AT&T to haul the cable up from the ocean floor.

Examining the recovered cable, AT&T engineers found shark teeth embedded in the outer plastic coating. Sharks had bitten the cable and left some of their easily detached teeth behind. The engineers were amazed; no one had ever seen sharks bite undersea coaxial cables, and they had no hint that sharks might develop a taste for fiber. The marine biologists whom they called also were amazed. The cable had been more than a kilometer (0.6 mile) deep, and no one realized that sharks swam that deep.

It wasn't the sort of thing that normally prompted AT&T press releases, but word inevitably leaked out. An engineer working on the problem[41] took a section of cable and a handful of sharks' teeth to a European conference, where he showed them to a few insiders. The story was too good for his fellow engineers to keep quiet. Once an enterprising reporter picked up the lead, it was all over the papers and the technical press.

Sharks had ignored coaxial cables. Why should they turn their razor-sharp teeth on fiber? Biologists suggested some experiments. They stretched a cable across a tank filled with captive sharks and watched the animals. Nothing happened until they switched on the current that normally powered repeaters, which prompted some sharks to investigate the cable. The electric field from the current attracted the predators, who apparently use it in hunting. Coaxial cable also carried a current, but its outer metal wrap blocked the electric field from reaching into the water. Nothing blocked the field from reaching the water around the fiber cable, where the sharks could sense it. Other factors also may have played a role. The Canaries cable hung above the surface in some areas, and also vibrated, which may have attracted sharks or simply gotten in their way as they swam.

Engineers took no chances when they returned the cable to the ocean floor. To deflect sharks' teeth and block at least part of the electric field, they wrapped the cable with a strong steel tape. They also extended the armor normally used to protect the cable in shallow water to depths well below where the sharks had bitten the cable. As they laid the cable, they made sure it rested on the bottom rather than hung in the water where it might tempt hungry sharks. "We haven't had a shark bite since," Runge told me with a grin.[42] The cable remains in regular service.

While shark bites grabbed the headlines, AT&T worried about semiconductor lasers. Bell Labs had problems making long-wavelength lasers; they had missed delivery schedules for the Bermuda trials,[43] and reliability was a continuing concern. Dreading the cost of hauling a repeater from the ocean floor to replace a dead laser, engineers designed repeaters with three spare lasers to back up each laser transmitting signals.[44] However, long-wavelength lasers improved as time passed, and by the time AT&T started building the final TAT-8 repeaters, they included just one backup laser per fiber.[45] The transmitters, cable, and receivers also worked better than had been expected in 1980, allowing AT&T to double repeater spacing from the original plan of 30 kilometers to beyond 65 kilometers. STL and Submarcom were more cautious, spacing repeaters about 40 kilometers in their parts of TAT-8.[46]

Turning the Tables on Satellites

In the mid-1970s, the communications satellite industry thought it might vanquish costly cables from the transatlantic market.[47] A decade later, fiber optics had turned the tables.[48] As TAT-8 moved toward completion, international consortia planned more submarine cables. In May 1986, AT&T, British Telecom, Teleglobe Canada, and the French and Spanish telephone authorities agreed to plan and build TAT-9, with service to start in 1991.

The new cable was designed around a fourth generation of fiber technology, operating at 1.55 micrometers, where fiber loss was lowest. Both Corning and British Telecom were pushing the advantages of fibers they had developed with minimum pulse dispersion shifted to 1.55 micrometers.[49] However, AT&T chose instead to use lasers that emitted an extremely narrow range of wavelengths, so pulses should suffer little dispersion at 1.55 micrometers. Their design let them double transmission speed to 560 million bits per second and stretch repeater spacing to a hundred kilometers in TAT-9. The same technology later reached 140 kilometers.[50]

The satellite industry fought back. The International Telecommunications Satellite Organization (Intelsat) promised to cut its rates and claimed that new echo-cancellation circuits made complaints about satellite delays "spurious."[51] As a frequent international caller, I knew better; satellites still had delays and bad circuits. Each side claimed its service was cheaper, but satellites were in retreat. As plans for new cables proliferated, the satellite industry quietly turned to routes not well served by fiber cables and services other than telephony, including video and data communications.

TAT-8 planners came remarkably close to their original plans of starting operation in July 1988. Ironically, the delays were due to relatively mundane electronic components, not fibers, cable, or lasers. AT&T had problems with one component; Submarcom had problems with another.[52] In August, AT&T's cable ship *Long Lines* laid its portion of the cable, stopping at the branch point off the French coast; French and British cable ships met it and laid the rest of the cable from the branch point to land.[53] After a few months of testing, the cable went on line in late December,[54] beating Alec Reeves's prediction of two decades earlier. By then, the *Long Lines* had almost finished laying the first transpacific fiber-optic cable.

The job was not finished. Strong currents off the French coast exposed the buried cable, and trawlers snagged it in February and March, damaging the wires carrying repeater power. In March, an electrical fault knocked out the British leg of the cable; calls had to be rerouted via satellite until repairs were finished and the cable buried deeper in early April.[55] Meanwhile, the transpacific cable began service to Japan.

The new cables did not revolutionize international communications overnight, but I could see the improvement. My calls to London were more likely to go through at the peak hours for transatlantic business. I got fewer bad circuits with maddening echoes or dead air from England. More and more calls were as clear as any long-distance call made over fiber-optic lines in America. The most reliable sign of a transatlantic call became an English accent on the other end.

Better Technology

Better technology brought higher cable capacity at lower cost per channel. The demand for international telecommunications exploded, filling TAT-8 to capacity soon after it began service. TAT-10 and-11 followed with the same 1.55-micrometer technology as TAT-9.

Meanwhile, a new technology arrived that dramatically improved submarine cables—optical amplifiers that increase the strength of a light signal without first converting it to some other form. Alec Reeves had seen the advantage of that approach back in the late 1960s, but he had thought semiconductor devices would do the job. The new technology, fiber amplifiers, grew from ideas Eli Snitzer developed in the early 1960s as he moved from fiber optics to lasers at American Optical.

Lasers are light amplifiers. Shine the right wavelength into a laser, and the light stimulates excited atoms to emit more light at that wavelength—precisely in step with the input light. Radio astronomers use masers, the microwave counterpart of lasers, to amplify weak microwave signals. Snitzer doped fiber cores with the rare earth neodymium and showed they could amplify light.[56] However, in the early days of lasers everybody wanted to generate their own beams, so they put mirrors on both ends to make oscillators.

By the late 1970s, Charlie Sandbank[57] and a few others were thinking about optical amplifiers but didn't know how to make them. Will Hicks came up with some novel schemes at 1984 Inc.,[58] but the winner proved to be a variant on Snitzer's original idea: a fiber with a small amount of a rare earth element in its core. The key was finding the right material to add to the fiber core so it would amplify light at one of the fiber windows, 1.3 or 1.55 micrometers. In the mid-1980s, Dave Payne at the University of Southampton came upon erbium, which amplified 1.55 micrometers nicely when excited by infrared light from a semiconductor laser[59].

The erbium-doped fiber is an almost ideal optical amplifier. It works over a range of wavelengths wide enough to amplify signals at dozens of different wavelengths at once. It responds quickly to pulses, so pulse spreading is not a problem. It has low noise. And fiber amplifiers fit easily into the torpedo-shaped cases designed to house undersea repeaters.

The latest generation of submarine cables use fiber amplifiers instead of repeaters. The Simplex Wire and Cable plant in Newington, New Hampshire, was humming with business when I visited in the fall of 1994. Cable ran endlessly through the plant. I saw fibers wrapped around the central steel king wire, encased in plastic, armored lightly, covered with copper conductor, then again encased in plastic. Heavy, sturdy, well-oiled machines wound the armor; machines that had spun at their tasks for years seemed ready to roll on forever.

The cable rolled through an overhead passageway onto a giant cable ship docked in back of the factory, where workers looped it into giant "tanks." They hung the amplifier casings above the tanks, solid metal cylinders, lumps in the thin coiled snake of white plastic cable. Today, that cable lies at the bottom of the Pacific, part of TPC-5, a 24,000-kilometer loop circling the ocean, linking Oregon, California, Hawaii, Japan, and Guam.[60] When I spoke to Charles Kao in Hong Kong, the pulses of light that carried our words may have been among the five billion bits per second that passed through each of the two fiber pairs in that cable.

No technology is perfect, and submarine cables are no exception. Cable breaks continue to be the main problem, despite air and sea patrols of cable landing points, but modern cables reroute signals almost instantly around a break, so callers hear only a small click as they talk. A few small cable imperfections have let salt water leak through, shorting out the electricity that powers repeaters or optical amplifiers. The optics are doing fine. Peter Runge proudly tells me: "We have seen no device failures," even among the lasers, which were their biggest worries.[61]

Thanks to submarine cables, calling overseas means just pushing a few more buttons on the dial. Words come clearly through the glass from places as exotic as Egypt and Japan; the problems are time zones and languages.

Runge didn't take transatlantic calls to his sister in their native Germany lightly when he came to America in the 1960s. Now, he says, "We don't think twice any more to pick up the phone even for little things, just to say hello." His sister talks fast; spoiled by the quick response of fibers; she hangs up when she gets a satellite circuit, knowing that dialing again will likely bring words as sharp and clear as glass. The impact has gone beyond their personal lives. "I don't know whether TAT-8 helped bring down the Berlin Wall or not," mused Runge as we sat in his Bell Labs office, "but communication in general has had a major impact on politics of the world."[62]

16

The Last Mile

An Elusive Vision

> These days when I watch television here at home, I have my choice of four channels that I can get with reasonable clearness and audibility. Even with only four channels at their disposal, however, the television moguls can supply me with a tremendous quantity of rubbish.
>
> Imagine what the keen minds of our entertainment industry could do if they realized they had a hundred million channels [the number of standard television channels that he calculated could fit in the visible spectrum] into which they could funnel new and undreamed-of varieties of trash.
>
> Maybe we should stop right now!"
>
> —Isaac Asimov, 1962[1]

The last mile has been the hardest for fiber optics. Optical fibers carry both telephone and cable television signals to your town or neighborhood, but except for a few experimental systems, the threads of glass do not reach homes. Somewhere within few miles of your home the fibers end in a box, which converts their optical signals into electrical form and sends them to copper wires. Telephone calls go over simple pairs of wires; cable television uses coaxial cable, a thin central wire surrounded by an insulator, which in turn is sheathed within a metal tube, then covered by protective plastic. Those wires connect to your phone and television.

Those wires also are bottlenecks. Telephone wires can connect you with any other phone on the planet but only at limited capacity. Cable systems have much higher capacity, but they connect you only to the cable company.

Why not use fibers? Visionaries have asked that question for a quarter of a century. The information capacity might not match the hundred million channels estimated by Isaac Asimov, but it would more than suffice for modern needs. Business managers counter that installing fibers would cost too much, and so far they have carried the day.

Wired Cities

Television pioneers hoped that their new medium would enlighten the public. C. Francis Jenkins and John Logie Baird did not live to see modern television, but it disappointed Vladimir Zworykin, the Russian-born engineer who led RCA's successful development team. "I hate what they've done to my child," he complained. "I would never let my own children watch it."[2]

Asimov was not the only one complaining about televised trash in the early 1960s. Newton Minow, freshly appointed to the Federal Communications Commission, earned sympathetic headlines by calling commercial television a "vast wasteland." Many critics sought to make television more interactive, so viewers could do something besides passively staring at the screen. The technological optimism of the era stimulated the dreamers who always lurk in the communications world. They outlined futuristic visions of "wired cities" where advanced telecommunications services would link homes, businesses, factories, and schools.[3]

They offered few specifics. "Advanced" video services often meant adding more channels in the naive hope they would attract more intelligent programs. "Two-way" television rarely went beyond viewer participation in game shows and televised debates on public policy. AT&T offered Picturephone. Utilities proposed automatic meter reading over phone lines; cities suggested automatic fire and burglar alarms. Others envisioned remote controls to adjust home thermostats or turn on ovens from work. A few wild-eyed visionaries proposed equipping home televisions with special controls and keyboards to access graphic computer-based information services over phone lines—a service called "videotex" that hinted at future development of the World Wide Web and the Internet.

Early proposals to link homes with high-capacity coaxial cables soon ran into the ugly reality of network design. Coaxial cables used in cable television systems were good pipes for delivering the same signals to everybody, but they worked poorly with switches that routed different signals to specific destinations. In fact, standard cable television systems were almost pure pipe, with virtually no switching capacity. The telephone network excelled at switching, but its aging wires were tiny pipes unable to carry video signals. Either the phone or the cable television network would require expensive rebuilding to provide two-way, high-capacity service with big pipes and good switching.

That wasn't an immediately fatal flaw in an era of urban planning and Lyndon Johnson's Great Society program. "New towns" projects, some government funded, sought to build new communities in virgin territory.[4] Visionaries hoped to make the new towns into "wired cities," promising a more equable, diverse, and democratic society. It was an enticing vision for a generation of idealists disturbed by pollution, the Vietnam War, and the nuclear arms race.

The Federal Communications Commission liked the idea of interactive television and decreed in 1972 that future cable systems must provide two-way service. No such systems existed at the time, but the National Science Foundation agreed to sponsor three experiments starting in early 1974. The technology worked, but high costs and unclear benefits later led the FCC to drop its two-way requirements.[5]

The Fibered City

In the heyday of "new town" planning, GTE Laboratories designed a network to carry the usual array of futuristic services to a couple thousand homes. The plan called for a separate coaxial cable to run from each home to a central facility. The logistics of handling all those cables left John Fulenwider aghast in 1972. Each cable was large, and a couple thousand would require serious, bulky plumbing to connect to the equivalent of a telephone switching office. A soft-spoken research engineer in his early forties with a passion for jogging years before it was fashionable, Fulenwider had an eye for new technology and had already spotted fiber optics. He knew fibers were thin and flexible—and that a lot would fit into one 3/4-inch (1.9-centimeter) cable.[6]

He sketched out a design for a fiber-optic network for the new town. Good semiconductor lasers were not available, so he proposed using LEDs to send signals to homes and bulkier solid-state lasers for the longer distances between regional distribution centers. Four multimode fibers would run to each home. Three would carry signals to the home, including color video-telephone, two color television channels, FM radio, digital data, control signals, and two ordinary voice phone lines. The fourth would carry signals in the other direction. Like the telephone network, the fiber system would be switched, sending users only signals they requested instead of all signals as in cable television.[7]

The fundamental problem Fulenwider faced was network topology. Early fiber systems ran between two points, but routing signals to many points was much more difficult—even worse than for coaxial cables. It required many transmitters, many receivers, many fiber connections, and ways to split signals between fibers that did not yet exist. Fulenwider did his best, but costs were discouraging. A GTE economist calculated his design would take seven years to earn back the investment; the developer wanted a two-year payback.

The fibers of the time were reasonable pipes, but they didn't fit very well with the switches. Although nobody was going to build the system, Fulenwider talked about it at the International Wire and Cable Symposium, December 5–7, 1972, in Atlantic City.[8] The audience was intrigued and asked probing questions about device lifetimes and fiber loss, concerns everywhere fiber was mentioned.

Back at the labs, Fulenwider persuaded GTE to buy a length of cable Corning had developed for the Navy. A pair of stiff Kevlar strength members kept the cable from bending in one direction, but he strung it between two conference rooms at opposite ends of the C-shaped GTE Labs complex. He hooked up a prototype GTE videoconferencing system and "all sorts of nifty things." Engineers used the cable as a testbed for 15 years, until it was removed during laboratory renovations.[9]

Meanwhile, wired-city development sputtered to a halt in America. Changes in the political climate ended government funding of new town projects. The government bolstered the regulatory wall separating the telephone and cable television industries, effectively barring integration of phone and cable service on a single network (except in sparsely populated rural areas). A handful of industry experiments remained, most notably Warner Communications' QUBE two-way cable trials from 1977 to 1984 in Columbus, Ohio. Built entirely with coaxial cables, it tested pay-per-view programming and MTV, helping shape modern cable television. However, the modest two-way services attracted little interest and were shut down when the parent Warner Company ran into financial trouble.[10]

Japan's Hi-OVIS Program

Japan's interest in wired cities grew as America's waned. In the early 1970s, the powerful Ministry for International Trade and Industry began planning a test of two-way video communications in a "new town" being built for Japan's growing professional class. Initial plans called for coaxial cable, but MITI shifted to optical fibers before announcing the final plan for Hi-OVIS, the Highly Interactive Optical Visual Information System, in March 1976.[11]

Hi-OVIS provided two-way service to 158 homes and 10 public institutions in the new town of Higashi-Ikoma. That choice mirrored a common vision of providing new services for the professional elite living in a new town. Yet Hi-OVIS also was a bold experiment for a traditionally cautious people, because it committed to running fibers to homes while AT&T was warily testing fibers in ducts under its Atlanta parking lot.

The technological differences between Hi-OVIS and Atlanta reflected their different missions. Bell designed the Atlanta system to carry many calls simultaneously between two points; the transmitters were expensive, but their cost was shared among many customers. Hi-OVIS needed separate links to each home, so they had to be inexpensive. AT&T could afford a few laser transmitters; Hi-OVIS needed many cheap LEDs. AT&T wanted maximum transmission distance, so it paid the premium for graded-index fibers. Hi-OVIS wanted to minimize connection costs, and they could get away with large-core multimode fibers, despite their large pulse spreading, because the home links were short.[12]

Video services went well beyond the usual retransmission of local television broadcasts. Hi-OVIS had its own programming center; homes had their own cameras and transmitters so residents could join in. The Japanese planners' standard example was having women demonstrate how they made their favorite recipes in their own kitchens. Other services included displaying information from computer files, such as weather forecasts, news highlights, or train timetables. A video-on-demand service let viewers request programs from the Hi-OVIS central library.[13]

As Japan built, Canada, Britain, France, and Germany planned their own home fiber trials. The American telecommunications establishment dithered, blocked by government regulations, the cable industry's limited capital, and the phone industry's timidity. In July 1978, the Visual Information System Development Association turned on Hi-OVIS. I watched from afar, fascinated by the experiment and worried America was falling behind.

Program director Masahiro Kawahata made Hi-OVIS a showcase, hosting a series of overseas visitors. Early results seemed good. The hardware worked, although it was primitive by modern standards. People used the system, although most were shy about getting in front of the camera, particularly the women. Children learned the system quickly and requested some programs over 500 times from the central library, literally wearing out the tapes.[14] Hi-OVIS officials apparently did not consider the children's choices wholesome enough, so they did not replace the worn-out tapes.

Carefully reading an English-language preliminary report on Hi-OVIS, I thought I saw the future of fiber to the home. The children of Higashi-Ikoma wanted to pick their own programs, and it was only logical to think adults would do the same. I envisioned a future where switched fiber networks would retrieve programs from on-line video libraries. I decided video-on-demand service could be the "killer application" that would pay for the new technology. It was clear video telephones wouldn't do the trick; Picturephone had crashed and burned. Videotex and similar home information services were struggling. I tried to stake my claim as a minor visionary with a talk at a fiber-marketing conference and an editorial in *Laser Focus*.[15]

In one sense I was right. The videotape market proves that people want to pick their own programs—but I expected them to select from on-line video libraries, not by renting tapes. Yet my forecast missed the mark because I did not realize the discouragingly high costs of both on-line video libraries and home fiber-optic networks. Hi-OVIS proved extremely expensive. The Japanese didn't talk much about their expenses at the time, but afterward Kawahata said they totaled about 10 billion yen, roughly $80 million, through the end of the experiment in early 1986.[16] That comes to a staggering $500,000 per household served! The fiber-optic hardware was expensive, but the steepest bills came from building and staffing the local operations center.[17]

Hi-OVIS had other weaknesses as well. While the tests showed what services people used, subscribers didn't have to pay the bills. In addition, telephone service was missing because Nippon Telegraph and Telephone would not yield its franchise. NTT and MITI were allied with rival bureaucracies in the Japanese government. Looking back, Kawahata points with pride to creating a sense of community among subscribers and fueling growth of Japan's fiber-optics industry, which has become a global powerhouse. He also claims credit for laying the groundwork that led NTT in 1990 to set stringing fiber to every Japanese home and business by 2015 as a long-term goal.[18]

However, NTT usually avoids mentioning Hi-OVIS, and has since backed away from that goal, which a top official said later was "not a business plan or an implementation schedule, just a wish list."[19] As of late 1997, NTT planned to install fiber-optic feeder cables to remote terminals *near* homes by 2010. With that network completed, the company says it "will be able to lay optical fiber to the home within a week or so."[20]

Fiber to the Farm: The Elie System

I never got to Hi-OVIS, but I did see the next trial of fiber-optic links to the home, in rural Canada. Canada wanted to enhance the quality of life on the big, productive farms that sprawl across the prairie. Farming is a major Canadian industry, but the best and brightest young people tend to leave rural areas. Canada's Department of Communications decided to test fibers for telephone, television, and data transmission in the tiny towns of Elie and Ste. Eustache about 50 kilometers (30 miles) west of Winnipeg.[21]

Elie was a little town, with small homes and a huge railside grain elevator that towered over the flattest landscape I have ever seen. Ste. Eustache was smaller, but had more trees and seemed less stark. They were places that desperately needed better communications. Party phones serving up to ten households were standard, so one call could tie up many other lines. Television was the four channels that tall antennas could pick up from Winnipeg. In the winter, temperatures hit 40 below, the point at which the Centigrade and Fahrenheit scales match and exposed skin freezes. The harsh weather outside would test the hardware; the people inside would test the services.

The Department of Communications and Manitoba Telephone teamed with Northern Telecom on a prototype system serving 150 homes, a third in the country and the rest in the two towns. They budgeted $7.5 million, about $50,000 per home, cheap compared to Hi-OVIS.[22] Steady technical advances gave Elie better fiber hardware than was available for Hi-OVIS. Service began in late 1981, and when I visited the following June, residents were still delighted with private phone lines. They also liked American television channels brought to their homes for the first time, somewhat to the annoyance of Canadian officials. The fibers also connected them to Canada's developmental videotex system. Think of it as a primitive, extremely limited version of the Internet accessed through an extremely slow modem and an antiquated computer that showed only boxy letters and pictures. To Elie in 1982, it was a marvel of modern technology. A man who farmed nearly three square miles used it to check grain futures and hog prices. Children mastered the new toy quickly, briefly overloading the system when they discovered video games.

The system was a humble wonder, serving everyday people. I talked with them in the local headquarters, a carpeted trailer partly filled with electronics, parked in back of the small block building that housed the local telephone office. They welcomed me, the visiting reporter from distant Boston, as part of the exotic world that had entered their ordinary lives. I relaxed, reflected on the future I thought was coming soon, and wrote an optimistic feature on the future of "Fiberopolis" for *Omni*.[23]

The Canadian communications department and Manitoba Telephone were less optimistic when they added up the numbers. Fiber-optic hardware worked, but it was expensive. Funding for a planned second stage fizzled, and they shut the system down after 18 months.[24]

A Grand Plan for Biarritz

Meanwhile, France was trying to assemble the world's biggest, boldest system, serving 1500 homes in the coastal resort of Biarritz. Politics played a big role in the decision. France was lagging badly in technology; in 1974, only 6.2 million phone lines served the country's 53 million people, and cable television was virtually nonexistent. The government pumped $30 billion into France Telecom for a badly needed modernization program, which added 10 million more phone lines by the end of 1980.[25] France became the only country to get videotex off the ground, by giving customers cheap terminals called Minitels in lieu of phone directories.

French planners also embraced fiber optics with an enthusiasm that made Hi-OVIS look almost timid. French engineers had returned to fiber optics after Charles Kao generated global interest, building up a small fiber industry.[26] Most observers thought France lagged well behind the United States, Britain and Japan. The Biarritz project, announced in September 1978, was intended to change that. Located on the Bay of Biscay near the Spanish border, Biarritz is a flashy resort with 28,000 year-round residents and an international reputation. Cliffs rise above warm ocean beaches that are crowded with sunbathers in the summer. The famous playground was a striking contrast to the upper-crust Japanese suburb and the tiny Canadian prairie towns.

The Biarritz system was designed to carry all the latest services—including videotex, picture telephone, standard telephone, and radio and television signals.[27] When initial contracts were signed at the end of 1980, the estimated bill was $100 million, more per home than the less-ambitious Elie system, but much less than Hi-OVIS.[28]

The initial goal was to connect 1500 homes in early 1983,[29] with possible later expansion to 5000 homes, but the schedule slipped because of laser problems. The first 50 homes went on line in 1984, and not until the summer of 1986 were 1500 homes connected.[30] The delays didn't discourage the French government, which in November 1982 announced plans to string fiber to 1.5 million homes in the next five years and expand service to 6 million homes (a third of the country) by 1992.[31]

Industry observers were unsure what to make of Biarritz. France couldn't match Japan, the United States, or Britain in high-speed, long-distance systems, low-loss fibers, and high-performance lasers,[32] but that wasn't the technology needed for fiber to the home. French officials made fiber optics a matter of national pride and invested massive sums in the demonstration, although today no one I could reach admits to knowing its total cost.

Once the system was installed, television service proved popular, particularly because Biarritz had poor broadcast reception. As in Elie, users had a choice of several channels, and the switching center routed the chosen channel to the home in about a second. Voice telephone and videotex signals also came through the fibers, but they also could have gone through ordinary phone wires. Video phones sent images in standard television format but were little used because service was limited to Biarritz.[33]

By the time results started coming in from Biarritz, the French government was committed to launching cable television over fiber, another global first. Video telephones were discarded; voice and videotex could go over phone lines. France Telecom kept analog LED transmitters and graded-index fibers, able to carry only a couple of user-selected television channels. They considered it a first-generation system, and if not as ambitious as Biarritz was when planned, it was bolder than the rest of the world, which had backed away from fiber to homes after Hi-OVIS and Elie. In 1988, France Telecom contracted for fiber-based cable systems in about a dozen communities, including Montpellier, Rennes, Serves, and some districts of Paris. Fibers now connect to some 200,000 homes in those areas and pass another half million who don't subscribe to the cable service.[34] You cannot find as many homes connected to optical fibers in all the rest of the world. However, the first-generation system was the last installed in France.

What is the legacy of Biarritz? "Nothing but a technical and economical disaster,"[35] e-mailed a French fiber developer. The fibers worked fine in Biarritz, but the laser transmitters suffered serious noise and reliability problems.[36] That led engineers to shift to LED transmitters for the cable system, but the LEDs left no room for expansion when customers wanted more channels and services. For its second-generation systems, France Telecom used single-mode fiber "trunks" to carry an array of many signals to neighborhood nodes but distributed signals to homes through coaxial cables. It's an approach now used in many other countries. The original fiber lines work, but they are technological orphans, incompatible with anything else anywhere else in the world. They can be upgraded only with new transmitters and special equipment.[37] The Biarritz system itself is gone; France Telecom stopped the experiment in early 1990, dismantled the network, and pulled the fiber cables from underground ducts.

Aftermaths of Early Experiments

Except for laser problems, hardware in the early trials worked reasonably well. Successful installations eased widespread worries that telephone construction crews couldn't handle fragile glass fibers. But it was expensive to adapt the pipes to the switches, driving costs up. "The issue wasn't technological, it was financial risk. Can you generate the revenues you need out of these kinds of systems?" recalls Rod Kachulak, head of network engineering at Manitoba Telephone who worked on Elie as a young engineer. "We couldn't make the economic case to get enough revenue from the subscriber base to justify the investment of millions and millions of dollars."[38]

American phone companies started trials of fiber to the home as Biarritz wound down in the late 1980s. They picked the most favorable possible sites—new subdivisions of expensive homes, where the subscribers had plenty of money and new services had to be installed anyway. Sometimes they battled with cable companies; sometimes they cooperated. But the plans were never very ambitious, and most quietly fizzled out. The ugly reality was that the market wasn't ready. Engineers were dazzled by the tremendous information capacity of optical fibers. The public was apathetic. The crystal clarity of digital fiber-optic lines was not going to persuade them to pay more for telephone service, although it could cut the phone company's costs for installing second and third lines to homes. Fiber-optic transmission likewise made little obvious difference to cable-television customers, although fibers cut costs in some parts of cable systems. The public paid no mind to videotex and automated meter reading. Fire and burglar alarms could be hooked up to regular phone lines. "Every trial was a technical success, but not a market success," says Paul Shumate, in charge of home transmission studies for Bellcore. "There was nothing subscribers were willing to pay for."[39]

In the summer of 1993, I visited a system in Cerritos, California, that GTE had heralded as "the largest and most advanced test of voice, video and data services in the U.S."[40] The phone company had teamed with the cable-television company serving the Los Angeles suburb for the first market tests of how subscribers reacted to new services. They hooked up most of the community with coaxial cables for television, and a few hundred homes with fibers for phone service. A few dozen got video through fibers.[41]

I sampled the new services in a showplace viewing room in GTE's Cerritos offices, dominated by a wall-size entertainment center. Sitting on a long, overstuffed couch I clicked through displays with a remote control. The technology had come a long way from Elie. The television was larger, with crisper pictures; the room was a long way from the inside of a trailer. GTE's Main Street service offered nicer pictures than the Canadian videotex service but didn't deliver much more information.

The man in charge of the showroom told me about video on demand, pay-per-view television, fiber-optic data transmission, and educational videoconferencing. Yet only a couple of homes were set up to receive video on demand, and their choice was among a mere ten programs. The town had 7500 cable subscribers, but only 250 paid $9.95 a month for Main Street. And GTE was keeping the details of its market trials proprietary because it was paying a lot for that data.

A bit disappointed, I cruised through Cerritos, looking for a sense of the planned community. I found private streets that greeted me as a trespasser, with high walls facing the public roads. The town had been dairy farms in the 1960s; now it seemed a fortress. I drove away to visit an old friend in another suburb where you could see yards, houses, and people from the street.

Kawahata had dreamed of building a futuristic community when he planned Hi-OVIS. The Canadians had tried to tailor their technology to the needs of a rural community in Elie. The French had wanted Biarritz to be a showplace. In Cerritos, GTE fit the mold of a subdivision, well planned and tastefully arranged, yet somehow sterile. Other American phone companies followed the Cerritos model, announcing home fiber trials in expensive new developments with great fanfare yet never reporting results. Such silence is the hallmark of failed experiments. The new technology had failed to grow because it offered nothing new to the customers.

A New Generation

Many survivors of the field trial wars have concluded that fibers are simply too expensive for home links. Bob Olshansky, now a top research manager at GTE Labs, sees the future as connecting fibers to copper telephone wires in your neighborhood. Digitize the signals, add new transmitters and receivers, clean up the wires, and phone lines called digital subscriber lines can carry millions of bits per second to the home, as long as it isn't too far from the phone company. That makes it attractive for Internet access, which he considers a more attractive market than delivering video programs.[42]

On the other side of the debate is the ever-optimistic Paul Shumate of Bellcore, who believes a new technology called Passive Optical Networks will bring costs down. Pioneered by British Telecom Labs, it saves on costly laser transmitters by splitting optical signals among 8 or 16 homes. New inexpensive lasers could meet the modest needs of home transmitters while easing the packaging requirements that account for most of the cost of modern lasers.[43] New data-compression technology could pack more digitized signals into fibers. Other savings come from careful attention to design details, like providing power for home telephones and other electronics and avoiding repairs caused by corrosion of metal cables. Put it all together, says Shumate, and fiber to the home already is cheaper than other technologies in rural areas where there are a dozen homes or fewer along each mile of roadway served. He expects the costs to keep coming down, but admits "the timetable is not real clear" for widespread fiber links.[44]

In Japan, Nippon Telegraph and Telephone plans to run fibers past every home by 2010, but that's more cautious than it sounds. So far, NTT has concentrated on the easy parts, stringing fiber to businesses in large cities.[45]

Although fibers will *pass* every home in Japan, the phone company won't promise to run them the last few hundred feet to connect homes. Instead, NTT plans to run wires from optical nodes along the street. That isn't too much different from American plans.

Will Hicks believes we have barely tapped the potential of fibers. He says that for $400 to $500 per user, he could build a fiber system to let people pick from among 5000 video channels for each of three televisions, access old television shows and a quarter of all the movies ever made, and call up a wealth of written material. "I know there aren't 5,000 TV programs. There aren't any at all worth seeing," Hicks says, echoing Isaac Asimov 30 years earlier. "But if there were, we could make them available."[46]

And there, in a sense, lies the central problem. Fiber optics may be too big a pipe; it can deliver more information than anyone knows what to do with. It can deliver an eight-lane freeway of information to our doorstep, when most of us want only a sidewalk. The Internet and the World Wide Web may change that. Try to download graphics-laden web pages and you quickly realize how slowly information trickles through the network. However, you often don't do too much better with a bigger pipe; the switches have also become bottlenecks. And despite all the wild-eyed optimists, the technology for optical switching is not yet ready.

That may change with time. The whole network is creaking with a massive and fast-growing load of traffic today. Keeping up with the growth is a major strain that has pushed a tremendous expansion in capacity of the fiber-optic backbone network. On the horizon are new generations of switches, some that route signals using optical technology instead of electronic circuits. Optical switching technology isn't easy; electrons are easier to manipulate because of their electric charges. Yet the demand for more bandwidth—for bigger switches as well as pipes—continues to escalate upward with little end in sight. Yesterday's fleet new modems are today's plodding dinosaurs. Engineers are working fast and furious to consign today's tired old telephone and cable systems to the same fate. Stay tuned for more thrilling episodes.

17

Reflections on the City of Light

> It is my firm belief that an almost ideal telecommunication network . . . going a long way toward eliminating the effects of mere distance altogether, can be made—but only, at least as now foreseeable, by optical means. I believe too that within and between the highly developed regions of the world such a technological revolution . . . will be demanded by the public by the start of the next century. It will be a real adventure, and lead too, to truly stimulating new challenges in human behavior—in all of which John Logie Baird, the pioneer in this venture whom we are commemorating tonight, would have been only too pleased, through his own eyes and hands, to take part.
>
> —Alec Reeves, John Logie Baird memorial lecture, May 1969[1]

Alec Reeves saw only the first steps toward fiber-optic communications, and probably had no idea that John Logie Baird himself had played a bit part in fiber-optic imaging. But he was prophetic in seeing the development of fiber communications as an adventure.

It was a marvelous adventure for John Midwinter. "Everything we touched turned to gold," he recalls. "It was unbelievable. From 1974 onwards, it was just one success after another. We had this wonderful group, with very strong backing." By 1984, British Telecom had effectively launched fiber optics, and it was time for him to move on to a new program, but he could see nothing else as fascinating at Martlesham Heath. Instead, he accepted British Telecom's offer to establish a professorship for him at University College, London. It was a plum position, leaving Midwinter ample room to pursue his own interests.

Over the years, his interests have evolved. He studies optical switching now, a complex and elegant technology, but confesses "the thrill of publishing yet another paper wears off." His new passion is guiding government policy on engineering, where he hopes to make greater contributions than in the laboratory. He showed me a miniature wooden coffin, a going-away present from Martlesham Heath, which his former colleagues filled with "technologies killed by fiber optics," bulky metal cables and a long-vanished section of millimeter waveguide. "It was a golden period," he said nostalgically as we sat in his office, seven floors above London on a rainy December Monday.[2]

Don Keck's office at Corning's Sullivan Park Research Center is only on the second floor, but it offers a commanding view of the Chemung River valley. In nearly three decades, he's climbed one floor up the glass-walled tower and a few levels up the corporate ladder to director of optics and photonics research. He thrives in the climate, like the plant that curls more than halfway around the ceiling of his office. I asked if he was nostalgic for the days of one breakthrough after another. "It's not over, Jeff," he replied, still full of energy and enthusiasm in his mid-fifties. I had to agree as he listed an impressive array of new optical wonders just presented at the Conference on Optical Fiber Communications in California.[3]

The Triumph of Fiber

British Telecom was the first phone company to commit to single-mode fiber, but the British Isles are tiny compared to America. MCI's decision to span the continent with single-mode fiber opened the floodgates to success. Deregulation of the long-distance telephone market created a tremendous demand for long-haul transmission that millimeter waveguides could never have met. Single-mode fibers were the right technology in the right place at the right time to become the national backbone of big digital pipelines. Sprint made fiber optics a household word by advertising that you could hear a pin drop through digital fiber lines. AT&T shifted gears and started installing single-mode fibers; smaller carriers followed.

The breakup of AT&T on January 1, 1984, shook Bell Labs far more than British Telecom's split from the Post Office affected Martlesham Heath. Stew Miller retired before the split of the labs where he had worked for 42 years.[4] One of his lieutenants, Jack Cook, started a company to make fiber-optic connectors using a design licensed from AT&T.[5] Some fiber developers went to Bellcore, the part of Bell Labs originally assigned to the seven regional operating companies, and Miller followed as a consultant. The rest remained with AT&T, most concentrating on long-distance, high-performance systems. As the atmosphere changed, some followed Miller into retirement, while others left for industry or academia.

The young fiber industry scaled up rapidly, but for a while fiber was in short supply. Even Corning hadn't been ready for the sudden flood of orders. AT&T, Northern Telecom, and Siecor cranked out cable by the mile; Lasertron and other companies sliced, diced, and packaged laser chips. New companies emerged as the market grew, while some old ones faltered.[6]

The telephone industry adapted a new way of doing business. As a monopoly carrier, AT&T tested each new technological step exhaustively, like a soldier crossing a suspected mine field. As competition grew and fiber technology matured, phone companies moved faster to deploy new hardware.

MCI planned its first systems to carry 400 million bits per second through each pair of fibers. The speed limit came from the transmitter electronics, not the capacity of the single-mode fibers. Faster transmitters came quickly, first at the European standard rate of 565 million bits per second, then at 800 million. By the time I finished the first edition of *Understanding Fiber Optics*[7] in 1986, AT&T had sent 1.7 billion bits per second through single-mode fibers installed to operate at 400 million bits per second. Nobody talked any more about the Northeast Corridor system; it was a techno-artifact as obsolete as a vacuum tube.

Single-mode fiber quickly conquered the market for cables running between switching centers as well as for long-distance systems. Single-mode offered the compelling attraction of easy upgradability even for modest needs. AT&T installed its first single-mode fibers in interoffice systems before using them for long distance.[8] Multimode survives only for short systems used within buildings, where ease of connections is critical and the bandwidth of graded-index fibers is adequate.

Fiber proved amazingly durable in the right kind of cable. With the hydrogen problem licked, glass was almost immune to environmental insults that corrode metal cables. The biggest hazard for fiber cables has been carelessness. Contractors and utilities who dig first and ask questions later have damaged more cables than all the sharks in the seven seas and all the hungry cable-gnawing gophers of the Great Plains put together.[9]

The web of high-capacity fiber pipes branched and spread across America, as similar backbone fiber webs spread in England, Japan, and western Europe. The market for fiber boomed as the network spread, then briefly slumped before sales of shorter systems pushed the volume upward again. Phone companies hauled old copper cables out of underground ducts, sold them for scrap, and put four higher-capacity fiber cables in their place. As phone companies installed new lines, they ran fiber cables from local switching centers to remote boxes, where electronics transferred the signals to old-fashioned phone wires. One pair of fibers carried 96 phone lines.

Developing countries built their own fiber networks, leapfrogging whole generations of technology, saving money and improving phone service. John Fulenwider circled the globe for the Arthur D. Little consulting firm, helping design fiber systems in China, Egypt, and Saudi Arabia.[10] By 1988, Corning was selling fiber to many developing countries, including India, Brazil, Argentina, Mexico, Thailand, and Indonesia.[11]

Even undersea cable engineers, the most cautious of the traditionally conservative telephone engineers, embraced fiber. Before TAT-8 was up and running at 1.3 micrometers, they signed contracts for more submarine cables operating at 1.55 micrometers. Satellite circuits were relegated to transmitting the growing volume of data and other signals not sensitive to delays. It seems years since I've recognized the delays of an overseas satellite circuit.

The Problem of the Future

This future has not been what the establishment expected. Thirty years ago, Picturephone and the millimeter waveguide were the future of telecommunications. The vision shifted continually. Fiber came and conquered, but the logical compromise of graded-index fibers turned out to be a dead end. Stew Miller looked back in retirement and wrote in the introduction to a massive 1988 book he edited on fiber optics, "Perhaps the least predicted trend in the last six years is the rapidity of the movement to single-mode fibers as the medium of choice."[12] Yet that shift was a compellingly logical one, too, once engineers solved the daunting but unglamorous problem of coupling light into single-mode fibers. The huge capacity of single-mode fiber pipes gave phone companies room for future upgrades without the high cost of installing new fibers. Using a single type of fiber throughout the whole network greatly simplified maintenance.

Miller was not an expressive man, but he must have been glad to see fibers succeed. Little legacy remains from other projects that occupied his 40 years at Bell Labs. Millimeter waveguides and hollow light pipes had gone down the tubes. His pet brainchild, integrated optics, has remained in the laboratory for decades. But fibers were a brilliant success. In February 1989, the Optical Society of America gave him the John Tyndall award for his contributions to fiber optics. He died a year later in New Jersey at 71.

As Miller's career shows, predicting the course of technology is no easy task. Fatal flaws can lurk deep inside the hearts of bright ideas. New concepts can outflank logical extrapolations of established knowledge, as the optical fiber beat out the millimeter waveguide. A pure and elegant idea can trump a technological compromise, as single-mode fiber proved far superior to graded-index fiber. Experts enamored of their own approaches can completely miss other possibilities, as Bell did when it ignored Charles Kao to concentrate on hollow optical waveguides. And as Kao showed when he asked about glass transmission, fundamental limits can be far more important than the state of the art.

Yet you can twist those conclusions the other way by citing other examples. Many fundamental limits are out of reach; many technological compromises yield workable and economic products. Most mistaken forecasts of trends in fiber technology made eminent sense at the time. Single-mode fiber offered no compelling attraction until the 1.3-micrometer window opened. Long-wavelength lasers *looked* like they should be more difficult to make than gallium arsenide lasers. Even the blind insistence that air *had* to be a better transmission medium than glass was logical given the state of knowledge in 1966. Most errors stemmed more from ignorance than from stupidity.

The understanding of glass and optics circa 1950 was sufficient to allow the development of fiber bundles for imaging. Communication fiber optics required pushing beyond that knowledge frontier into the unexplored territory of ultratransparent materials. The breakthroughs came when people like Kao asked the right questions and teams like Bob Maurer, Don Keck, and Peter Schultz used their collective expertise to steer the right course across new terrain. They took risks, evaluated their course, then corrected their direction to aim at the most promising areas. Corning's first low-loss fiber would have been merely an interesting data point if taken by itself. Corning's true success came from expanding on that work, replacing the troublesome titanium dopant with germanium to make a lower-loss, more durable fiber. They might never have reached that end point without seeing promising results in their earlier experiments with titanium.

Research is full of false starts and experiments that don't work. It wouldn't be research if everything worked; you hope to learn from experiments. It's nice if the experiments confirm your ideas, but if not, you try to learn from your mistakes. Even when Maurer thought the odds were against success, he hoped to learn something useful.

In many ways, Bell Labs was a surprising also-ran, a world-class laboratory that scored few conceptual breakthroughs and spent untold millions chasing dead ends. Bell suffered from being fat and happy and too quick to reject outside ideas in favor of its own. The labs were slow to recognize problems in its favorite ideas, notably the gas-lens optical waveguide and graded-index fibers, but that is not surprising. One of the hardest jobs for research managers is to kill their own faltering projects before they grow into money pits consuming departmental budgets. It is harder to explain why the Bell environment only rarely sparked the creativity that ignited new ideas at Standard Telecommunication Labs, British Post Office/British Telecom, Corning, Lincoln Lab, and Nippon Telegraph and Telephone. Had the prestigious lab become stuck in the mud?

Nonetheless, Bell doggedly stayed in the fiber-optic race, often finishing a respectable second and sometimes winning. Large and talented teams relentlessly beat problems to death. Zhores Alferov's Russian group was only weeks ahead of Mort Panish and Izuo Hayashi at Bell in making the first room-temperature semiconductor laser, and the Russians lacked the resources to overcome the tough problem of laser reliability. It was the Bell Labs team led by Barney DeLoach that had the skills and resources to make million-hour lasers. Bell engineers did a superb job in designing and building the crucial Atlanta and Chicago systems, which decisively made the case for fiber optics. Peter Runge's group succeeded in the singular challenge of adapting fiber technology to the difficult world of submarine cables. Today the labs, now part of Lucent Technologies after the second breakup of AT&T, are top competitors at the forefront of fiber technology.

The Fiber-Optic Performance Olympics

Bell is usually in the thick of the "hero experiments" that seek to demonstrate ever more spectacular feats of sending more information over greater distances. Teams of specialists boast of their achievements at the two big annual meetings, the winter Conference on Optical Fiber Communications in America and the fall European Conference on Optical Communications. They work long hours fine-tuning their experiments before dashing off to catch their flights, then report their latest and greatest results at special sessions of papers that came in after the normal conference deadline.

There are no formal rules and no referees to declare the winners, but the fiber-optic performance Olympics test the cutting edge of the technology. The usual entrants come from the major industrial labs, including Bell Labs and its spin-offs in America, NTT and some of the big electronics companies in Japan, and British Telecom Labs in England. In the 1980s, hero experiments laid the groundwork for long-distance systems on land and under the sea, which now carry billions of bits per second through each fiber. They tested ways of reducing the pulse dispersion that limited transmission speed at 1.55 micrometers. They stretched the spacing between repeaters and eventually showed that optical amplifiers could replace repeaters.

They pushed the speed and distance frontiers further in the early 1990s. For Bell Labs, a major goal was to develop better technology for a new generation of submarine cables. Neal Bergano at Crawford Hill and Linn Mollenauer at Holmdel took different approaches, using different types of light pulses. Bergano used the standard pulses that spread gradually with distance, at the mercy of the pulse dispersion inherent in fibers. Mollenauer took more exotic pulses, called solitons or solitary waves, which do not spread in the same way.

Soliton transmission is among the boldest ideas to emerge from Bell Labs. The allure of the soliton is its peculiar stability, first seen in water waves that retained their shapes for long distances along nineteenth-century canals. Theoretical physicists explained the anomaly as a kind of balancing act, possible only in certain materials when certain conditions were met. In 1973, Akira Hasegawa, a Japanese theorist working at Bell Labs, found that optical fibers could carry soliton light pulses at wavelengths longer than 1.2 micrometers. However, high fiber attenuation quickly dimmed them, making the idea seem impractical.

That changed when fiber developers opened the long-wavelength window, and Mollenauer soon produced solitons in fibers. That renewed Hasegawa's interest, and he calculated that solitons could travel thousands of miles if the pulses were amplified. Mollenauer started a new round of experiments but bogged down in developing special equipment for the tests. Although he preached the gospel of solitons with missionary zeal, a couple of years passed with little progress. Bell Labs managers gave Mollenauer a deadline to stop playing and get back to work. Fortunately, he was already getting close and soon sent solitons around and around a loop of fiber until they traveled more than 4000 kilometers.[13]

Those experiments used an exotic process called Raman amplification[14] that Will Hicks had tested at 1984 Inc. It wasn't very efficient, but it was spread along the length of the fiber, and it worked.[15] Then erbium-doped fiber amplifiers arrived, and they worked even better. Mollenauer kept refining the equipment, shifting to more practical lasers and to fibers with near-zero dispersion at 1.55 micrometers. He added more refinements to eliminate tiny residual effects that built up over thousands of kilometers.

I was among the audience of 250 people who watched Mollenauer at the 1993 fiber-optic performance Olympics. It was a Thursday night at the San Jose Convention Center, the traditional night for late papers at the Conference on Optical Fiber Communications. Mollenauer had competition in the soliton race, Masataka Nakazawa, a bespectacled clean-cut scientist from NTT Laboratories. Nakazawa spoke with a thick accent that I had to strain to understand, but his title made a bold claim: "soliton transmission over unlimited distances."

He offered no optical perpetual motion machine. Two years earlier, he had sent solitons through a million kilometers of fiber—more than enough to reach the moon and return—but he could not show that the signals were free of noise. Now he could. He sent 10 billion bits per second through a series of ten separate 50-kilometer loops of fiber, with fiber amplifiers between each pair of loops. Then he passed the output through a modulator to regenerate the signal before going back through the same loops. This left the optical signal clear and clean after 180 million kilometers, far enough to reach the sun. It could go on indefinitely, he claimed.[16]

Mollenauer, sitting in the front row, fired back a question. Then it was his turn. A tall man of strong voice and opinion, he could not match 180 million kilometers. Instead, he offered a simpler approach without a complex signal regenerator or other components that required active controls. His method sufficed to send 10 billion bits per second through 20,000 kilometers of fiber—the equivalent of halfway around the world. And when he added a second series of pulses, at a slightly different wavelength, the combined 20 billion bits per second could travel through 13,000 kilometers of fiber.[17]

The numbers told one story: Nakazawa had won the distance race. Yet no one needs a fiber-optic cable to the sun, and to reach that incredible distance the NTT team needed complex optical regenerators to clean up the signal every 500 kilometers. Mollenauer's approach was simpler and could reach halfway around the globe, adequate for all earthly communication systems.

Later, when I visited Mollenauer at Holmdel, he told me he had stretched 10 billion bit transmission to circle the earth, 40,000 kilometers or 25,000 miles. Racks of equipment stood around the walls of his 20 × 30 foot laboratory; they included eight or ten soliton transmitters operating at different wavelengths, for experiments through a single fiber. In the middle were a pair of massive black optical tables—built like pool tables covered with flat aluminum plates painted a dull, reflectionless black. Fiber-optic cables dangled from overhead runways, linking precision optics on the tables and equipment racks to the basement, where Bell stored reels and reels and reels of fiber forming a loop 2400 kilometers (1500 miles) long.[18]

A Trillion Bits a Second

At Crawford Hill, I visited a lab where Andy Chraplyvy and Bob Tkach set a different target, packing a trillion bits per second through a single fiber. They don't use solitons and they don't seek to span the globe. Several hundred kilometers is adequate for their mission, new technology for long-distance service on land.

A trillion bits per second is a tremendous pipeline. Tanned and enthusiastic, Chraplyvy told me it's more than all the long-distance traffic that usually flows in the whole American telephone network.[19] It's the equivalent of ten million voice telephone circuits, or eight percent of all Americans talking to each other at the same instant. Depending on how efficiently digital video signals can be compressed, a trillion-bit fiber could carry a mind-numbing hundreds of thousands of standard television channels, or perhaps tens of thousands in high-definition format. You might wonder why anyone would need that gigantic an information pipeline, but that was what people said about fibers carrying a billion bits per second a dozen years earlier.

You can't turn a laser on and off a trillion times a second. The record is around a hundred billion pulses a second, but that's tough to achieve, even at Bell Labs. What you want are many lasers at separate wavelengths, each modulated very fast, to send signals through the same fiber. It's the same concept of wavelength-division multiplexing that Bell tried nearly two decades ago in the Northeast Corridor system, but now it's vastly more sophisticated and easier. Erbium-doped fiber amplifiers can amplify all the wavelengths between about 1.52 and 1.58 micrometers. The problem is how close can you pack the wavelengths without the signals interfering with each other.

When I visited in July 1995, Chraplyvy and Tkach had 17 laser transmitters, each running at 20 billion bits per second. That comes to 340 billion bits per second,[20] more than all the long-distance traffic in North America, going through the 9-micrometer (0.009 millimeter) core of a single fiber.

It wasn't an easy task. "The cost of these experiments is outrageous," said Chraplyvy. He showed me a metal laboratory rack about the size of a refrigerator holding eight laser transmitters. Their experiments demand lasers with wavelengths that can be adjusted very, very precisely, to better than one part in 100,000. Each precision laser costs $60,000. To measure the power at each wavelength used in the experiments you need several optical spectrum analyzers, boxes that run about $50,000 each.

Generating 20 billion pulses a second is a stretch. The fastest test sets produce only 10 billion per second and cost $400,000. It takes a separate $100,000 box to double the speed to 20 billion bits. That's half a million dollars, so separate instruments for each channel would blow even Bell Labs' budget. The Crawford Hill team instead runs all the channels through the same test set, then splits them apart and adjusts them to make each signal seem independent.

More costly and delicate instruments are scattered about the lab. A high-speed oscilloscope to examine the shapes of the pulses runs $50,000. Optical amplifiers are $20,000. Another rack has six shelves, each holding reels with 120 kilometers of fiber; three more racks bring the total to 3000 kilometers in the lab. My eyes scanned the boxes and my mind ran a quick total: $2 million worth of instruments in a room about 20 feet (6 meters) square. The staggering bill means that fewer and fewer groups can run system experiments. It may be good for Bell Labs, said Chraplyvy, "but it's bad for scientists, because progress comes from a lot of people racing each other."[21]

Seven months later, Bell Labs reached the trillion-bit milestone. Chraplyvy's group scraped together eight more lasers to make a total of 25. They split each laser output into beams with two different polarizations, and modulated each of the 50 signals at 20 billion bits per second. Then they sent it through 55 kilometers of fiber and reported the results at the annual hero experiments session at the Optical Fiber Communications conference.[22] By then, the split of Lucent Technologies from AT&T had divided the group between AT&T Research and Bell Labs, although all still worked at the Crawford Hill building.

However, the high cost of the experiments had not kept the Japanese out of the race. In fact, Fujitsu Laboratories won it, by combining the signals from 55 lasers modulated at 20 billion bits per second. That yielded 1.1 trillion bits per second, and Fujitsu sent the signals through 150 kilometers of fiber—nearly three times as much as Bell Labs.[23] It was an impressive, headline-setting demonstration.

Nippon Telegraph and Telephone Laboratories was not to be left out. At the same conference, they reported sending a trillion bits per second through 40 kilometers of fiber using a different technique. They generated ten wavelengths from a single light source, and modulated each one at 100 billion bits per second.[24]

To fans of the fiber-optic performance Olympics, those are elegant and awe-inspiring demonstrations. A dozen years earlier, people dismissed Will Hicks as a wild-eyed dreamer for suggesting such speeds might be possible. Today they require extraordinary effort, and some techniques used to set records may never prove practical. Yet others point the way to future developments, and commerical versions of the equipment may be only a few years away.

Today's Technology

The latest generation of submarine cables operating as I write carry ten billion bits per second through each fiber—2.5 billion bits at each of four wavelengths. Optical amplifiers simultaneously boost the strengths of all four wavelength signals in each fiber. Because they use optical amplifiers, that transmission rate can be multiplied by adding more laser transmitters at additional wavelengths on each end. The most ambitious system in the works is called Project Oxygen. It's a $10 billion global network that is supposed to carry 160 billion bits per second on each fiber—10 billion bits per second at each of 16 wavelengths. With four fiber pairs, each length of submarine cable will carry up to 640 billion bits per second. That's over a thousand times the capacity TAT-8 offered a decade ago. When it's up and running in 2003, Project Oxygen's 168,000 kilometers (100,000 miles) of cable will connect 99 points in 78 countries.[25]

On land. telecommunications companies are multiplying the capacity of their long-distance systems by adding extra wavelengths. The first systems used four wavelengths, each operating at 2.5 billion bits per second, to sencd a total of 10 billion bits per second. [26] In 1997, MCI started field tests using four wavelengths each carrying 10 billion bits per second, a total capacity of 40 billion bits per second. [27] By the end of 1998, Lucent Technologies promises to deliver a system that can pack a staggering 40 billion bits per second—10 billion bits at each of 40 wavelengths—through a single fiber.[28] Other companies are not far behind.

In early 1997, I visited FiberFest, an exhibition run by the New England Fiberoptics Council at a suburban Boston hotel, to keep up with the industry and chat with old friends. It was just a little local show, but I counted 92 exhibitors—triple the size of the first fiber industry shows I attended back in 1978. The technology ranged from the exotic to the mundane. The Optical Corporation of America showed its latest devices for wavelength-division multiplexing, using ideas Will Hicks patented in 1987.[29] Nearby, other tables showed plastic trays for routing and organizing fiber-optic cables, products that are useful, but hardly high-tech.

Bundles of imaging fibers are still around. I stopped to look at one on a table labeled Schott-CML Fiberoptics Inc. It's a joint venture of the corporate descendants of American Optical and Jim Godbey's original fiber-bundle venture, Electro-Fiberoptics. The bundle is a prototype of a fiberscope for the do-it-yourself market. You can use it to see what went down the drain or to guide wires through holes in the wall. It's just what Clarence Hansell was looking for 70 years ago when he started writing the notes that became his patent application. If all goes well, you should be able to buy one for under a hundred dollars soon after this book is in print.

The City of Light has sprawled across the world, vast and complex. I'm surprised how many people I encounter who know about fiber optics, or even work with them. By one of the odd coincidences of life, one of C. W. Hansell's granddaughters is a fiber-optic engineer who once worked for Charles Kao; she had no idea of her grandfather's contribution to her field.

The Global Fiber Network

I have spoken to people on six continents through glass; my faxes and e-mail follow the same routes around the globe, but sometimes via satellite links. Only when I spoke to a scientist working in Antarctica was I sure the words made part of their trip via satellite. The City of Light has become global.

Fiber-optic communications has spread beyond the fixed telecommunications network. Compact, lightweight, durable, and high in capacity, optical fibers are standard equipment for on-the-spot television reporting of sports and other events. Rich Cerny's latest company, Telecast Fiber Systems, supplies fiber systems for demanding jobs like covering the Olympics; he delights in the adventures as well as the business. Scientist and explorer Robert Ballard extols the virtues of fiber-optic cables for remote control of robotic vehicles investigating the ocean floor by scientists on surface ships.

Fibers go beyond communications to deliver laser energy inside the body for delicate orthoscopic surgery on joints. Optical fibers molded into concrete or plastic composites can monitor the strengths of structures from bridges to aircraft. Subtle changes in light waves passing through fibers looped many times around a spool can measure rotation, serving as gyroscopes with no moving parts. The wonders go on and on.

The next wave in fiber communications is likely to be networks that route signals by their wavelength. Light is moving into switches. Africa One, a submarine cable planned to circle Africa, will use that technology to drop selected signals at different points. You can think of it as violet goes to Senegal, blue to Zaire, green to Ghana, yellow to Nigeria, orange to Angola, and red to South Africa, although all the wavelengths will be near 1.55 micrometers. Peter Runge says the system will be built, but negotiations among the more than 40 parties in its international consortium have taken a long time.[30] Bob Tkach at Crawford Hill is developing similar technology for use on land, where the color routing would be to cities like Cleveland, Detroit, Pittsburgh, and Cincinnati.

In Japan, Europe, and America, developers still talk about bringing fibers all the way to homes some day. The key is reducing the costs of fiber installations to match those of copper wires. NTT planners think the key is passive optical networks, which contain few transmitters.[31] Paul Shumate at Bellcore tells me passive optical networks are the key to bringing fiber to American homes as well.[32]

It's about time. I live in an older suburb of Boston, where aging cables make telephone repair trucks a common sight. Technicians spend a lot of time down the manhole across the street and riding cherry pickers to work on the overhead cables. Each time I see them working, I ask hopefully if they're putting in fiber, but they invariably reply "no."

Erratic, intermittent noise plagues my phone line. It hit when I was talking to Jason Stark, a Bell Labs scientist who devised an ingenious way to generate 206 wavelengths from a single laser source. He started with a pulse lasting just a tenth of a trillionth of a second, which contained a wide range of wavelengths. Passed through a 20-kilometer fiber, it stretched 200,000 times, with the longer wavelengths leading the shorter ones. Then a modulator chopped the long rainbow pulse into a couple hundred short ones, each of a different color. Optics can sort the light, routing each color pulse to a different destination. Thus, one laser transmitter could deliver signals to a couple hundred homes.

The technology is not ready yet. Right now, it's a big box sitting on a bulky optical table in his lab. Stark has plans to modify it, squeezing the components down to semiconductor devices that fit on an integrated wafer. He hopes that would help solve the problem of bringing fibers to homes economically.[33]

I could use it right now. One Friday morning, something knocked out a cable that runs down my street, and two of my three phone lines died. As a journalist, I live by the phone; I paced the floor because I couldn't get urgent calls returned. After a thoroughly frustrating day, I was surprised to hear the phone ring about 5:30. Don Keck was returning a call from a few days earlier. I apologized for my phone problems and complained about the poor service. He said the phone company should have been working the light harder. "No, Don," I replied, "They're still using obsolete old electrons." We both laughed.

Appendix A

Dramatis Personae: Cast of Characters

Alferov, Zhores: Russian semiconductor laser pioneer; made first room-temperature diode lasers at Ioffe Physico-Technical Institute.

American Optical Corporation: Manufacturer of spectacles and optical instruments founded in nineteenth century that became an early developer of imaging fiber optics.

Armistead, Bill: Research director at Corning Glass Works.

AT&T: American Telephone and Telegraph Corporation, the company founded by Alexander Graham Bell that was a monopoly in most of America until it divested local telephone service in 1984. It split off its equipment business in 1996 as Lucent Technologies, leaving AT&T a long-distance phone company.

Babinet, Jacques (1794–1872): French physicist, first to report light guiding in bent glass rods in 1842.

Baird, John Logie (1888–1946): Scottish inventor of mechanical television, patented sending an image through array of glass rods in 1920s.

Barlow, Harold: Professor at University College, London; leader in developing millimeter waveguides.

Bechmann, G.: Chief water engineer for Paris, designer of illuminated fountains for 1889 Universal Exposition in Paris.

Bell, Alexander Graham (1847–1922): Inventor of the telephone in 1876, and the Photophone—which sent voices through the air optically—in 1880.

Bell Laboratories: Originally Bell Telephone Laboratories, the research arm of AT&T when it was America's telephone monopoly. In 1984, part was divested as Bellcore, at first owned by the seven regional Bell operating companies. In 1996, AT&T divested Bell Labs as part of Lucent Technologies, keeping some operations as AT&T Laboratories.

Bergano, Neal: Bell Labs developer of high-speed fiber-optic systems.

Berreman, Dwight: Bell Labs physicist who invented the gas lens.

Black, Phil: Fiber developer at Standard Telecommunication Labs.

Bolton, Sir Francis (1830–1887): English engineer, London city water inspector, and designer of illuminated fountains for 1884 London exhibition.

Boys, Charles Vernon (1855–1944): British physicist who made strong quartz fibers for mechanical measurements.

Bray, John: Research director at British Post Office Research Station from 1966 to 1975.

British Post Office Research Station: Research arm of the British Post Office, which operated the country's telephone system until British Telecom was formed around 1980. Initially at Dollis Hill in London, the research group moved to Martlesham Heath in the early 1970s and became British Telecom Research Laboratories, now British Telecom Laboratories.

British Telecom: British telephone company, originally part of the British Post Office.

Brouwer, Willem: Student of van Heel's who worked on fiber bundles and proposed image scramblers; immigrated to America in 1953 and became noted optical designer.

Cerny, Rich: Fiber-optic marketeer and entrepreneur at Valtec and other companies.

Chown, Martin: Engineer who developed early fiber systems under Charles Kao at Standard Telecommunication Labs.

Chraplyvy, Andy: Bell Labs developer of high-speed fiber-optic systems.

Chynoweth, Alan: Director of materials research at Bell Labs in Murray Hill in 1970s and 1980s.

Colladon, J. Daniel (1802–1891): Swiss physicist and engineer, demonstrated light guiding in a jet of water in 1841, a principle later used in illuminated fountains.

Cook, Jack: Fiber-optic group manager working for Stew Miller at Bell Labs; later founded Dorran Photonics.

Courtney-Pratt, Jeofry: British physicist who patented the fiber-optic faceplate; later worked at Bell Labs.

Curtiss, Lawrence: Made first practical glass-clad fibers while an undergraduate student at the University of Michigan; co-developer of first practical fiber-optic endoscope.

Daglish, Hugh: Early fiber developer at British Post Office.

DeLoach, Barney: Bell Labs semiconductor specialist; led group that made million-hour diode lasers.

de Réaumur, Réné: French industrialist and engineer who made spun glass fibers for artificial heron feathers in early 1700s.

Dobson, Paul: Engineer who started communication fiber optics program at Valtec.

Drendel, Frank: Head of Comm/Scope, later chairman of Valtec.

Duke, David: Manager of Corning's fiber business from 1976 through 1980s; retired as Corning vice chairman in 1996.

Dyment, Jack: Diode laser developer who made first narrow-stripe lasers at Bell Labs.

Dyott, Richard B.: Early fiber developer at British Post Office; left to join Andrew Corporation in America in 1975.

Eaglesfield, Charles C.: Electronic engineer at Standard Telecommunication Labs; proposed transmission through reflective pipes in 1960s.

Epworth, Richard: Engineer at Standard Telecommunication Labs; discoverer of modal noise.

French, William: Early fiber developer at Bell Labs.

Fulenwider, John: GTE Laboratories research engineer; first to suggest fiber optics to homes.

Gambling, William Alec: Professor at University of Southampton, England, and early fiber developer.

Geneen, Harold: Chairman of ITT; built company into a powerful conglomerate.

Gloge, Detlef: German-born Bell Labs communications researcher; studied confocal lens waveguides and fiber optics.

Godbey, Jim (1935–1978): Founder of Electro-Fiberoptics, and later president of Valtec who launched the company into the fiber-optic communications business.

Goubau, Georg: German-born engineer and waveguide developer at Army Electronics Command.

Gould, Gordon: Received a series of patents on lasers based on notebooks he wrote in 1957 while a graduate student at Columbia; later co-founded a small fiber-optics company, Optelecom.

Hall, Robert: Made first semiconductor laser at General Electric Research and Development Center in Schenectady, New York.

Hammesfahr, Herman: German glass blower who made glass fiber fabrics for 1892 World's Fair in Chicago.

Hansell, Clarence W. (1898–1967): American electronic engineer who received over 300 patents while working for RCA, including one in 1930 for image transmission through bundles of optical fibers.

Hartman, Robert: Bell Labs semiconductor specialist; helped develop million-hour diode lasers.

Hayashi, Izuo: Japanese physicist who co-developed first American room-temperature diode laser while working at Bell Labs.

Heckingbottom, Roger: Fiber-optic system developer at British Post Office.

Hicks, J. Wilbur "Will": Independent-minded physicist, inventor, entrepreneur, and visionary who led development of glass-clad fibers and image scram blers at American Optical, then left to found Mosaic Fabrications in 1958. Later patented designs and components for high-capacity telecommunications.

Hirschowitz, Basil (1925–): South-African born gastroenterologist; developed first practical fiber-optic endoscope at University of Michigan.

Hockham, George: Engineer at Standard Telecommunication Laboratories; co-author with Charles Kao of paper concluding fiber loss could be reduced below 20 decibels per kilometer.

Holonyak, Nick, Jr.: Pioneer in LEDs and diode lasers at General Electric and University of Illinois.

Hopkins, Harold H. (1918–1994): British physicist and optical designer who proposed and developed imaging bundles in 1950s.

Hopkins, Robert: Succeeded Brian O'Brien as head of the University of Rochester's Institute of Optics.

Horiguchi, Masahara: Nippon Telegraph and Telephone co-developer of first fibers with low loss at 1.3 and 1.55 micrometers.

Hsieh, J. Jim: Chinese-born American physicist who made first long-wavelength diode lasers at MIT Lincoln Lab, then founded Lasertron to manufacture them.

Hudson, Marshall: Fiber engineer at Corning and Valtec.

Hyde, Frank: Corning Glass Works chemist who developed flame hydrolysis process to make fused silica in 1930s.

ITT: Originally International Telephone and Telegraph Corporation, which became a conglomerate in the 1960s under Harold Geneen, but later sold its communications business.

Jacobs, Ira: Director of digital transmission at Bell Labs, Holmdel; headed development of early fiber systems.

Jaeger, Ray: Early fiber developer at Bell Labs; later founded Spectran Corporation.

Javan, Ali: Invented helium-neon laser at Bell Labs.

Jenkins, C. Francis (1867–1934): American inventor of movie projector and mechanical television; used glass rods to guide light in one patented television receiver.

Kahn, Irving (1917–1994): Cable-television promoter and fiber-optic entrepreneur.

Kaiser, Peter: Bell Labs specialist in gas lenses and fiber optics.

Kao, Charles Kuen (1933–): Engineer who proposed that fiber-optic loss could be reduced below 20 decibels per kilometer, launching modern fiber-optic communications. Led first fiber-optic communications research program at Standard Telecommunication Laboratories. Now president of Chinese University of Hong Kong.

Kapany, Narinder S. (1927–): Indian-born physicist and entrepreneur who earned first doctorate in fiber optics for research directed by Harold H. Hopkins; a prolific author who popularized fiber optics.

Kapron, Felix: Theoretical physicist who calculated and measured properties of Corning's first low-loss fiber.

Karbowiak, Antoni E.: Engineering manager at Standard Telecommunication Labs; worked on millimeter waveguides and early thin-film and fiber-optic communications.

Kawahata, Masahiro: Director of Japan's Hi-OVIS project for fiber to the home.

Kawakami, Shojiro: Fiber developer at Tohoku University; improved graded-index fibers.

Keck, Donald: Physicist on Corning team that made first low-loss fiber; now director of optics and photonics research at Corning.

Kessler, John: Wrote first article on fiber optics for *Electronics* magazine; later founded market-research firm.

Keyes, Robert: Early semiconductor laser researcher at MIT Lincoln Laboratory.

Kiyasu, Zen-ichi: Engineering professor at Tohoku University; early Japanese fiber developer.

Kompfner, Rudolf (1909–1977): Manager of transmission research and head of Bell Labs in Crawford Hill.

Kressel, Henry: Head of RCA's semiconductor laser development group.

Lamm, Heinrich (1908–1974): First person to transmit images through a bundle of optical fibers in 1930, while a medical student at University of Munich. Became a surgeon in Texas after fleeing Nazi Germany.

Lazay, Paul: Bell Labs physicist who set up first fiber measurement lab at Murray Hill; later an executive who managed Valtec group for ITT.

Lewin, Len: Division manager at Standard Telecommunication Laboratories, responsible for early optics research.

Li, Tingye: Fiber-optic group manager under Stew Miller at Bell Labs.

Lucy, Chuck: Business manager of Corning's early fiber-optics program.

MacChesney, John: Developed fiber deposition process at Bell Labs in Murray Hill.

MacNeille, Steve: High-level research manager at American Optical.

Maiman, Theodore: Built first working laser in 1960 at Hughes Research Labs.

Marcatili, Enrique "Henry": Communications theorist and manager at Bell Labs under Stew Miller.

Marsh, Jock: Managing director of Standard Telecommunication Laboratories.

Maurer, Robert: Corning Glass Works physicist; directed development of first low-loss fibers as manager of a small glass research group.

McGowan, Bill: Chairman of MCI.

Midwinter, John: Managed British Telecom Research Labs fiber-optics program from 1977 to 1984, succeeding F. F. Roberts.

Miller, Stewart E. (1919–1990): Managed guided-wave transmission research at Bell Labs in Crawford Hill, including millimeter waveguides, hollow optical light pipes, and fiber optics.

Mollenauer, Linn: Bell Labs physicist who developed soliton transmission systems.

Møller Hansen, Holger (1915–): Danish inventor who proposed and demonstrated imaging bundles and recognized the need for clad fibers, but failed to secure a Danish patent in 1951.

Mosaic Fabrications: Founded by Will Hicks in 1958 to make imaging fiber optics; now Galileo Corporation.

Nelson, Herb: RCA engineer who developed liquid-phase epitaxy for semiconductor lasers.

Newman, David: Semiconductor laser developer at British Post Office.

Newns, George R.: Early fiber developer at British Post Office.

Nichizawa, Jun-ichi: Early Japanese fiber developer at Tohoku University.

Nill, Ken: Co-founder of Lasertron with J. Jim Hsieh.

Norton, Frederick H. "Ted": Retired MIT glass scientist who consulted with American Optical.

NTT: Nippon Telegraph and Telephone, Japan's national telephone company.

O'Brien, Brian (1898–1992): Eminent American optical physicist who proposed cladding optical fibers to improve image transmission; left University of Rochester to head research at American Optical Corporation.

O'Hara, Sidney: British Post Office semiconductor laser developer and fiber-optic project manager.

Ogilvie, Graeme: Australian metallurgist who developed liquid-core fibers.

Olshansky, Robert: Theoretical physicist who joined Corning fiber group in 1973; now at GTE Laboratories.

Osanai, Horoshi: Fujikura Cable Works co-developer of first fibers with low loss at 1.3 and 1.55 micrometers.

Panish, Mort: Bell Labs chemist and co-developer of first American room-temperature diode laser.

Payne, David: Engineer at University of Southampton; developed early solid-and liquid-core fibers with Alec Gambling; also developed fiber amplifiers.

Pearson, David: First fiber developer at Bell Labs.

Peters, C. William "Pete" (1919–1989): Optics professor at University of Michigan; co-developer of practical fiber-optic endoscope.

Pierce, John R.: Bell Labs engineering manager and developer of millimeter waveguide and communications satellites. Headed the Holmdel lab; later became a professor at Caltech and Stanford.

Potter, Robert: Earned first American doctorate in fiber optics from University of Rochester; developed measurement techniques and fiber-optic reader of punched computer cards.

Quist, Tom: Early semiconductor laser researcher at MIT Lincoln Laboratory.

Ramsay, Murray: Physicist at Standard Telecommunication Labs; managed fiber optic communication development after Charles Kao.

Randall, Eric: Fiber developer at Corning, Valtec, and other companies.

RCA: Radio Corporation of America, founded in 1920 as a radio patent trust. Became a leader in electronics and radio and television broadcasting; later bought by General Electric.

Rediker, Robert: Early semiconductor laser researcher at MIT Lincoln Laboratory.

Reeves, Alec Harley (1902–1971): English engineer and inventor of pulse-code modulation for digital communications; godfather of fiber-optic communications as a research adviser at Standard Telecommunication Laboratories.

Ritchie, Simon: Semiconductor laser developer at British Post Office.

Roberts, Frederick Francis "F. F." (1917–1977): Crusty engineering manager at British Post Office who pushed fiber communications research; headed BPO program from 1965 to 1977.

Runge, Peter: Director of submarine fiber-optic cable development at Bell Labs.

Saint-René, Henry C.: Teacher at French agriculture school in Crezancy and would-be inventor of television, who in 1895 proposed sending an image through a bundle of bent glass rods.

Sandbank, Charles: Division manager at Standard Telecommunications Laboratories; responsible for optical communications research from mid-1960s.

Schindler, Rudolf: German-born developer of semiflexible gastroscope replaced by fiber optics.

Schultz, Peter: Glass chemistry specialist on Corning team that made first low-loss fiber, now president of Heraeus Amersil Inc.

Schwalow, Arthur L.: Co-author of key paper on laser theory with Charles Townes; later Nobel Laureate for development of laser spectroscopy.

Shumate, Paul: Bellcore engineer who directs development of broad-band home systems, including fiber optics.

Siegmund, Walter: Former student of Brian O'Brien's who headed fiber optic development at American Optical from the late 1950s.

Simon, Jean-Claude: Director of CSF central research laboratory in France.

Smith, David D.: Patented bent glass rod as dental illuminator in 1890s.

Snitzer, Elias: American Optical scientist who formulated theory of single-mode fibers and developed fiber-optic lasers. Later worked at Polaroid and Rutgers University.

Spitz, Eric: Head of microwave lab at CSF Central Research Laboratory in France.

Standard Telecommunication Laboratories: Harlow, England-based research arm of Standard Telephones and Cables, a British subsidiary of ITT. Now part of Northern Telecom (Nortel).

Steventon, Alan: Semiconductor laser developer at British Post Office.

Stewart, Walter: President of American Optical Corporation.

Stone, Julius: Bell Labs physicist; developer of liquid-core fibers.

Tillman, Jack: Deputy research director at British Post Office during early fiber development.

Tillotson, Leroy: Managed research on communications through the atmosphere at Bell Labs in Crawford Hill.

Tkach, Bob: Bell Labs developer of high-speed fiber-optic systems.

Todd, Mike (1907–1958): American entertainment impresario and promoter of wide-screen movies.

Townes, Charles H.: Nobel laureate in physics for inventing the fundamental principles of maser and laser operation while at Columbia University.

Tyndall, John (1820–1893): Irish-born British physicist who demonstrated light guiding in a water jet in 1854.

Uesuge, Naoshi: Nippon Telegraph and Telephone engineer who identified hydrogen problem in submarine cables.

Upton, Lee: Glass specialist at American Optical.

Valpey, Ted: Chairman of Valtec.

van Heel, Abraham C. S. (1899–1966): Professor of optics at the Technical University of Delft in the Netherlands and leader of Dutch optical community, who proposed imaging bundles and made first bundles of clad optical fibers.

Werts, Alain: French engineer at CSF central research laboratory who published proposal for fiber communications with loss of 20 decibels per kilometer two months after Kao and Hockham.

Western Electric Company: Former name for the manufacturing operations of AT&T, which were divested along with most of Bell Labs as Lucent Technologies.

Wheeler, William (1851–1932): American engineer who patented light pipes to distribute light from a central electric arc throughout homes.

White, Robert Williamson: Manager of a waveguide development section at British Post Office Research Station.

Williams, Don: Early sponsor of fiber-optics research at Royal Signals Research and Development Establishment at Christchurch, England.

Zimar, Frank: Corning scientist with furnace for melting fused silica.

Appendix B

A Fiber-Optic Chronology

Circa 2500 *BC*: Earliest known glass.
Roman Times: Glass is drawn into fibers.
1713: Réné de Réaumur makes spun glass fibers.
1790s: Claude Chappe invents "optical telegraph" in France.
1841: Daniel Colladon demonstrates light guiding in jet of water in Geneva; it also is demonstrated in London and Paris.
1842: Daniel Colladon publishes report on light guiding in *Comptes Rendus*; Jacques Babinet also reports light guiding in water jets and bent glass rods.
1853: Paris Opera uses Colladon's water jet in the opera *Faust*.
1854: John Tyndall demonstrates light guiding in water jets at the suggestion of Michael Faraday, duplicating but not acknowledging Colladon.
1873: Jules de Brunfaut makes glass fibers that can be woven into cloth.
1880: Alexander Graham Bell invents Photophone.
1880: William Wheeler invents system of light pipes to illuminate homes from an electric arc lamp in basement, Concord, Mass.
1884: International Health Exhibition in South Kensington district of London has first fountains with illuminated water jets, designed by Sir Francis Bolton. Colladon republishes his 1842 paper to show the idea was his.

1887: Charles Vernon Boys draws quartz fibers for mechanical measurements.

1887: Royal Jubilee Exhibition in Manchester has illuminated "Fairy Fountains" designed by W. and J. Galloway and Sons.

1888: Dr. Roth and Prof. Reuss of Vienna use bent glass rods to illuminate body cavities for dentistry and surgery.

1889: Universal Exhibition in Paris shows refined illuminated fountains designed by G. Bechmann.

1892: Herman Hammesfahr shows glass dress at Chicago World's Fair.

1895: Henry C. Saint-René designs a system of bent glass rods for guiding light in an early television scheme (Crezancy, France).

1898: David D. Smith of Indianapolis applies for patent on bent glass rod as a surgical lamp.

1920s: Bent glass rods common for microscope illumination.

June 2, 1926: C. Francis Jenkins applies for US patent on a mechanical television receiver in which light passes along quartz rods in a rotating drum to form an image.

October 15, 1926: John Logie Baird applies for British patent on an array of parallel glass rods or hollow tubes to carry image in a mechanical television. He later built an array of hollow tubes.

December 30, 1926: Clarence W. Hansell proposes a fiber-optic imaging bundle in his notebook at the RCA Rocky Point Laboratory on Long Island. He later receives American and British patents.

1930: Heinrich Lamm, a medical student, assembles first bundle of transparent fibers to carry an image (of an electric lamp filament) in Munich. His effort to file a patent is denied because of Hansell's British patent.

December 1931: Owens-Illinois mass-produces glass fibers for Fiberglas.

August 20, 1932: Norman French of Bell Labs applies for patent on an "optical telephone system" using quartz rods.

Mid-1930s: Frank Hyde develops flame hydrolysis to make fused silica at Corning Glass Works.

1939: Curvlite Sales offers illuminated tongue depressor and dental illuminators made of Lucite, a transparent plastic invented by DuPont.

October 31, 1945: Ray D. Kell and George Sziklai apply for patent on transmitting signals through quartz or glass rods, issued May 9, 1950.

Circa 1949: Holger Møller Hansen in Denmark and Abraham C. S. van Heel at the Technical University of Delft begin investigating image transmission through bundles of parallel glass fibers.

April 11, 1951: Holger Møller Hansen applies for a Danish patent on fiber-optic imaging in which he proposes cladding glass or plastic fibers with a transparent low-index material. Patent claim is denied because of Hansell patent.

October 1951: Brian O'Brien (University of Rochester) suggests to van Heel that applying a transparent cladding would improve transmission of fibers in his imaging bundle.

July 1952: Harold Horace Hopkins applies for a grant from the Royal Society to develop bundles of glass fibers for use as an endoscope at Imperial College of Science and Technology. Hires Narinder S. Kapany as an assistant after he receives grant.

Early 1953: O'Brien joins American Optical as vice president and research director. His top priority is developing a wide-screen movie system for promoter Mike Todd; fiber optics is sidetracked.

Spring 1953: Hopkins tells Fritz Zernicke his idea of fiber bundles; Zernicke tells van Heel, who decides to publish quickly.

May 21, 1953: *Nature* receives brief paper by van Heel on simple bundles of clad fibers.

June 12, 1953: Dutch-language weekly *De Ingeneur* publishes van Heel's first report of clad fiber.

November 22, 1953: *Nature* receives paper on bundles of unclad fibers for imaging written by Hopkins and Kapany.

January 2, 1954: *Nature* publishes papers by Hopkins and Kapany and by van Heel. The long delay of the van Heel paper has never been explained.

1954: Basil Hirschowitz visits Hopkins and Kapany in London from the University of Michigan.

September 1954: American Optical hires Will Hicks to develop fiber-optic image scramblers, proposed by O'Brien to the Central Intelligence Agency.

Summer 1955: Kapany completes doctoral thesis on fiber optics under Hopkins, moves to University of Rochester.

Summer 1955: Hirschowitz and C. Wilbur Peters hire undergraduate student Larry Curtiss to work on their fiber-optic endoscope project.

1956:	First transatlantic telephone cable, TAT-1, goes into operation. It uses coaxial cable to carry 36 voice circuits.
Summer 1956:	Curtiss suggests making glass-clad fibers by melting a tube onto a rod of higher-index glass. Peters and other Michigan physicists push plastic-clad fibers, which Curtiss makes instead.
October 1956:	Frederick H. Norton starts consulting with American Optical on fiber development. Later he suggests ways to make glass cladding.
October 1956:	Curtiss and Peters describe plastic-clad fibers at Optical Society of America meeting in Lake Placid, New York. Kapany also presents a paper. Hicks attends but does not give a talk.
December 8, 1956:	Curtiss makes first glass-clad fibers by rod-in-tube method; they are much clearer than plastic-clad fibers.
February 18, 1957:	Hirschowitz tests first fiber-optic endoscope in a patient.
Early 1957:	Hicks experiments with glass-clad fibers and fusing many fibers into a rigid bundle, an idea suggested by Norton.
May 1957:	Hirschowitz demonstrates fiber endoscope to American Gastroscopic Society.
Mid-1957:	Kapany leaves Rochester to head group at Illinois Institute of Technology Research Institute in Chicago.
Mid-1957:	Image scrambler project ends after Hicks tells CIA the code is easy to break. American Optical shifts to developing faceplates, adding more people as Todd-AO wide-screen movie project fades.
1957:	Hirschowitz, Peters, and Curtiss license gastroscope technology to American Cystoscope Manufacturers Inc.
Late 1957:	Charles Townes and Arthur Schawlow outline principles of laser operation. Gordon Gould starts work on his own laser proposal.
Early 1958:	Hicks develops practical fiber-optic faceplates for military imaging systems.
1958:	Hicks, Paul Kiritsy, and Chet Thompson leave American Optical to form Mosaic Fabrications in Southbridge, Mass., the first fiber-optics company.
1958:	Alec Reeves begins investigating optical communications at Standard Telecommunication Laboratories.
1959:	Working with Hicks, American Optical draws fibers

so fine they transmit only a single mode of light. Elias Snitzer recognizes the fibers as single-mode waveguides and applies for a patent (with Hicks) in 1960.

May 16, 1960: Theodore Maiman demonstrates the first laser at Hughes Research Laboratories in Malibu.

December 12, 1960: Ali Javan makes first helium-neon laser at Bell Labs, the first laser to emit a steady beam.

Circa 1960: George Goubau at Army Electronics Command Laboratory, Stew Miller of Bell Telephone Laboratories, and Murray Ramsay of Standard Telecommunication Laboratories begin investigating confocal optical waveguides with regularly spaced lenses.

January 1961: Charles C. Eaglesfield of STL proposes hollow optical pipeline made of reflective pipes.

May 1961: Eli Snitzer of American Optical publishes theoretical description of single-mode fibers.

1961: Narinder Kapany founds Optics Technology Inc.

1962: Experiments at STL show high loss in Eaglesfield's hollow optical pipeline.

1962: AT&T starts converting to digital telephone transmission.

September-October 1962: Four groups nearly simultaneously make first semiconductor diode lasers, which emit pulses at liquid-nitrogen temperature. Robert N. Hall's group at General Electric is first.

1962: Dwight Berreman of Bell Labs proposes gas lens waveguide.

1962–1963: STL abandons millimeter waveguide development. Alec Reeves pushes optical waveguides but sees problems with confocal lens waveguides.

1962–1963: Experiments show high loss when sending laser beams through atmosphere.

1963: Heterostructures proposed for semiconductor lasers.

1963–1964: Antoni E. Karbowiak of STL realizes that unclad transparent optical waveguides would have to be impractically thin. He considers clad optical fibers, but thinks a flexible thin-film waveguide would have lower loss.

October 1964: Charles Koester and Eli Snitzer describe first optical amplifier, using neodymium-doped glass.

December 1964: Charles K. Kao takes over STL optical communication program when Karbowiak leaves to become chair of electrical engineering at the University of New South Wales. Kao and George Hockham

	soon abandon thin-film waveguide in favor of single-mode clad optical fiber.
February 1965:	Stewart Miller of Bell Labs applies for patent on graded-index waveguides for light and millimeter waves.
Autumn 1965:	Kao concludes that the fundamental limit on glass transparency is below 20 decibels per kilometer, which would be practical for communications. Hockham calculates that clad fibers should not radiate much light. They prepare a paper proposing fiber-optic communications.
January 1966:	Kao tells Institution of Electrical Engineers in London that glass fibers could be made with loss below 20 decibels per kilometer for communications.
Early 1966:	F. F. Roberts starts fiber-optic communications research at British Post Office Research Laboratories.
July 1966:	Kao and Hockham publish paper outlining their proposal in *Proceedings of the Institution of Electrical Engineers.*
July 1966:	John Galt at Bell Labs asks Mort Panish and Izuo Hayashi to figure out why diode lasers have high thresholds at room temperature.
September 1966:	Alain Werts, a young engineer at CSF in France, publishes proposal similar to Kao's in French-language journal *L'Onde Electronique*, but CSF does nothing further for lack of funding.
1966:	Roberts tells William Shaver, a visitor from the Corning Glass Works, about interest in fiber communications. This leads Robert Maurer to start a small research project on fused-silica fibers.
1966:	Kao travels to America early in year but fails to interest Bell Labs. He later finds more interest in Japan.
Early 1967:	British Post Office allocates an extra £12 million to research; some goes to fiber optics.
Early 1967:	Shojiro Kawakami of Tohoku University in Japan proposes graded-index optical fibers.
Summer 1967:	Corning summer intern Cliff Fonstad makes fibers. Loss is high, but Maurer decides to continue the research using titania-doped cores and pure-silica cladding.
October 1967:	Clarence Hansell dies at 68.
Late 1967:	Robert Maurer recruits Peter Schultz from Corning's glass chemistry department to help make pure glasses.

January 1968: Donald Keck starts work for Maurer as the first full-time fiber developer at Corning.

August 1968: Dick Dyott of British Post Office picks up suggestion for pulling clad optical fibers from molten glass in a double crucible.

1968: Kao and M. W. Jones measure intrinsic loss of bulk fused silica at 4 decibels per kilometer, the first evidence of ultratransparent glass, prompting Bell Labs to seriously consider fiber optics.

1969: Martin Chown of Standard Telecommunication Labs demonstrates fiber-optic repeater at Physical Society exhibition.

April 1970: STL demonstrates fiber-optic transmission at Physics Exhibition in London.

Spring 1970: First continuous-wave room-temperature semiconductor lasers made in early May by Zhores Alferov's group at the Ioffe Physical Institute in Leningrad (now St. Petersburg) and on June 1 by Mort Panish and Izuo Hayashi at Bell Labs.

June 30, 1970: AT&T introduces Picturephone in Pittsburgh. The telephone monopoly plans to install millimeter waveguides to provide the needed extra capacity.

Summer 1970: Maurer, Donald Keck, and Peter Schultz at Corning make a single-mode fiber with loss of 16 decibels per kilometer at 633 nanometers by doping titanium into fiber core.

September 30, 1970: Maurer announces Corning's fiber results at London conference devoted mainly to progress in millimeter waveguides.

November 1970: Measurements at British Post Office and STL confirm Corning results.

Late Fall 1970: Charles Kao leaves STL to teach at Chinese University of Hong Kong; Murray Ramsay heads STL fiber group.

1970–1971: Dick Dyott at British Post Office and Felix Kapron of Corning separately find pulse spreading is lowest at 1.2 to 1.3 micrometers.

May 1971: Murray Ramsay of STL demonstrates digital video transmission over fiber to Queen Elizabeth at the Centenary of the Institution of Electrical Engineers.

October 13, 1971: Alec Reeves dies in London.

1971–1972: Unable to duplicate Corning's low loss, Bell Labs, the University of Southampton, and CSIRO in Australia experiment with liquid-core fibers.

1971–1972:	Focus shifts to graded-index fibers because single-mode offers few advantages and many problems at 850 nanometers.
June 1972:	Maurer, Keck, and Schultz make multimode germania-doped fiber with 4 decibel per kilometer loss and much greater strength than titania-doped fiber.
Late 1972:	STL modulates diode laser at 1 billion bits per second. Bell Labs stops its last work on hollow light pipes.
December 1972:	John Fulenwider proposes a fiber-optic communication network to carry video and other signals to homes at International Wire and Cable Symposium.
1973:	John MacChesney develops modified chemical vapor deposition process for making fiber at Bell Labs.
Mid-1973:	Diode laser lifetime reaches 1000 hours at Bell Labs.
Spring 1974:	Bell Labs settles on graded-index fibers with 50 to 100 micrometer cores.
December 7, 1974:	Heinrich Lamm dies at 66.
February 1975:	Bell completes installation of 14 kilometers of millimeter waveguide in New Jersey. After tests, Bell declares victory and abandons the technology.
June 1975:	First commercial continuous-wave semiconductor laser operating at room temperature offered by Laser Diode Labs.
September 1975:	First nonexperimental fiber-optic link installed by Dorset (UK) police after lightning knocks out their communication system.
October 1975:	British Post Office begins tests of millimeter waveguide; like Bell it declares the tests successful, but never installs any.
1975:	Dave Payne and Alex Gambling at University of Southampton calculate pulse spreading should be zero at 1.27 micrometers.
January 13, 1976:	Bell Labs starts tests of graded-index fiber-optic system transmitting 45 million bits per second at its plant in Norcross, Georgia. Laser lifetime is main problem.
Early 1976:	Valtec launches Communications Fiberoptics division.
Early 1976:	Masahara Horiguchi (Nippon Telegraph Telephone Ibaraki Lab) and Hiroshi Osanai (Fujikura Cable) make first fibers with low loss—0.47 decibel per kilometer—at long wavelengths (1.2 micrometers).

March 1976: Japan's Ministry for International Trade and Industry announces plans for Hi-OVIS fiber-optic "wired city" experiment involving 150 homes.

Spring 1976: Lifetime of best laboratory lasers at Bell Labs reaches 100,000 hours (10 years) at room temperature.

Summer 1976: Horigushi and Osanai discover third fiber-optic transmission window at 1.55 micrometers.

July 1976: Corning sues ITT alleging infringement of American patents on communication fibers.

Late 1976: J. Jim Hsieh makes indium-gallium arsenide-phosphide (InGaAsP) lasers emitting continuously at 1.25 micrometers.

Spring 1977: F. F. Roberts reaches mandatory retirement age of 60; John Midwinter becomes head of fiber-optic group at British Post Office.

April 1, 1977: AT&T sends first test signals through field test system in Chicago's Loop district.

April 22, 1977: General Telephone and Electronics sends first live telephone traffic through fiber optics (6 million bits per second) in Long Beach, Calif.

May 1977: Bell System starts sending live telephone traffic through fibers at 45 million bits per second fiber link in downtown Chicago.

June 1977: British Post Office begins sending live telephone traffic through fibers in underground ducts near Martlesham Heath.

June 29, 1977: Bell Labs announces one million hour (100 year) extrapolated lifetime for diode lasers.

Summer 1977: F. F. Roberts dies of heart attack.

October 1977: Valtec "acquires" Comm/Scope, but Comm/Scope owners soon gain control of Valtec.

Late 1977: AT&T and other telephone companies settle on 850-nanometer gallium arsenide light sources and graded-index fibers for commercial systems operating at 45 million bits per second.

1977–1978: Low loss at long wavelengths renews research interest in single-mode fiber.

May 22–23, 1978: Fiber Optic Con, first fiber-optic trade show.

July 1978: Optical fibers begin carrying signals to homes in Japan's Hi-OVIS project.

August 1978: Nippon Telegraph and Telephone transmits 32 million bits per second through a record 53 kilometers of graded-index fiber at 1.3 micrometers.

September 1978: Richard Epworth reports modal noise problems in graded-index fibers.

September 1978: France Telecom announces plans for fiber to the home demonstration in Biarritz, connecting 1500 homes in early 1983.

1978: AT&T, British Post Office, and Standard Telephones and Cables commit to developing a single-mode transatlantic fiber cable, using the new 1.3-micrometer window, to be operational by 1988. By the end of the year, Bell Labs abandons development of new coaxial cables for submarine systems.

Late 1978: NTT Ibaraki lab makes single-mode fiber with record 0.2 decibel per kilometer loss at 1.55 micrometers.

January 1980: AT&T asks Federal Communications Commission to approve Northeast Corridor system from Boston to Washington, designed to carry three different wavelengths through graded-index fiber at 45 million bits per second.

February 1980: STL and British Post Office lay 9.5-kilometer submarine cable in Loch Fyne, Scotland, including single-mode and graded-idex fibers.

Winter 1980: Graded-index fiber system carries video signals for 1980 Winter Olympics in Lake Placid, New York, at 850 nanometers.

September 1980: With fiber optics hot on the stock market, M/A Com buys Valtec for $224 million in stock.

1980: Bell Labs publicly commits to single-mode 1.3-micrometer technology for the first transatlantic fiber-optic cable, TAT-8.

July 27, 1981: ITT signs consent agreement to pay Corning and license Corning communication fiber patents.

1981: Commercial second-generation systems emerge, operating at 1.3 micrometers through graded-index fibers.

1981: British Telecom transmits 140 million bits per second through 49 kilometers of single-mode fiber at 1.3 micrometers, starts shifting to single-mode.

Late 1981: Canada begins trial of fiber optics to homes in Elie, Manitoba.

1982: British Telecom performs field trial of single-mode fiber, abandons graded-index in favor of single-mode.

December 1982: MCI leases right of way to install single-mode fiber from New York to Washington. The system will operate at 400 million bits per second at 1.3 micrometers. This starts the shift to single-mode fiber in America.

Late 1983: Stew Miller retires as head of Bell Labs fiber development group.

January 1, 1984: AT&T undergoes first divestiture, splitting off its seven regional operating companies but keeping long-distance transmission and equipment manufacture.

1984: British Telecom lays first submarine fiber cable to carry regular traffic, to the Isle of Wight.

1985: Single-mode fiber spreads across America to carry long-distance telephone signals at 400 million bits per second and up.

Summer 1986: All 1500 Biarritz homes connected to fiber to the home system.

October 30, 1986: First fiber-optic cable across the English Channel begins service.

1986: AT&T sends 1.7 billion bits per second through single-mode fibers originally installed to carry 400 million bits per second.

1987: Dave Payne at University of Southampton develops erbium-doped fiber amplifier operating at 1.55 micrometers.

1988: Linn Mollenauer of Bell Labs demonstrates soliton transmission through 4000 kilometers of single-mode fiber.

December 1988: TAT-8, first transatlantic fiber-optic cable, begins service using 1.3-micrometer lasers and single-mode fiber.

February 1991: Masataka Nakazawa of NTT reports sending soliton signals through a million kilometers of fiber.

February 1993: Nakazawa sends soliton signals 180 million kilometers, claiming "soliton transmission over unlimited distances."

February 1993: Linn Mollenauer of Bell Labs sends 10 billion bits through 20,000 kilometers of fibers using a simpler soliton system.

February 1996: Fujitsu, NTT Labs, and Bell Labs all report sending one trillion bits per second through single optical fibers in separate experiments using different techniques.

Notes

Chapter 1

1. Daniel Colladon, "On the reflections of a ray of light inside a parabolic liquid stream," *Comptes Rendus 15*, pp. 800–802 (Oct. 24, 1842), translated by Julian A. Carey, Apr. 1, 1995.

2. The number keeps increasing and is likely to be outdated by the time you read this. Phone companies do not yet need all that capacity; the highest speeds in use in late 1998 were about 80 billion bits per second. In practice, the signals consist of bit streams sent simultaneously using different colors of light. A single fiber can carry 80 billion bits per second as eight different wavelengths each carrying 10 billion bits, or 32 wavelengths each carrying 2.5 billion bits.

3. Tiles installed on the inbound side of the Kendall Square station, which serves MIT, chronicle a century of inventions and the growth of MIT. *The 1997 World Almanac* says the same thing in its list of inventions.

Chapter 2

1. David Napoli, "The luminous fountains at the French Exposition," *Scientific American*, Dec. 14, 1889, pp. 376–377, translated from the French *La Nature*.

2. French embassy, in response to telephone query.

3. Napoli, "The luminous fountains."

4. This account draws heavily on unpublished research by Kaye Weedon, a Norwegian engineer who collected information on Colladon's and Babinet's work and gave several talks on the origins of fiber optics around 1970. Jeofry Courtney-Pratt kindly gave me copies of Weedon's unpublished manuscripts. Sadly, Weedon died in 1992, shortly before I began research on this book.

5. Savart died Mar. 16, 1841, shortly before Colladon repeated his experiment.

"Felix Savart" entry in Charles Coulston Gillespie, ed., *Dictionary of Scientific Biography (Vol. XII*; Scribner's, New York, 1976, pp. 129–130).

6. Daniel Colladon, "On the reflections of a ray of light inside a parabolic liquid stream," *Comptes Rendus 15*, (Oct. 24, 1842), pp. 800–802, translated by Julian A. Carey, Apr. 1, 1995.

7. René Sigrist, cites an unpublished manuscript by Pierre Speziali; (personal communication, Oct. 10, 1995), Colladon's *Comptes Rendus* paper cites a London demonstration but does not say who conducted it.

8. Colladon cited experiments by Joseph Plateau, who steered light around curves in a different way. See Joseph Plateau, "On a curious consequence of the laws of light's reflection," *Bulletins de l'Academie Royale des Sciences et Belles Lettres de Bruxelles IX* 2d, Partie (1842), pp. 10–14, read July 4, 1842, translated by Jean-Louis Trudel.

9. A device he invented to measure polarization, called a Babinet Compensator, is still widely used by optical specialists. *Dictionary of Scientific Biography (Vol. I*, "Jacques Babinet," pp. 357–358).

10. Jacques Babinet, "Note on the transmission of light by sinuous canals," *Comptes Rendus #15*, (Oct. 24, 1842) p. 802, translated by Julian A. Carey, Apr. 1, 1995; also Kaye Weedon, unpublished manuscript.

11. T. K. Derry and Trevor I. Williams, *A Short History of Technology: From the Earliest Times to A.D. 1900* (Dover, New York, 1993 (reprint of 1960 edition), pp. 84–85).

12. Specimen in Corning Glass Museum.

13. The colors of the rainbow come from a combination of two effects: total internal reflection and differences in the refractive index of water with wavelength. Light rays that enter a water droplet are reflected back around the sphere, emerging back toward the sun (which is why we see the rainbow when the sun is behind us). Water refracts different wavelengths at different angles, so the angle at which the light rays emerge depends on their wavelength, creating the rainbow.

14. René Sigrist, personal communications, Oct. 10, 1995, and Nov. 13, 1995, citing Pierre Speziali, "La physique," in Jacques Trembley, ed., *Les Savants Genevois dans l'Europe intellectuelle* (Journal de Geneva, Geneva, 1987, p. 154).

15. David E. Nye, *Electrifying America: Social Meanings of New Technology* (MIT Press, Cambridge, Mass., 1990); see chap. 2, "The Great White Way."

16. Gösta M. Bergman, *Lighting in the Theatre* (Rowman and Littlefield, Totowa, N.J.; and Almqvist & Wiskell International, Stockholm, 1977, pp. 278–280).

17. "Preparations for the holding of the international health exposition," *Nature*, Feb. 21, 1884, pp. 388–389; Leslie Stephen and Sidney Lee, eds., *Dictionary of National Biography: Vol. 22 Supplement* (Oxford University Press, London, pp. 230–231).

18. Bolton was not a specialist in optics, and the importance of total internal reflection may have eluded him. It definitely eluded the writers of contemporary accounts, and it takes careful analysis of published drawings to deduce how the fountains guided light. "Illumination of fountains by the electric light," *Scientific American Supplement #847*, p. 7774 (May 2, 1885).

19. "The fountains at the health exhibition," *The Illustrated London News*, Aug. 2, 1884, p. 106.

20. "Health Exhibition-X: the electrically illuminated fountain," *The Electrician*

13, pp. 456–457 (Sept. 27, 1884); "The illuminated fountains at the Healtheries," *Nature*, Nov. 6, 1884, pp. 11–12.

21. D. Colladon, "La Fontaine Colladon," *La Nature*, 2nd half year 1884, pp. 325–326. In a brief introduction, editor Gaston Tissandier says he requested the paper. Translated as Daniel Colladon, "The Colladon Fountain," *Scientific American*, Dec. 6, 1884, p. 359.

22. That was a large sum in the 1880s. Andrew A. Gillies, ed., *Report of the Executive Committee, Royal Jubilee Exhibition Manchester 1887* (John Heywood, Manchester, 1888, p. 189).

23. Charles John Galloway and John Henry Beckwith, British Patent 1460, "An improvement in illuminated fountains," filed Jan. 29, 1887, accepted Nov. 8, 1887.

24. *Official Guide to the Royal Jubilee Exhibition, Manchester 1887* (John Heywood, Manchester, 1887, p. 9).

25. Perilla Kinchin and Juliet Kinchin, *Glasgow's Great Exhibitions* (White Cockade Publishers, 1988); quoted in letter to author from Glasgow City Library dated Dec. 5, 1995.

26. "Glasgow International Exhibition," *The Electrician 21*, June 29, 1888, pp. 239–241.

27. Although he obviously paid attention to the illuminated fountains, he lists the wrong dates for the British exhibitions. He puts Glasgow in 1884, followed by London and Manchester, and does not mention other European fairs with great fountains, such as one held in Barcelona in 1888. Jean-Daniel Colladon, *Souvenirs et Memoires, Autobiographie* Aubert-Schuchardt, Geneva, 1893, p. 289).

28. Napoli, "The luminous fountains."

29. G. Bechmann, "Fontaine Lumineuse," *Le Grand Encyclopedie: Vol. 17 Fanum-Franco* Lamirault, Paris, 1886–1902 p. 733). It includes the clearest drawings I have found of vertical and parabolic fountains.

30. Nye, *Electrifying America* (pp. 38–39).

31. Woodward Hudson, "William Wheeler, December 6, 1851–July 1, 1932," *Social Circle Memoirs* (Social Circle, Concord, Mass., pp. 331ff). Clark was a colorful and controversial figure whose shorter stay left him a legend in Japan. See Robert H. Guest, "The rise and fall of an Amherst immortal," *Amherst*, Summer 1983, pp. 66–67, 78–79.

32. William Wheeler, US Patent 247,229, "Apparatus for lighting dwellings or other structures," filed Dec. 10, 1880, granted Sept. 20, 1881.

33. Ibid., p. 1, ll. 27–40.

34. Wheeler served as company president, per state records; he also was an engineering consultant. A leading citizen of Concord, he moved in the same social groups as the intellectual Emerson family, and was deeply involved in town government, until shortly before his death in 1932. *The Story of Massachusetts: Personal and Family History* (Vol. 4: The American Historical Society, New York, 1938, pp. 129–132). Wheeler Reflector was still advertising in the Oct. 1958 issue of *The American City* (p. 143).

35. "A new method of illuminating internal organs," *The Lancet*, Jan. 5, 1889, p. 52; the same brief note was reprinted in *Scientific American 60*, p. 14 (Apr. 6, 1889). The original *Lancet* piece is by-lined Vienna, Dec. 1888. The two men were identified only as "Dr. Roth and Professor Reuss."

36. David D. Smith, US Patent 624,392, "Surgical lamp," filed Apr. 25, 1898, issued May 2, 1899. At least two similar patents were issued many years later:

Isaac J. Smit, US Patent 1,246,338, "Illuminated transparent retractor," filed Aug. 21, 1916, issued Nov. 13, 1917; Frank G. Young, Jr., US Patent 1,246,338, "Light projector," filed Mar. 26, 1926, issued Sept. 13, 1927.

37. It was known as Lucite in America and Perspex in England; its proper name is polymethyl methacrylate. It also was used in tongue depressors. "Cold light (Lucite) surgical instruments," *Scientific American*, Feb. 1939, p. 99; see also "Piped light aids surgeons and dentists," *Popular Science*, Mar. 1939, p. 108.

38. *Dictionary of Scientific Biography (Vol. XIII,* "John Tyndall," pp. 521–524).

39. John Tyndall, notebook preserved at Royal Institution, London, dated Friday evening, May 19, 1854.

40. John Tyndall, "On some phenomena connected with the motion of liquids," *Proceedings of the Royal Institution of Great Britain 1*, pp. 446–448 (1854). He says his demonstration repeats Savart's work on fluid flow.

41. Faraday's memory failed gradually, beginning around 1840; the problem was episodic, so he was able to work much of the time but could not concentrate at other times. He became quite senile before his death in 1868 and probably suffered from what we now call Alzheimer's disease. L. Pearce Williams, *Michael Faraday: A Biography* (Basic Books, New York, 1965).

42. Colladon in *Comptes Rendus* says the water-jet apparatus was demonstrated in London, probably in 1841, but does not say where or by whom. Faraday gave two or three lectures at the Royal Institution in 1841, which was the worst year of his first breakdown in health. His topics almost certainly did not include light guiding. There is no evidence that either Colladon or de la Rive spoke at the Royal Institution at the time, although they could have lectured elsewhere in London, such as at the Adelaide Galleries or the London Institution. Faraday spent much of his time that year in Switzerland away from other scientists, as he tried to recover his faculties, but even if he did not see the demonstration, he must have heard of it. (Frank James, Royal Institution, e-mail Sep. 14, 1998).

43. The mistaken crediting of Tyndall with first guiding light in a water jet is a reminder of how easily and widely mistakes can spread. Tyndall's original 1854 account (Tyndall, "On some phenomena") does not claim the idea of light guiding is new, but also does not explicitly attribute it to someone else. He wound up credited with the idea half a century after his death largely because he described the experiment in one of his widely-circulated popular books, where it was rediscovered in the 1950s (an American edition is John Tyndall, *Light and Electricity*, Appleton & Co., New York, 1871, pp. 41–43). It was natural to credit the eminent Tyndall with the discovery. The first publication I have found to credit Tyndall is Narinder S. Kapany, "Fiber optics," *Scientific American 203* (5), 72–81, November 1960, and Kapany told me he found the reference (telephone interview, Feb. 13, 1996). Kapany's thesis advisor Harold H. Hopkins may have played a role in finding the Tyndall reference; he was well read and had a large library of old science books. As fiber optics spread, most people—including myself—accepted Tyndall as the originator of the idea. The Optical Society of America later named its major fiber-optics award after Tyndall. The late Kaye Weedon uncovered Colladon's and Babinet's 1842 reports in the late 1960s, but although he gave several talks on their work, he never published an account, and few others took notice.

Chapter 3

1. Charles Vernon Boys, "Quartz fibres," *Nature*, July 11, 1889, pp. 247–251.

2. Oszkar Knapp, *Glasfasern* (Glass Fibers) (Akademiai Kaido, Budapest, 1966, p. 9; translation by Max J. Riedl).

3. The oldest mention of glass fibers is in a book by Antonio Neri called *L'Arte Vetraria*, published in Florence in 1612. David G. Mettes, "Glass fibers," in George Lubin, ed., *Handbook of Fiberglass and Advanced Plastics Composites* (Krieger, Huntington, N.Y., 1975, p. 143).

4. Knapp, *Glasfasern* (pp. 9–12); Mettes, "Glass fibers" (p. 143).

5. K. L. Lowenstein, *The Manufacturing Technology of Continuous Glass Fibres* (Elsevier, Amsterdam, 1983, p. 1).

6. Johann Georg Krünitz, "Perrücke, Glas," *Ökonomischtechnologische Encyclopädie* (Berlin, 1808), cited in Hanne Frøsig, "A glass wig," *Journal of Glass Studies 16*, pp. 92–94 (1974).

7. An English silk weaver named Louis Schwabe in 1842 drew fibers this way and wove them into fabric on a spinning machine. Knapp, *Glasfasern* (pp. 9–12).

8. "The uses of spun glass," *Scientific American, Supplement 1706*, Sept. 12, 1908, p. 163.

9. "Glass spinning," *Art Workman 15*, p. 240 (1873). The best fibers were made by Jules de Brunfaut; see a clipping dated July 10, 1895 (p. 17) quoting the "Pottery Gazette," obtained from the Corning Glass Museum, possibly from a magazine called *China, Glass, and Lamps*. Other sources, also from the files of the Corning Glass Museum, are "Cloth made of glass," *Health and Home 4*, No. 51, Dec. 21, 1872, and "Glass spinning."

10. Charles C. Gillispie, ed., *Dictionary of Scientific Bibliography* (Vol. XV, Suppl. 1: Scribner's, New York, 1978, pp. 59–61). Apologists add that Boys later becam"an able expositor," but Boys was hardly the only eminent scientist whose classroom manner was less than effective.

11. Boys, "Quartz fibres."

12. Charles Vernon Boys, "On the production, properties, and some suggested uses of the finest threads," *Philosophical Magazine 23*, No. 145, pp. 489–499, (June 1887). (p. 492)

13. Ibid., (p. 493).

14. Knapp, *Glasfasern* (pp. 9–12).

15. Dorothy Stafford, "Glass dress, now at art museum, was World Fair sensation in 1892," *Toledo Blade*, Jan. 21, 1951, p. 7; also *Can This Be Glass?* booklet published for 1939 World's Fair by Owens-Corning Fiberglas Corp., Toledo, Ohio.

16. H. S. Souttar, "Demonstration of a method of making capillary filaments," *Proceedings of the Physical Society of London 24*, No. 3, pp. 166–167 (1912).

17. C. C. Hutchins, "How to make quartz fiber," *Scientific American Supplement*, Aug. 17, 1912, p. 100.

18. The odd spelling allowed them to trademark the word "fiberglas."

19. *Can This Be Glass?*

Chapter 4

1. Quoted in Milton Wright, "Successful inventors VIII: they seek more than money, says one of them," *Scientific American*, Aug. 1927, pp. 140–142.

2. Albert Abramson, *The History of Television, 1880–1941* (McFarland & Co., London, 1987, p. 6).

3. "Seeing by electricity," *Scientific American 42*, p. 373 (June 12, 1880). The inventor was George R. Carey.

4. H. C. Saint-René, "On a solution to the problem of remote viewing," *Comptes Rendus 150*, pp. 446–447 (1910), translation by Jean-Louis Trudel, Oct. 1994.

5. Kaye Weedon found this detail when he examined the packet that Saint-René had sent to the French Academy; it was not published in *Comptes Rendus*. Kaye Weedon, unpublished manuscript.

6. Hubert Masson, letter to author, May 10, 1996. Masson is vice president of the agriculture school and deputy mayor of Crezancy.

7. Michael Ritchie, *Please Stand By: A Prehistory of Television* (Overlook Press, Woodstock, N.Y., 1994, p. 24).

8. See, e.g., R. W. Burns, "J. L. Baird: success and failure," *Proceedings IEE 126*, No. 9, pp. 921–928 (Sept. 1979); Archer S. Taylor, "Origins of British television," *CED: Communications Engineering and Design*, June 1995, p. 162.

9. J. L. Baird, British Patent 285,738, "An improved method of and means for producing optical images," filed Oct. 15, 1926, issued Feb. 15, 1928.

10. Ray Herbert, letter to author, Aug. 17, 1994. Herbert worked with Baird.

11. David E. Fisher and Marshall Jon Fisher, *Tube: The Invention of Television* (Counterpoint, Washington, D.C., 1996).

12. Jenkins's importance in early motion-picture development is controversial. He and Thomas Armat patented a projector called the Phantoscope (US Patent 586,953, issued July 20, 1897) but later split, with Armat obtaining a later patent on a variant projector. Jenkins insisted he projected the first motion pictures, but his claims were never verified. Armat claimed he deserved sole credit for inventing the projector, because the Jenkins-Armat patent was impractical. Armat licensed his patents to Thomas Edison, leading to the common perception that Edison invented motion pictures. Edison's contribution was the camera. See Raymond Fielding, ed., *A Technological History of Motion Pictures and Television* (University of California Press, Berkeley, 1967), especially F. H. Richardson, "What happened in the beginning" (pp. 23–41). Jenkins's gift for self-promotion earned him flattering magazine profiles: "The original movie man and his first 'show,' " *The Literary Digest*, Apr. 30, 1921, pp. 38–39; Homer Croy, "The infant prodigy of our industries: the birth and growth of motion pictures," *Harper's Monthly 135*, pp. 349–357, (Aug. 1917).

13. Wright, "Successful inventors,"

14. Hugo Gernsback of *Radio News* (who would soon publish the world's first science-fiction magazine) and Watson Davis of *Popular Radio*.

15. Background on Jenkins and his mechanical television systems comes from Albert Abramson, "Pioneers of television—Charles Francis Jenkins," *Journal of the Society of Motion Picture and Television Engineers*, Feb. 1986, pp. 224–238.

16. C. Francis Jenkins, US Patent 1,683,137, "Method of and apparatus for converting light impulses into enlarged graphic representations," filed June 2, 1926, issued Sept. 4, 1928.

17. Technical details from Abramson, "Pioneers of television."

18. "Broadcasts pictures," *New York Times*, May 6, 1928, p. 3.

19. Ritchie, *Please Stand By*; (pp. 22–31); for a fuller account, see Fisher and Fisher, *Tube*.

20. Frank Melville Jr. Memorial Library, State University of New York, Stony Brook, Special Collections Dept.: Clarence Weston Hansell Collection, Collection 209, biographical sketch of Hansell.

21. Orrin E. Dunlap, Jr., *Radio's 100 Men of Science* (Harper & Brothers Publishers, New York, 1944, pp. 269–272).

22. James Hillier, telephone interview, May 19, 1994.

23. Clarence W. Hansell, US Patent 1,751,584, "Picture transmission," filed Aug. 13, 1927, issued Mar. 25, 1930.

24. Hansell notebook dated Apr. 11, 1925–Sep. 12, 1930, Clarence Weston Hansell Collection, Special Collections Dept., State University of New York, Stony Brook.

25. Daniel M. Costigan, *Electronic Delivery of Documents and Graphics* (Van Nostrand Reinhold, New York, 1978), p. 5.

26. Howard Rosenthal, telephone interview, May 19, 1994.

27. Patricia (Hansell), Sisler, telephone interview, Dec. 29, 1995.

28. William P. Vogel, Jr., "Inventing is vision plus work," *Popular Science*, Oct. 1947, pp. 97–101.

29. The physician was Adolf Kussmaul. Basil I. Hirschowitz, "Development and application of endoscopy," *Gastroenterology 104*, pp. 337–342 (1993).

30. Unnamed source, quoted in H. H. Hopkins, "Optics in clinical medicine," presidential address, physics and mathematics section, British Association meeting, Sept. 1977, typed manuscript.

31. Audrey B. Davis, "Rudolf Schindler's role in the development of gastroscopy," *Bulletin of the History of Medicine 46*, No. 2, pp. 150–170 (Mar.–Apr. 1972).

32. Hirschowitz, "Development."

33. Sam Carter, MD, telephone interview, Sept. 9, 1994; he remembers taking a course from a Prof. Rose who in the 1930s had a solid "bundle that looked like a garden house," 8 to 10 feet long. "He put a light source at one end of it . . . and by golly . . . the light would show up at the end of the bundle."

34. Heinrich Lamm, "Biegsame optische Geräte" (Flexible optical instruments), *Zeitschrift fur Instrumentenkunde 50*, pp. 579–581 (1930), translated by Lamm many years later.

35. Lamm, "Biegsame optische Geräte," p. 580.

36. The British patent was a duplicate of Hansell's American patent; RCA assigned the British rights to the Marconi Company as part of a broader patent agreement. The British patent was issued before the American one. Clarence Weston Hansell, British Patent 295,601, "Improvements in or relating to means for transmitting radiant energy such as light and to apparatus for use therewith," filed Aug. 13, 1928, accepted Feb. 21, 1929 (US Patent 1,751,584, "Picture transmission," filed Aug. 13, 1927, issued Mar. 25, 1930).

37. Lamm, "Biegsame optische Geräte," p. 581.

38. Almost 30 years later, Schindler was in the audience at a small conference where the first successful gastroscope was described. He stood up after the talk, and recalled that Lamm had suggested the idea in 1928 or 1929. "We put bundles of glass fibers together, but we failed. . . . There was not yet any coating and, more important, we lacked the advice of physicists . . ." (Basil I. Hirschowitz et al, "Demonstration of a new gastroscope, the 'fiberscope,' " *Gastroenterology 35* pp. 50–53, July 1953, Schindler quoted in discussion section, p. 52). Unless Schindler had forgotten Gerlach's help, that indicates Lamm worked with the physicist indepen-

dently. Lamm may have turned to Gerlach after early experiments with Schindler failed. Schindler was quite busy with other activities at the time, including perfecting his own semi-rigid gastroscope and conducting the Physicians' Orchestra of Munich (Martin Carey, personal communication). Interestingly, Schindler knew where Lamm was working in the late 1950s, but the two were not in regular contact, and there is no evidence Schindler ever told Lamm about the later successful gastroscope.

39. Lamm, "Biegsame optische Geräte," p. 581.

40. Schindler, whose father was Jewish, spent six months in jail before emigrating to America. Davis, "Rudolf Schindler's role."

41. Rudolf Hecht, telephone interview, Sept. 27, 1994.

42. Michael Lamm, telephone interview, Sept. 24, 1994.

43. Hecht interview; "Dr. H. Lamm," obituary, *Texas Medicine*, Mar. 1975, p. 120.

Chapter 5

1. Date from Lewis Hyde (interview, Feb. 15, 1993), who says van Heel was in Rochester Oct. 18–28, 1951.

2. O'Brien's son, an optical engineer who had a security clearance and sometimes worked with his father, was also present. Brian O'Brien, Jr., telephone interview, Sept. 12, 1994.

3. Willem Brouwer, interview, Jan. 12, 1994.

4. Lewis Hyde, telephone interview, Feb. 28, 1994.

5. Walter P. Siegmund and F. Dow Smith, "Brian O'Brien—pioneer in optics," *Optics & Photonics News*, Mar. 1993, pp. 48–51.

6. Brian O'Brien, Jr., interview, Feb. 4, 1994.

7. I calculated the number assuming light entering at a 5 degree angle to the axis of a 50-micrometer glass fiber with refractive index of 1.5.

8. O'Brien, Jr., interview, Feb. 4, 1994, telephone interview, Apr. 5, 1994.

9. The smaller the refractive-index difference, the steeper the critical angle for total internal reflection. However, even a one percent difference leaves a critical angle of eight degrees in glass, and that was enough for practical purposes.

10. W. S. Stiles and B. H. Crawford, "The luminous efficiency of rays entering the eye pupil at different points," *Proceedings Royal Society of London B112*, pp. 428–450 (1933).

11. Ernst W. von Brücke, "Die physiologische Bedeutung des stabförmigen Körper und der Zwillingszapfen in den Augen der Wirbelthiere" ("The physiological meaning of rods and cones in the eyes of vertebrates"), *Müller's Archiv fur Anatomie und Physiologie 11*, pp. 444–451 (1844), translated by Susanna Lammert. Von Brücke was trying to explain why some animal eyes look bright. He did not cite Colladon's light guiding demonstration, but the timing suggests it could have influenced him.

12. Brian O'Brien, "Vision and resolution in the central retina," *Journal of the Optical Society of America 41*, No. 12, pp. 882–894 (Dec. 1951).

13. "Brian O'Brien, Frederic Ives Medalist for 1951," *Journal of the Optical Society of America 41*, No. 12, pp. 879–881 (Dec. 1951).

14. O'Brien, Jr., interview, Feb. 4, 1994; an engineer, he was visiting his father at the time and had a clearance because he had worked with him on several military programs.

15. H. Møller Hansen, Danish patent application 1094/51, "Flexible picture transport cable," Apr. 11, 1951 (in Danish, translated by Jonathan D. Beard).

16. Olav Hergel, "Geni-og taber" ("A genius and a loser"), *Berlingske Tidende*, Jul. 23, 1989 Section 5, p. 1 translated by Jonathan D. Beard.

17. H. Møller Hansen, private communication, Feb. 10, 1995.

18. H. Møller Hansen, telephone interview, Apr. 3, 1995.

19. Møller Hansen thought the cladding material should have a refractive index much lower than that of glass, and close to the index of air. Solid coatings like the instrument lacquer O'Brien used have an index only slightly lower than glass, so Møller Hansen experimented with oils, which have lower indexes. H. Møller Hansen, letter to author, received July 1996.

20. Møller Hansen, "Flexible picture transport cable."

21. Reuters, "Danish engineer invents 'eye' that can see around corners," *Los Angeles Times*, June 11, 1951. The article clearly demonstrates that Møller Hansen was using optical fibers for imaging, although the terminology is antiquated. It is not clear where else the Reuters article appeared. The wire service was not widely distributed in America at the time, and a search of the *New York Times* failed to find any mention of the invention.

22. Thore Sandell, "Han ser sig Själv I örat," *Teknikens Varld 14/51*, 1951, p. 12.

23. Møller Hansen telephone interview.

24. Møller Hansen telephone interview.

25. Adriaan Walther, "A. C. S. van Heel," obituary, *Journal of the Optical Society of America* 56, pp. 1411–1412 (Oct. 1966).

26. Daniel Boorstin, *The Discoverers* (Vintage, New York, 1985, p. 314).

27. Brouwer interview.

28. Walther, "A. C. S. van Heel."

29. The computer is now in the Delft Museum; Brouwer interview.

30. G. J. Beernink, "The Delft cradle of glass fiber optics" (in Dutch), *Nederlands Tijdschrift voor Fotonica*, Dec. 1981, pp. 7–10, translated by Adriaan Walther.

31. Brouwer interview.

32. Beernink, "The Delft cradle."

33. W. L. Hyde and C. M. Caldwell, "The National Defense Research Council and the Defense Research Physical Laboratory of the Netherlands," Office of Naval Research, London, Technical Report ONRL-99-52, Sept. 22, 1952, p. 3 (originally secret, declassified Jan. 10, 1984).

34. The students were G. J. Beernink and H. de Vries.

35. Zernicke is identified in H. H. Hopkins, letter to Martin C. Carey, Feb. 8, 1989.

36. Beernink, "The Delft cradle."

37. Walter Siegmund, telephone interview, Mar. 1, 1994.

38. Siegmund and Smith, "Brian O'Brien."

39. Brian O'Brien, letter to Martin Carey, May 2, 1990.

40. F. Dow Smith, telephone interview, Apr. 4, 1994.

41. Walter Siegmund, interview June 29, 1983.

42. Milton Silverman, "The man with the invisible light," *Saturday Evening Post*, Sept. 14, 1946, p. 22.

43. O'Brien, Jr., interview, Feb. 4, 1994.

44. For his description, see O'Brien, "Vision and resolution."

45. Walter Siegmund, telephone interview, Jan. 4, 1994.

46. Robert E. Hopkins, telephone interview, May 29, 1996.

47. Lee DuBridge, who went on to become president of Caltech and President Nixon's science adviser, is quoted in Silverman, "The man with the invisible light."

48. For more details on the Todd-AO affair, see Jeff Hecht, "The amazing optical adventures of Todd-AO," *Optics & Photonics News*, Oct. 1996.

49. Walter Siegmund, telephone interview, Apr. 7, 1994.

50. "Todd, Mike," *Current Biography 1955 (H. W. Wilson* New York) pp. 608–610), quoting John Chapman, *Colliers*, May 12, 1945.

51. Michael Todd, Jr., and Susan McCarthy Todd, *A Valuable Property: The Life Story of Michael Todd* (Arbor House, New York, 1983, p. 244).

52. David A. Cook, *A History of Narrative Film* (Norton, New York, 1981, p. 415).

53. Todd, Jr., and McCarthy Todd, *A Valuable Property* (p. 245).

54. O'Brien, Jr., interview, Feb. 4, 1994.

55. Ibid.

56. Walter Siegmund telephone interview, Jan. 5, 1994. W. Lewis Hyde also recalls seeing a letter from van Heel to O'Brien, but O'Brien's son doubts his father received one.

57. Fluent in English, van Heel probably wrote the English abstract so the words are his. A. C. S. van Heel, "Optische afbeelding zonder lenzen of Abfeeldingsspielgels," *De Ingenieur* No. 24, June 12, 1953; other information based on English translation by Bill Ornstein for American Optical, dated Feb. 23, 1954.

58. A plastic fiber with refractive index of 1.52 was coated with another plastic having an index of 1.47 and painted with black lacquer to block light leakage between fibers.

59. Van Heel lists Willem Brouwer, G. J. Beernink, and H. de Vries as the people who performed the experiments.

60. A. C. S. van Heel, "A new method of transporting optical images without aberrations," *Nature 173*, p. 39 (Jan. 2, 1954).

61. Brouwer interview.

62. H. H. Hopkins, telephone interview, Mar. 1, 1994.

63. Kelvin Hopkins, interview, Dec. 3, 1994.

64. Hopkins credits Hugh Gainsborough of St. George's Hospital in London with asking the question; H. H. Hopkins, letter to the editor of *Photonics Spectra*, Aug. 26, 1982 [unedited version given to me by L. Hyde; edited version appeared in November 1982 *Photonics Spectra*].

65. A search of the major London papers failed to find any mention of Møller Hansen in May or June 1951. Nikki Smith, letter to author, Oct. 14, 1995.

66. H. H. Hopkins, transcript of interview by Dr. Martin Gordon, New York City, May 19, 1981, p. 1.

67. H. H. Hopkins, letter to *Photonics Spectra*.

68. Narinder S. Kapany, telephone interview, Feb. 13, 1996.

69. H. H. Hopkins telephone interview.

70. H. H. Hopkins and N. S. Kapany, "A flexible fibrescope, using static scanning," *Nature 173*, pp. 39–41 (Jan. 2, 1954).

71. Willem Brouwer, note to author, Feb. 24, 1995.

72. *Nature*, letter to author, Apr. 1995.

73. H. H. Hopkins telephone interview.

74. van Heel, "Optische afbeelding."

75. Brian O'Brien, US Patent 2,825,260, "Optical image forming devices," filed Nov. 19, 1954, issued Mar. 4, 1958.

76. H. H. Hopkins interview. In fact, Hansell's patent would have been a more serious obstacle than Baird's.

77. H. H. Hopkins and N. S. Kapany, "Transparent fibres for the transmission of optical images," *Optica Acta 1*, No. 4, pp. 164–170, Feb. 1955.

78. H. H. Hopkins telephone interview.

79. See, e.g., "Narinder Kapany: photonics polymath," *Photonics Spectra*, Aug. 1982, p. 59.

80. H. H. Hopkins, "Fiber optic origins," and reply from N. S. Kapany, *Photonics Spectra*, letters column, Nov. 1982. In his reply, Kapany wrote, "I lay no paternity claim to the field of fiber optics." The published version of Hopkins's letter was edited to moderate its tone.

81. N. S. Kapany, curriculum vitae, p. 1.

82. H. H. Hopkins, telephone interview.

83. Kapany, telephone interview.

Chapter 6

1. Basil I. Hirschowitz, "A personal history of the fiberscope," *Gastroenterology* 76, pp. 864–869 (1979) p. 864.

2. H. H. Hopkins and N. S. Kapany, "A flexible fiberscope, using static scanning," *Nature 173*, pp. 39–41 (Jan. 2, 1954).

3. H. H. Hopkins, telephone interview, Mar. 1, 1994. His feud with Kapany also may have discouraged him.

4. Henk J. Raterink says it stopped by 1956 or 1957 (letter to author, Feb. 14, 1994); Beernink says it stopped in 1954 (G. J. Beernink, "The Delft cradle of glass fiber optics" (in Dutch), *Nederlands Tijdschrift voor Fotonica*, Dec. 1981, pp. 7–10, (translated by Adriaan Walther).

5. Beernink, "The Delft cradle."

6. He was alerted to these papers by his mentor at Central Middlesex Hospital in London, Sir Francis Avery Jones. Avery Jones, letter to Martin Carey, July 21, 1989.

7. Strictly speaking, endoscopes can inspect any part of the body, while gastroscopes are intended only to inspect the stomach. However, gastroscopes are often called endoscopes.

8. Hirschowitz, "A personal history."

9. Lawrence E. Curtiss, telephone interview, Feb. 7, 1994.

10. Curtiss, telephone interview, Feb. 7, 1994.

11. Hirschowitz, "A personal history."

12. Lawrence E. Curtiss, letter to author, Dec. 9, 1994.

13. Mary Jo Peters, telephone interview, Jan. 26, 1996.

14. Credit for the invention of the glass-clad fiber eventually became the subject of a long and bitter court fight, which Curtiss won. Board of Patent Interferences, U.S. Patent Office, *Patent Interference No. 93,002, Norton v. Curtiss: Fiber Optical Components* Final Hearing Jun. 1, 1967, paper No. 101.

15. A. C. S. van Heel, "A new method of transporting optical images without aberrations," *Nature 173*, p. 39 (Jan. 2, 1954)

16. Will Hicks interview, Feb. 4, 1994.

17. Ibid.

18. Court papers indicate Hicks continued some plastic-cladding experiments into 1957. (Board of Patent Interferences, *Norton v. Curtiss*).

19. Hicks interview. Although Hicks didn't realize it at the time, a German named Armand Lamesch had patented the idea a decade earlier as a way of strengthening glass fibers. See Armand Lamesch, US Patent 2,313,296; "Fiber or filament of glass," filed Sept. 23, 1937 (in Germany Sept. 30, 1936), issued Mar. 9, 1943.

20. Curtiss telephone interview. Feb. 7, 1994.

21. Lawrence E. Curtiss, interview, Jan. 20, 1995.

22. Curtiss letter; Hirschowitz, "A personal history."

23. Basil I. Hirschowitz, US Patent 3,010,357, "Flexible light transmitting tube," filed Dec. 28, 1956; issued Nov. 28, 1961.

24. Lawrence E. Curtiss, US Patent 3,589,793, "Glass fiber optical devices," filed May 6, 1957, issued June 29, 1971. Norton and American Optical bitterly fought this patent in court, delaying it for well over a decade and setting a record for most voluminous patent litigation!

25. Hirschowitz, "A personal history."

26. Curtiss telephone interview, Feb. 7, 1994.

27. Avery Jones letter.

28. Hirschowitz had to settle for a post at the less prestigious University of Alabama at Huntsville. Martin Carey, telephone interview, May 7, 1996.

29. That instrument is now at the Smithsonian Institution. Basil I. Hirschowitz, "Demonstration of a new gastroscope, the 'fiberscope,' " *Gastroenterology 35*, pp. 50–53 (July 1958).

30. Hirschowitz, "A personal history." p. 867.

31. Hirschowitz, "Demonstration."

32. The exact timing was the subject of lengthy patent litigation when American Optical tried unsuccessfully to block Curtiss's patent. Although Norton claimed to have made his suggestion in October 1956, the court ruled he could not document invention before early 1957. See Board of Patent Interferences, *Norton v. Curtiss*.

33. Hicks interview.

34. Ibid.

35. Bill Wetherell, conversation, January or February 1994.

36. Robert Greenler, telephone interview, Sept. 18, 1995.

37. The series started with Narinder S. Kapany, "Fiber optics. Part I, Optical properties of certain dielectric cylinders," *Journal of the Optical Society of America 47*, pp. 413–422 (May 1957), and eventually reached part XI in 1965.

38. A bibliography published when he won the Ives award lists 51 papers, but some were abstracts or co-authored and many were published in the 1930s. "Brian O'Brien, Frederic Ives Medalist for 1951," *Journal of the Optical Society of America 41*, No. 12, pp. 879–881 (Dec. 1951). People who knew him later recall he never got around to publishing much.

39. Walter Siegmund, interview, June 29, 1993.

40. Lawrence E. Curtiss, telephone interview, Aug. 6, 1997.

41. Figures derived by author from Robert J. Potter and Cecelia E. Beasor, "The history and evolution of fiber optics," paper presented at the SPIE Fiber Optics Seminar in Depth, Apr. 29, 1968, Baltimore.

42. Narinder S. Kapany, "Fiber optics," *Scientific American 203*, No. 5, pp. 72–81 (Nov. 1960).

43. Narinder S. Kapany, *Fiber Optics: Principles and Applications* (Academic Press, New York, 1967).

44. N. S. Kapany, curriculum vitae, p. 1.

45. Numbers compiled from *2500 Fiber Optics Patent Abstracts 1881–1979* (Patent Data Publications, Wheaton, Ill., 1980).

46. Narinder S. Kapany telephone interview, Feb. 13, 1996.

47. Walter Siegmund, telephone interview, Aug. 1, 1996.

48. Curtiss telephone interview, Feb. 7, 1994.

49. Hirschowitz, "A personal history," p. 867.

50. He spent 22 years at American Cystoscope, leaving only in 1982, a year after American Hospital Supply bought the company.

51. Basil I. Hirschowitz, "Development and application of endoscopy," *Gastroenterology 104*, pp. 337–342 (1993).

52. Hicks interview.

53. William Gardner, telephone interview, Jan. 19, 1996.

54. Jeofry S. Courtney-Pratt, a British specialist in high-speed photography, came up with the same idea independently, while trying to get enough light to photograph the early stages of nitroglycerine explosions. He promptly filed for British and American patents, which he later licensed to American Optical. Jeofry S. Courtney-Pratt, interview, Jan. 14, 1994, and letter to author, Jeofry S. Courtney-Pratt, British Patent 841,200 Dec. 3, 1994; "Improvements in or relating to electronic image forming tubes," filed Sept. 17, 1956; issued July 13, 1960 (US Patent 3,321,658 issued May 1967).

55. Hicks interview.

56. The friends from American Optical were Paul Kiritsy and Chet Thompson. A third coworker, Bart Frey, left with Hicks but stayed less than a year at Mosaic Fabrications. Hicks interview.

57. Ibid.

58. Siegmund telephone interview, Aug. 1, 1996.

59. O'Brien was chairman of NRC's division of physical sciences from 1953 to 1961, chaired the National Academy of Sciences' Air Force Study board from 1962 to 1974, and headed the NASA panel that recommended building the Space Shuttle.

60. The company survived by specializing in sound production, and remains in business.

61. Brian O'Brien, US Patent 2,825,260, "Optical image forming devices," filed Nov. 19, 1954, issued Mar. 4, 1958.

62. Van Heel, "Optische afbeelding zonder lenzen of Abfeeldingsspielgels," reprint from "*De Ingenieur* No. 24, 1953." The issue is dated June 12, 1953, but the only date on the copies sent to AO was "12/6/53"; English translation by Bill Ornstein for American Optical, dated Feb. 23, 1954.

63. Walter Siegmund telephone interview, Jan. 4, 1994.

64. Curtiss interview, Jan. 20, 1995. Inventors sometimes can benefit from a long delay in issuing a patent if their invention comes into wide use. Sales of fiber-optic endoscopes had grown by 1971, so Curtiss, Peters, and Hirschowitz earned much more during the patent's 17-year lifetime than they would have if it had been issued promptly a decade earlier. They would have earned even more if their lawyers had not worded the patent so narrowly that it did not cover the fibers used for communications.

65. Walt Siegmund (personal communication) showed me a photo of Garro-

way taken from the side of the studio. It's eerie to realize how many people saw this window into future technology through morning-bleary eyes and immediately forgot about it.

66. Working under Robert Hopkins, O'Brien's successor at Rochester, Potter analyzed fiber properties in detail and explained some previously puzzling features of fiber transmission. Robert Joseph Potter, *A Theoretical and Experimental Study of Optical Fibers* (University of Rochester, 1960).

67. Siegmund telephone interview, Aug. 1, 1996.

68. Hicks interview.

Chapter 7

1. Alec Harley Reeves, "Future prospects for optical communication," John Logie Baird Memorial Lecture, University of Strathclyde, May 30, 1969.

2. Many years were needed to perfect practical switches after Strowger applied for a patent in 1889, but the simple and durable equipment remained in use for decades. Peter Young, *Power of Speech* (George Allen & Unwin, London, 1983).

3. Irwin Lebow, *Information Highways & Byways* (IEEE Press, New York, 1995).

4. While testing his scheme, Bell heard audio tones that the current had carried through the wires. As a teacher of the deaf, he realized that the wires which could carry tones also could carry speech. Arthur C. Clarke, *How the World Was One: Beyond the Global Village* (Bantam, New York, 1992, p. 112).

5. The modern telephone system carries only frequencies of 300 to 3000 hertz, much smaller than the nominal range of human hearing from 20 to 20,000 hertz but adequate for most people to understand conversations. The limit has been retained from the early days of telephone electronics.

6. They also can be seen as particles, clumps of energy called "photons." Sometimes they behave like waves, other times like particles.

7. The story of the optical telegraph is a fascinating example of the rise and fall of a nineteenth-century technology, and reminds us how fast once-vital systems can be forgotten. For a fascinating account, see Gerard J. Holzmann and Björn Pehrson, *The Early History of Data Networks* (IEEE Computer Society Press, Los Alamitos, Calif., 1995).

8. Robert V. Bruce, *Alexander Graham Bell and the Conquest of Solitude* (Little Brown and Co., Boston, 1973, p. 336).

9. He sketched an optical telephone in one of his copious notebooks but never worked out the details of how one might operate. Neil Baldwin, "The laboratory notebooks of Thomas Edison," *Scientific American*, Oct. 1995, pp. 160–160C.

10. Bruce, *Alexander Graham Bell* (p. 337).

11. Alexander Graham Bell, "The photophone," *Scientific American, Supplement 246*, Sept. 18, 1880, pp. 3921–3923.

12. Forgotten, that is, except by a few curious souls who thought the "photophone" Bell had sealed in the Smithsonian archives might be an early television. They were disappointed when the box was opened after Bell's death.

13. America and Britain tested systems during both world wars; Germany used an infrared system during World War II. See N. C. Beese, "Light sources for optical communication" (*Infrared Physics* 1, pp. 5–6 (1961)) for technical details.

14. Earlier transmitters had worked only sporadically or transmitted only telegraph signals. H. D. Arnold and Lloyd Espenschied, "Transatlantic radio telephony," *Journal American Institute of Electrical Engineers*, Aug. 1923, pp. 815–826.

15. Reflection dropped during daytime, so transmission was possible only at night.

16. Orrin E. Dunlap, Jr., *Radio's 100 Men of Science* (Harper & Brothers Publishers, New York, 1944, pp. 269–272 on Hansell).

17. Clarence W. Hansell, "Radio-relay systems development by the Radio Corporation of America," *Proceedings of the Institute of Radio Engineers*, Mar. 1945, pp. 156–168.

18. Richard J. O'Rorke, Jr., *1990 World's Submarine Telephone Cable Systems* (US Government Printing Office, Washington, D.C., 1991; see pp. 112–113, 123).

19. Murray Ramsay, telephone interview, Sept. 12, 1994.

20. Kenneth W. Cattermole, "A. H. Reeves: the man behind the engineer," *IEE Review*, Nov. 1990, pp. 383–386.

21. Ibid.

22. Richard Epworth, interview, Dec. 2, 1994.

23. His notebooks on the paranormal were so thoroughly mingled with his other work that the management of BNR-Europe inadvertently put one on display in a case at Harlow. This wasn't discovered until I visited in late 1994, when Richard Epworth went to check the notebook and examined its contents. Epworth interview.

24. R. V. Jones, *The Wizard War: British Scientific Intelligence 1939–1945* (Coward, McCann & Geohegan, New York, 1978, p. 276).

25. Reeves experimented with "two-terminal" devices called diodes, where two wires make contact with different parts of a semiconductor chip. Transistors have three terminals, a design that lets a voltage applied to one terminal modulate the current flowing between the other two. Cattermole, "A. H. Reeves."

26. A. G. Fox and W. D. Warters, "Waveguide research," in S. Millman, ed., *A History of Engineering and Science in the Bell System: Vol. 5. Communications Science 1925–1980* (AT&T Bell Laboratories, Indianapolis, 1984, p. 264).

27. Richard Dyott, telephone interview, June 23, 1994.

28. E. F. O'Neill, ed., *A History of Engineering and Science in the Bell System: Vol. 7. Transmission Technology 1925–1975*, (AT&T Bell Laboratories, New York, 1985, pp. 624–627).

29. Clarke actually proposed adding a transmitter to a manned space station, not launching a satellite dedicated only to automatic communications (Arthur C. Clarke, "Extra-terrestrial relays," *Wireless World*, Oct. 1945, pp. 305–308). Four decades later, Clarke remembered that he had mentioned an early version of the idea in a brief letter to the magazine in February of that year, and in a letter that he circulated to friends in May. See Arthur C. Clarke, *Ascent to Orbit* (Wiley-Interscience, New York, 1984). He cites a paper by C. W. Hansell on ground-based radio relays.

30. Pierce published fiction as J. J. Coupling; he remains a member of the Science Fiction and Fantasy Writers of America.

31. Ramsay telephone interview.

32. Antoni E. Karbowiak telephone interview, Feb. 5, 1995.

33. Reeves tested hollow-cathode discharge lamps. W. A. Atherton, "Pioneers 16: Charles Kuen Kao, father of optical communications," *Electronics & Wireless World*, Apr. 1988, pp. 406–407.

34. The others were John Lytollis, Ian Turner, Bernard Fairchild, and Ron Lomax. Murray Ramsay, letter to author, Jan. 12, 1997.

35. Antoni E. Karbowiak, vita and telephone interview, Feb. 5, 1995.

36. George Hockham, interview, Dec. 6, 1994.

37. Norman R. French, US Patent 1,981,999, "Optical telephone system," filed Aug. 20, 1932, issued Nov. 27, 1934.

38. Ray D. Kell and George C. Sziklai, US Patent 2,506,672, "Signal transmission system," filed Oct. 31, 1945, issued May 9, 1950. They envisioned applications in television transmission.

39. Neither Bell Labs nor the David Sarnoff Research Center—the former RCA Laboratories—has any records indicating anything was ever built. The technical difficulties would have been formidable, and signals would not have gone far.

40. This appeared only in an internal memo, cited in R. Kompfner, "Optics at Bell Laboratories—optical communications," *Applied Optics 11*, pp. 2412–2425 (Nov. 1972).

41. C. C. Eaglesfield, "Optical pipeline: a tentative assessment," *Proceedings of the IEE 109B*, pp. 26–32 (Jan. 1962).

42. I could not find any of Wheeler's calculations, and believe none survive. I estimated theoretical transmission of light pipes using his design from standard optical formulas, assuming ideal straight pipes and perfect optical surfaces. Theoretical transmission is better for the metal-coated pipes Eaglesfield proposed than for Wheeler's glass pipes because glancing-angle reflectivity is lower from a glass surface than from an ideal metal reflector.

43. R. W. Lomax, B. D. Fairchild, and G. I. Turner, "Attenuation measurement of an optical pipeline," Standard Telecommunications Laboratories, report 1001/162/0051, undated but probably 1962.

44. The Czechs measured loss of 57 decibels per kilometer (about 90 decibels per mile) in 100 meters of aluminum-coated glass pipe hung from the ceiling in halls of a reinforced concrete building, but even they concluded the idea was impractical. M. Prochazka, J. Pachman, and J. Muzik, "Experimental investigation of a pipeline for optical communications," *Electronics Letters 3*, pp. 73–74 (Feb. 1967).

45. Murray Ramsay, interview, Dec. 2, 1994.

46. C. H. Chandler, "An investigation of dielectric rod as wave guide," *Journal of Applied Physics 20*, pp. 1188–1192 (Dec. 1949).

47. Elias Snitzer interview, Mar. 4, 1994. The firing led to a legal battle between the American Association of University Professors and Lowell, now the University of Massachusetts at Lowell.

48. Snitzer described the results at a Rochester optics conference in mid-1960, then published two detailed studies: Elias Snitzer, "Cylindrical dielectric waveguide modes," *Journal of the Optical Society of America 51*, No. 5, pp. 491–498 (May 1961); Elias Snitzer and Harold Osterberg, "Observed dielectric waveguide modes in the visible spectrum," *Journal of the Optical Society of America 51*, No. 5, pp. 499–505 (May 1961).

49. John W. Hicks, Jr., Elias Snitzer, and Harold Osterberg, US Patent 3, 157,726, "Optical energy transmitting devices and method of making same," filed Mar. 1, 1960, issued Nov. 17, 1964.

50. Narinder S. Kapany, "Fiber optics," *Scientific American 203*, No. 5, pp. 72–81 (Nov. 1960).

51. Will Hicks, interview, Feb. 4, 1994.

Chapter 8

1. "Light amplifier extends spectrum," *Electronics*, July 22, 1960, p. 43.

2. Strictly speaking, some other sources generate light that is coherent over very short distances, and some lasers are not very coherent. However, lasers remain the most practical sources of coherent light.

3. Internal memo by W. A. Tyrell, quoted in Rudolf Kompfner, "Optics at Bell Laboratories—optical communications," *Applied Optics 11*, No. 11, pp. 2412–2425 (Nov. 1972).

4. For more details, see Jeff Hecht, *Laser Pioneers* (Academic Press, Boston, 1991).

5. Townes grew up in Greenville, South Carolina, the home town of Will Hicks, and like Hicks attended Furman University before leaving for graduate school. At Furman, Hicks was a classmate of Townes's younger sister, Aurelia, who later married Schawlow. While they share sharp minds, fertile technical imaginations, and the same small-city southern background, the two have contrasting styles. Townes is a gentleman of the old school, scholarly and dignified; Hicks is casual and folksy, with an unmistakably southern tone to his voice. Townes has spent his life in academia, publishing widely in scholarly journals and serving on many government science-policy panels. Hicks has spent most of his life as an entrepreneur and inventor, collecting over a hundred patents and starting a series of companies.

6. Arthur L. Schawlow and Charles H. Townes, "Infrared and optical masers," *Physical Review 112*, p. 1940 (1958).

7. The story of laser development is a rich tale in itself. For highlights, see Hecht, *Laser Pioneers*, and Joan Bromberg, *The Laser in America* (MIT Press, Cambridge, 1991).

8. The journal's eminent and opinionated founding editor, Samuel Goudsmit, was annoyed by a series of papers reporting minor advances, and Maiman had published a related paper in the journal just weeks before. Hecht, *Laser Pioneers*; Bromberg, *The Laser in America*" (p. 92).

9. Theodore H. Maiman, "Stimulated optical radiation in ruby," *Nature 187*, p. 493 (Aug. 6, 1960).

10. "Light amplifier."

11. "Scientists demonstrate optical maser," *Electronics*, Oct. 21, 1960, p. 38. (Bell Labs stubbornly continued to call the laser an "optical maser" although others quickly adopted the name laser.)

12. R. N. Schwartz and Charles H. Townes, "Interstellar and interplanetary communication by optical masers," *Nature 190*, pp. 205–208 (Apr. 15, 1961).

13. Hecht, *Laser Pioneers* (interview with Ali Javan).

14. Richard Smith, telephone interview, June 11, 1997.

15. Ivan Kaminow, interview, July 19, 1995.

16. James P. Gordon, telephone interview, July 5, 1994.

17. S. E. Miller and L. C. Tillotson, "Optical transmission research," *Applied Optics* 5, pp. 1538–1549 (Oct. 1966).

18. Strictly speaking, by over 60 decibels (decibels are explained on pages 115–116). Miller and Tillotson, "Optical transmission research."

19. R. Meredith, Royal Radar and Signals Establishment, "Unguided optical propagation in the atmosphere and under-sea," paper presented at the Conference

on Lasers and Their Applications, Sept. 1964, Institution of Electrical Engineers, London.

20. Kaminow interview.

21. Jena was in the Russian-occupied sector that became East Germany. Felix Schwering, telephone interview, May 15, 1996.

22. Georg Goubau, Felix Schwering, J. Robert Christian, and James W. Mink, "Comments on: optics at Bell Laboratories—optical communications," *Applied Optics 12*, No. 5, pp. 936–937 (May 1973). This responds to a long paper by Rudolf Kompfner that implied Bell Labs was first; in a reply to Goubau's letter, Kompfner denies he claimed priority.

23. G. Goubau and J. R. Christian, "Investigations on a beam waveguide for optical frequencies," paper presented at the Army Science Conference, June 1962, copy supplied by R. Epworth.

24. Ibid.

25. Georg Goubau and J. R. Christian, "Loss measurements with a beam waveguide for long distance transmission at optical frequencies," *Proceedings of the IEEE*, Dec. 1964, p. 1739.

26. Georg Goubau, "Lenses guide optical frequencies to low-loss transmission," *Electronics*, May 16, 1966, pp. 83–89.

27. Paul Lazay, interview, Nov. 12, 1996.

28. Kompfner "was a seed of many many thoughts that people thought were their own but in reality had grown out of their interactions with this extraordinary man." Enrique (Henry) Marcatili, telephone interview, July 19, 1994.

29. E. F. O'Neill, ed., *A History of Engineering and Science in the Bell System: Vol. 7. Transmission Technology 1925–1975* (AT&T Bell Labs, 1985, New York, pp. 623–647).

30. Detlef Gloge, telephone interview, June 27, 1994.

31. Kompfner, "Optics." It isn't clear exactly what those assumptions were because the numbers Kompfner published on p. 2414 are wrong. His paper gives loss of two percent or "1/13 decibel" for each of a series of lenses spaced 100 meters apart, and assumes a repeater is needed to amplify the signal after total loss reaches 50 decibels. However, those numbers would yield a repeater spacing of 65 kilometers, not 650 kilometers. Reaching Kompfner's goal of 650 kilometer transmission between repeaters would have required loss a factor of ten lower, or 0.2 percent (about 0.0087 decibel) per lens.

32. The loss estimate comes from Takanori Okoshi, *Optical Fibers* (Academic Press, New York, 1982, pp. 5–6), citing D. Gloge and W. H. Steier, "Pulse shuttling in a half-mile optical lens guide," *Bell System Technical Journal 47*, pp. 767–782 (1968). Gloge and Steier did not directly measure or indicate loss in their paper.

33. D. Gloge and D. Weiner, "The capacity of multiple beam waveguides and optical delay lines," *Bell System Technical Journal 47*, pp. 2095–2109 (1968).

34. Okoshi, *Optical Fibers* (pp. 6–7).

35. The manager was Andy Hutson. Dwight W. Berreman, telephone interview, Dec. 8, 1995.

36. Dwight W. Berreman, "A lens or light guide using convectively distorted thermal gradients in gases," *Bell System Technical Journal 43*, pp. 1469–1475. (1964)

37. Berreman telephone interview.

38. Both papers appeared in the same journal issue: Dwight W. Berreman,

"Growth of oscillations around the irregular wavy axis of a light guide," *Bell System Technical Journal 44*, pp. 2117–2132 (Nov. 1965); Dietrich Marcuse, "Properties of periodic gas lenses," *Bell System Technical Journal 44* pp. 2083–2116, (Nov. 1965).

39. Peter Kaiser, telephone interview, May 17, 1995.

40. Irwin Dorros, "The Picturephone system: the network," *Bell System Technical Journal 50*, No. 2, pp. 221–233 (Feb. 1971); this entire thick issue of Bell's official technical journal is devoted to plans for the Picturephone, which managers thought would shape the phone system.

41. Marcatili telephone interview; also A. G. Fox and Ivan P. Kaminow "Lightwave communications," in S. Millman ed., *A History of Engineering and Science in the Bell System: Vol. 5 Communications Sciences* (AT&T, Bell Laboratories, Indianapolis, 1984, pp. 282–283).

42. Stewart E. Miller, "Communication by laser," *Scientific American 214*, No. 1, pp. 19–27 (Jan. 1966).

43. Kompfner, "Optics."

44. Probably Narinder S. Kapany and James Burke, "Fiber optics. Part IX, waveguide effects," *Journal of the Optical Society of America 51*, No. 10, pp. 1067–1078 (Oct. 1961). This appeared five months after Elias Snitzer published a paper on waveguide effects in the same journal.

45. Jeofry S. Courtney-Pratt, telephone interview, Jan. 5, 1995.

46. Kaminow interview. Kaminow says he suggested optical fibers to Kompfner; other people must have done the same.

47. Rudolf Kompfner, "Optical communications," *Science 150*, pp. 149–155 (Oct. 8, 1965).

48. Miller and Tillotson, "Optical transmission research."

49. Stewart E. Miller, "Possible broadband systems of the future," *Signal*, Aug. 1966, pp. 52–56.

50. Will Hicks interview, Feb. 4, 1994.

51. Ibid.

52. Lawrence Curtiss, interview, Jan. 20, 1995.

53. Heinrich Lamm, letter to Norbone B. Powell, Aug. 26, 1965, copy supplied by Michael Lamm. Although Rudolf Schindler knew about fiber-optic endoscopes eight years earlier, and knew where Lamm was working, there is no evidence Schindler told Lamm about the development.

54. Per telephone interviews with the Hansell family, who were surprised to hear about his fiber-optics patent.

55. "Clarence W. Hansell, 69, dead; electrical engineer for RCA," *New York Times*, Oct. 24, 1967.

56. Richard Cunningham, "Applications pioneer interview: Narinder Kapany," *Lasers & Applications*, July 1986, pp. 65–68.

57. "2 ex-giants stop making lasers," *Laser Focus*, Mar. 1973, pp. 30–32.

58. Kapany was forced out as chief executive as the company drifted toward bankruptcy. By 1974, Optics Technology had vanished, one of several early laser companies hit hard by the slumping economy ("2 ex-giants"). The company never did venture into fibers for communications, which Kapany did not mention in a 1969, interview. "Fiber optics, growth field of the 1970s: an interview with Narinder S. Kapany," *Laser Focus*, July 1969, pp. 30–31.

Chapter 9

1. A. E. Karbowiak, "Guided propagation at optical frequencies," paper 33 in *Conference on Lasers and Their Applications* (Institution of Electrical Engineers, London, 1964).

2. "There could, however, be some future in large-capacity systems using laser communication between balloons suspended above the weather-affected layer of the troposphere. The service reliability of such systems could be very good indeed." A. E. Karbowiak, "Communication systems in the visible and infrared spectra: present and future," paper 30 in *Conference on Lasers and Their Applications*. (p. 30–5).

3. R. W. Lomax, B. D. Fairchild, and G. I. Turner, "Attenuation measurement of an optical pipeline," Standard Telecommunication Laboratories Ltd., report 1001/162/00F1, updated (probably 1962).

4. Charles K. Kao, interview, Apr. 9, 1995.

5. Karbowiak, "Guided propagation."

6. Antoni E. Karbowiak, telephone interview, Feb. 5, 1995.

7. K. C. Kao, "The theory and practice of quasi-optical waveguide components," *Symposium Proceedings on Quasi-Optics* (Polytechnic Institute of Brooklyn, 1964, pp. 497–514).

8. The company's French name translates roughly as General Company for Telegraphy Without Wires; like RCA, it specialized in radio transmission.

9. Eric Spitz, telephone interview, Feb. 14, 1996; Alain Werts, telephone interview, Feb. 13, 1996.

10. J. C. Simon and E. Spitz, "Propagation guidee de lumiere coherente" (The guided propagation of coherent light), *Communications a la Societe Francaise de Physique 24*, No. 2, pp. 149–149 (1963), translation by Jean Louis Trudel; Karbowiak, "Guided propagation."

11. C. H. Chandler, "An investigation of dielectric rod as wave guide," *Journal of Applied Physics 20*, pp. 1188–1192 (Dec. 1949).

12. George Hockham, interview, Dec. 6, 1994.

13. In microwave terms, this means the outer material had to have a smaller dielectric constant.

14. Karbowiak, "Guided propagation."

15. Werts telephone interview; Spitz telephone interview. Spitz says the project started in 1963; Werts remembered 1964.

16. Simon and Spitz, "Propagation guide."

17. A. E. Karbowiak, "New type of waveguide for light and infrared waves," *Electronics Letters 1*, No. 2, pp. 47–48 (Apr. 1965).

18. This is often called the near to mid-infrared range. Visible light has wavelengths of 0.4 to 0.7 micrometer. Antoni E. Karbowiak, "Optical waveguides," in Leo Young, ed., *Advances in Microwaves* (Vol. 1; Academic Press, New York, 1966, pp. 75–113).

19. Karbowiak telephone interview.

20. Antoni E. Karbowiak, *Trunk Waveguide Communication* (Chapman and Hall, London, 1965).

21. Karbowiak telephone interview; Hockham telephone interview.

22. Family names are written first in Chinese; his family name is Kao.

23. Kao interview.

24. Hockham interview.

25. Karbowiak "Optical waveguides" (see note p. 103).

26. Charles K. Kao, "Historical notes on optical fiber communications," unpublished manuscript, p. 13.

27. Kao interview.

28. Hockham interview.

29. He attributed the estimate to a Mme. A. Winter, evidently a glass specialist. Kao, "Historical notes" (p. 11).

30. Robert D. Maurer, "Light scattering by glasses," *Journal of Chemical Physics 25*, p. 1206 (1956).

31. Kao, "Historical notes" (p. 13).

32. Ibid. (p. 16).

33. Murray Ramsay, interview, Dec. 2, 1994; Ramsay, letter to author, Jan. 12, 1997.

34. Kao, "Historical notes" (p. 16).

35. K. C. Kao and G. A. Hockham, "Dielectric-fibre surface waveguides for optical frequencies," *Proceedings IEE 113* pp. 1151–1158, July 1966.

36. Standard Telecommunication Laboratories, "STL develops techniques aimed at communication by guided light," press release dated Jan. 27, 1966.

37. "Optical waveguides: single-mode propagation in optical fibers and films," *Wireless World 72*, p. 194 (Apr. 1966).

38. ITT Communications also put out a press release, and the editor evidently did not know ITT owned STL, so he thought that two companies had developed the same, technology separately. "Two British firms build tiny optical waveguides," *Laser Focus 1*, Apr. 1966, pp. 3–4.

39. Searches by author.

40. Stewart E. Miller, "Communication by laser," *Scientific American 214*, No. 1, pp. 19–27 (Jan. 1966).

41. Alain Werts, "Propagation de la lumiere coherente dans les fibres optiques," *L'Onde Electronique 46*, pp. 967–980 (Sept. 1966), cites a paper by Kao, "Dielectric fiber surface waveguides for optical frequencies," evidently an internal STL paper or a preprint. Kao used the same title for both his Delft paper and the paper in *Proceedings of the IEE*.

42. Werts telephone interview.

43. Werts, "Propagation de la lumiere." The Journal does not indicate a submission date, but it probably was much later than the Kao and Hockham paper. One reason the paper has been largely forgotten was its publication in French.

44. Spitz telephone interview.

Chapter 10

1. Charles K. Kao, interview, Apr. 9, 1995.

2. The light source he spoke of was the semiconductor laser. Charles K. Kao, telephone interview, July 14, 1994.

3. Richard B. Dyott, telephone interview, Jan. 11, 1994.

4. Richard B. Dyott, "Some memories of the early years with optical fibers at the British Post Office: a personal account," *IEE Proceedings 133 J*, No. 3, pp. 199–201 (June 1986); also telephone interview, June 23, 1994.

5. Jack Tillman, telephone interview, Feb. 9, 1995.

6. John Bray, telephone interview, June 22, 1994.

7. Prestel was the archetypal videotex system, which used electronic adapters

that provided an interface between phone lines (hooked to remote computers) and home television screens. Videotex arrived at the same time as the personal computer, which rendered it obsolete. Ironically, some features of the graphically oriented World Wide Web clearly derive from videotex.

8. "It failed to take off so decisively," recalls Roger Heckingbottom, interview, Dec. 1, 1994.

9. Charles Kao, "Historical notes on optical fiber communications," unpublished manuscript, with notes.

10. Full name provided by Institution of Electrical Engineers, Archives dept.

11. Don Williams (telephone interview, Sept. 9, 1994) said he "looked after all sorts of odd topics."

12. Clive Day, interview, Dec. 1, 1994. The project was a novel scheme for storing data at the sites of atomic dislocations in crystals.

13. Dyott, telephone interview, Jan. 23, 1995; Clive Day, interview, Dec. 1, 1994.

14. The Post Office lab took pride in beating the better-funded Bell Labs. The property is called Faraday rotation, in which applying a magnetic field to a material called a ferrite rotates the plane of polarization of the microwaves (information provided by IEE Archives dept.).

15. Heckingbottom, interview.

16. Charles Sandbank telephone interview, Sept. 7, 1994.

17. Robert Maurer, interview, Mar. 7, 1995.

18. Williams telephone interview.

19. W. A. Gambling, "Optical fibres: the Southampton scene," *IEE Proceedings 133J* No. 3, pp. 205–210 (June 1986).

20. Kao, "Historical notes" (p. 20).

21. Martin Chown, interview, Dec. 2, 1994, and telephone interview Nov. 15, 1994.

22. Eric Spitz, telephone interview, Feb. 14, 1996. Spitz thinks the visit came in 1964 or 1965, but the involvement of Roberts indicates it could not have been before 1966.

23. Shojiro Kawakami, "The early days of optical fiber communication research," *Optoelectronics—Devices and Technologies 1*, No. 2, pp. 125–136 (Dec. 1986).

24. Miller's idea was a theoretical extension of work on millimeter and confocal waveguides, which also applied to fibers. From a theoretical standpoint, a graded-index waveguide is merely the limiting case of infinitely thin lenses spaced infinitesimally close together. Stewart E. Miller, US Patent 3,434,774, "Waveguide for millimeter and optical waves," filed Feb. 2, 1965, as division of application filed Feb. 25, 1964; issued Mar. 25, 1969.

25. Kao, "Historical notes" (pp. 20–21).

26. Richard B. Dyott, "Some memories."

27. Heckingbottom, interview, Dec. 1, 1994.

28. Richard B. Dyott, "Some memories."

29. Ibid.

30. Dyott telephone interview, Jan. 23, 1995.

31. Chown interview; Laszlo Solymar, personal communication. The precise chain of command at STL varied over the years when fibers were being developed, and Reeves functioned more as a technical consultant than a line manager, es-

pecially as he neared retirement. Managers not responsible for the fiber program tended to be the most skeptical.

32. Charles Sandbank, interview, Dec. 6, 1994; this probably was the 1967 budget, but he was not certain.

33. Murray Ramsay, letter to author, Jan. 12, 1997.

34. He used the fiber work in his doctoral thesis. Hockham ran the antenna group until he left STL at the end of 1974. George Hockham, interview, Dec. 6, 1994.

35. Richard Epworth, interview, Dec. 2, 1994.

36. K. C. Kao and Turner W. Davies, "Spectrophotometric studies of ultra low loss optical glasses 1: single-beam method," *Journal of Scientific Instruments 1 (Journal of Physics E)*, pp. 1063–1068 (Nov. 1968).

37. Kao, "Historical notes."

38. M. W. Jones and K. C. Kao, "Spectrophotometric studies of ultra low loss optical glasses 2: double-beam method," *Journal of Scientific Instruments 2 (Journal of Physics E)*, pp. 331–335 (Apr. 1969).

39. Murray Ramsay, interview, Dec. 2, 1994.

40. David Pearson, telephone interview, Feb. 16, 1996.

41. Robert C. Weast, ed. *CRC Handbook of Chemistry and Physics* 62nd ed., (CRC Press, Boca Raton, Fla., 1981) other sources say higher temperatures, up to 2000°C.

42. Ramsay interview.

43. Klaus Loewenstein, technical director of Deeglass Fibres Ltd. in Camberley, a small town west of London, suggested the idea to Dyott when he visited the plant on Aug. 30, 1968.

44. Armand Lamesch, US Patent 2,313,296, "Fiber or filament of glass," filed Sept. 23, 1937, issued Mar. 9, 1943 (German patent filed Sept. 30, 1936).

45. Will Hicks, interview, Feb. 4, 1994.

46. Dyott telephone interviews, Jan. 23, 1995, and Feb. 2, 1996. He is not sure if Roberts ever knew Newns used sugar.

47. Clive Day, interview, Dec. 1, 1994.

48. Pearson added William French and Eric Rawson from Murray Hill, and Art Tynes and D. L. Bisbee from Crawford Hill.

49. Pearson telephone interview.

50. They made the measurements in different units, converted by author. A. R. Tynes, A. David Pearson, and D. L. Bisbee, "Loss mechanisms and measurements in clad glass fibers and bulk glass," *Journal of the Optical Society of America 61*, pp. 143–153 (Feb. 1971).

51. John Noble Wilford, "Soviet scientist reports successful trials in using laser light beams for telephoning," *New York Times*, Mar. 29, 1970, p. 53.

52. Kawakami, "The early days."

53. Teiji Uchida et al. "Optical characteristics of a light-focusing fiber guide and its applications," *IEEE Journal of Quantum Electronics QE-6*, pp. 606–612 (Oct. 1970).

54. Ramsay letter. It was later upgraded and operated at 100 million bits per second. Chown interview, Dec. 2, 1994.

55. Ramsay and Chown interviews, Dec. 2, 1994.

56. Ibid.

57. Kao, "Historical notes."

58. Sandbank interview, Dec. 6, 1994.

59. See Detlef Gloge, "Optical waveguide transmission," *Proceedings of the IEEE 58*, pp. 1513–1522 (Oct. 1970). He cites D. Marcuse and R. M. Derosier, "Mode conversion caused by diameter changes of a round dielectric waveguide," *Bell System Technical Journal 48*, pp. 2103–2132 (Sept. 1969).

60. Stewart E. Miller, "Optical communications research progress," *Science 170*, pp. 685–695 (Nov. 13, 1970).

Chapter 11

1. Robert Maurer, interview, Mar. 7, 1995.
2. Ibid.
3. David E. Fisher and Marshall Jon Fisher, *Tube: The Invention of Television* (Counterpoint, Washington, D.C., 1996).
4. Peter Schultz, interview, May 17, 1995.
5. Robert Kammer Cassetti, Corning Inc., "Pure silica glass: vitreous silica," memo dated Oct. 2, 1996, based on interviews with Hyde and others.
6. David Pearson, telephone interview, Feb. 16, 1996.
7. Schultz interview; Cassetti, "Pure silica glass."
8. "Fiber Optics: Corning enters intriguing field," *Corning Gaffer*, June 1965, pp. 1–2.
9. Charles J. Lucy, telephone interview, Dec. 3, 1996.
10. Maurer interview.
11. Ibid.
12. Clifton Fonstad, telephone interview, June 5, 1996.
13. Donald Keck, interview, Mar. 7, 1995.
14. Kriedl was an unusual character, an Austrian who had run a family glass factory in Czechoslovakia and married a princess, only to barely escape with a suitcase (and his wife) from the invading Nazis in 1938. He joined Bausch and Lomb, one of the old giants of traditional American optics, in time to head optical glass development during World War II. Full of energy when he retired from the company at 60, he was hired by Rutgers as the first glass specialist in the university's ceramics program (Schultz, interview).
15. Schultz, interview.
16. Keck, interview.
17. Felix Kapron, interview, June 12, 1995.
18. Keck, interview.
19. Lucy, telephone interview. Corning had a philosophy of forming industrial partnerships with companies with expertise in areas that complemented Corning's glass technology. Fiber-optic communication systems would require cabled fiber, so cable companies were logical partners.
20. Keck interview.
21. Further heating crystallized the whole fiber, and crystalline quartz is neither as clear nor as flexible as disordered glass. Keck interview.
22. Ibid.
23. Ibid.
24. Maurer interview.
25. Keck interview.
26. Jack Cook, telephone interview, Jul. 9, 1997.
27. The published abstract does not mention the low-loss fiber. The sentence reporting the record low loss is at the end of the second paragraph in the published

paper. F. P. Kapron, D. B. Keck, and R. D. Maurer, "Radiation losses in glass optical waveguides," *Applied Physics Letters 17* pp. 423–425, Nov. 15, 1970, p. 423.

28. Keck interview.

29. F. P. Kapron, D. B. Keck, and R. D. Maurer, "Radiation losses in glass optical waveguides," *Conference on Trunk Telecommunications by Guided Waves* (Institution of Electrical Engineers, London, 1970, pp. 148–153).

30. Kapron, et al "Radiation losses in glass optical waveguides."

31. Harold E. M. Barlow, "Introductory address," *Conference on Trunk Telecommunications* (pp. xi–xiv).

32. Maurer interview.

33. Cook telephone interview.

34. Richard Dyott, telephone interview, Jan. 11, 1994.

35. Lucy telephone interview.

36. Maurer interview. Although Hopkins and Kapany did important pioneering work in England, the country never developed its own medical-fiber industry.

37. Richard B. Dyott, "Some memories of the early years with optical fibres at the British Post Office: a personal account," *IEE Proceedings 133 J*, No. 3, pp. 199–201 (June 1986).

38. Dyott telephone interview.

39. Ibid.

40. Murray Ramsay, interview, Dec. 2, 1994.

41. Corning's low-loss fiber was a singularly unreported breakthrough. The only contemporary account I could find was in *New Scientist* ("Clearer future for fibre optic communications," *New Scientist*, 8 Oct. 1970, p. 77). The magazine three months earlier had published a feature on fiber communications written by an STL engineer (Martin Chown, "Light: the long-distance answer," *New Scientist*, 16 July 1970, pp. 14–19 in supplement "Telecommunications: the expanding Spectrum"). Even major trade magazines such as *Electronics* and *Industrial Research* ignored the breakthrough for months, although they reported lesser developments in late 1970 and early 1971. ("Glass laser fibers help transmit and amplify beams," *Electronics* Sep. 28, 1970, pp. 129–130; "Transmission system uses coated glass for laser beam," *Electronics* Nov. 9, 1970, "International Newsletter" page; "British use R&D on fiber telecom link," *Industrial Research*, Jan. 1971, p. 33). An important reason for the oversight probably was that Corning did not issue a press release.

42. Schultz, interview.

43. Pearson, telephone interview.

44. The idea that glass flows very slowly at normal temperatures is an urban myth of science. Although glass panes in some medieval cathedral windows may be thicker at the bottom than at the top, that is an artifact of old glass-making techniques which did not yield flat sheets. Much older glass artifacts show no evidence of flow, and measurements at higher temperatures indicate glass should be stable at room temperature for the life of the universe. (Edgar D. Zanotto, "Do cathedral glasses flow?" *American Journal of Physics 66*, No. 5, pp. 392–395, May 1998).

45. Keck interview.

46. Schultz interview.

47. Ibid.

48. D. B. Keck, R. D. Maurer, and P. C. Schultz, "On the ultimate lower limit

of attenuation in glass optical waveguides," *Applied Physics Letters 22*, No. 7, pp. 307–309 (Apr. 1, 1973).

Chapter 12

1. David Newman, interview, Dec. 1, 1994.

2. The name diode comes from a two-terminal vacuum tube, which like a semiconductor diode conducts current in only one direction.

3. Applying the voltage in the direction that makes the current flow is called forward-biasing the diode; applying it in the opposite direction is called reverse-biasing. A high enough reverse-bias voltage will tear valance electrons free from atoms, making current flow in a process called "breakdown," but we won't worry about that.

4. Robert N. Hall, quoted in Jeff Hecht, *Laser Pioneers*, (Academic Press, Boston, 1991, p. 181).

5. Strictly speaking, the second law holds that in a closed system, any irreversible process must inevitably increase the entropy, a measure of disorder. Real processes are irreversible, so entropy invariably increases. The quote is from Robert Rediker, telephone interview, Dec. 11, 1996. Many details also are in Robert H. Rediker, "Research at Lincoln Laboratory leading up to the development of the injection laser in 1962," *IEEE Journal of Quantum Electronics QE-23*, No. 6, pp. 692–695, June 1987.

6. Hall, quoted in Hecht, *Laser Pioneers*.

7. Keyes gave his paper in early July 1962; the journal received Hall's paper 77 days later, on Sept. 24. R. N. Hall, Gunther E. Fenner, J. D. Kingsley, T. J. Soltys, and R. O. Carlson, "Coherent light emission from GaAs junctions," *Physical Review Letters 9*, pp. 366–368 (Nov. 1, 1962).

8. Marshall I. Nathan, "Invention of the injection laser at IBM," *IEEE Journal of Quantum Electronics QE-23*, No. 6, pp. 679–683 (June 1987); Marshall I. Nathan, W. P. Dumke, G. Burns, F. H. Dill, Jr., and G. Lasher, "Stimulated emission of radiation from GaAs p-n junctions," *Applied Physics Letters 1*, pp. 62–64 (Nov. 1, 1962).

9. Nick Holonyak, Jr., "Semiconductor alloy lasers—1962," *IEEE Journal of Quantum Electronics QE-23*, No. 6, pp. 684–691 (June 1987); Nick Holonyak, Jr., and S. F. Bevacqua, "Coherent (visible) light emission from $Ga(As_{1-x}P_x)$ junctions," *Applied Physics Letters 1*, pp. 82–83 (Dec. 1962).

10. T. M. Quist, R. H. Rediker, R. J. Keyes, W. E. Krag, B. Lax, A. L. McWhorter, and H. J. Zeiger, "Semiconductor maser of GaAs," *Applied Physics Letters 1*, pp. 91–92 (Dec. 1962).

11. G. F. Dalrymple, B. S. Goldstein, and T. M. Quist, "A solid-state room-temperature operated GaAs laser transmitter," *Proceedings of the IEEE 52*, No. 12, pp. 1742–1743 (Dec. 1964). To be fair, diode lasers ran at much lower voltage and thus lower power, but that was a tremendous current to put through a crystal as big as a grain of sand.

12. H. Kroemer, "A proposed class of heterojunction injection lasers," *Proceedings of the IEEE 51*, pp. 1782–1783 (Dec. 1963).

13. Nick Holonyak, Jr., telephone interview, July 24, 1996.

14. Zhores Alferov, "The history and future of the semiconductor heterostructures from the point of view of a Russian scientist," Physica Scripta T68, pp. 32–

45 (1996) (*Proceedings of the Nobel Symposium* "Heterostructures in Semiconductors), forthcoming.

15. Alferov, "The history and future."

16. Herb Nelson, "Epitaxial growth from the liquid state and its application to the fabrication of tunnel and laser diodes," *RCA Review 24* pp. 603–615 (Dec. 1963).

17. Alferov, "The history and future"; he refers to Z. I. Alferov et al., "High-voltage p-n junctions in GaxAl$_{1-x}$As crystals," *Fizika i Tekhnika Poluprovodnikov 1*, p. 1579 (1967) (submitted May 18, 1967); published in English in *Soviet Physics—Semiconductors 1*, p. 1313 (1968).

18. A team at the IBM T. J. Watson Research Center made gallium aluminum arsenide heterostructures on gallium arsenide substrates. H. Rupprecht, J. M. Woodall, and G. D. Pettit, "Efficient visible electroluminescence at 300°K from Ga$_{1-x}$A$_{1-x}$As *p-n* junctions grown by liquid-phase epitaxy," *Applied Physics Letters 11*, pp. 81–83 (Aug. 1, 1967).

19. H. Kressel and H. Nelson, "Close confinement gallium arsenide pn junction lasers with reduced optical loss at room temperature," *RCA Review 30*, pp. 106–113 (Mar. 1969).

20. "What is in a name? Plenty—when it refers to new diodes," *Laser Focus 5*, pp. 20–21 (Aug. 1969).

21. Izuo Hayashi, "Heterostructure lasers," *IEEE Transactions on Electron Devices ED-31*, pp. 1630–1642 (Nov. 1984).

22. Morton Panish, interview, July 19, 1995.

23. Hayashi, "Heterostructure lasers."

24. Morton Panish, telephone interview, June 30, 1995.

25. Hayashi, "Heterostructure lasers."

26. I. Hayashi, M. Panish, and P. W. Foy, "A low-threshold room-temperature injection laser," *IEEE Journal of Quantum Electronics QE-5*, pp. 211–213 (Apr. 1969).

27. Kressel and Nelson, "Close-confinement."

28. See fig. 4 in Z. I. Alferov et al., "AlAs-GaAs Heterojunction injection lasers with a low room-temperature threshold," *Fizika i Tekhnika Poluprovodnikov 3*, No. 9, pp. 1328–1332 (Sept. 1969); published in English in *Soviet Physics—Semiconductors 3*, No. 9, pp. 1107–1110 (Mar. 1970).

29. Zhores I. Alferov, letter to author, June 14, 1996.

30. Hayashi, "Heterostructure lasers."

31. Ibid. Memorial Day was celebrated on May 30, which was a Saturday in 1970. Bell employees had the following Monday off.

32. The executive was Bill Boyle. Panish telephone interview, June 30, 1995.

33. I. Hayashi, M. B. Panish, P. W. Foy, and S. Sumski, "Junction lasers which operate continuously at room temperature," *Applied Physics Letters 17*, No. 3, pp. 109–111 (Aug. 1, 1970).

34. Alferov, "The history and future."

35. Jack C. Dyment, telephone interview, Dec. 18, 1996; J. C. Dyment, "Hermite-Gaussian mode patterns in GaAs junction lasers," *Applied Physics Letters 10*, No. 3, pp. 84–86 (Feb. 1, 1967).

36. Peter Selway, telephone interview, Jan. 31, 1995.

37. Zhores I. Alferov et al., "Effect of the heterostructure parameters on the laser threshold current and the realization of continuous generation at room temperature," *Fizika i Tekhnika Poluprovodnikov 4*, p. 1826 (1970) (submitted May 6,

1970); published in English in *Soviet Physics—Semiconductors 4*, p. 1573 (1971), cited in Alferov, "The history and future."

38. Holonyak telephone interview.

39. Alferov letter.

40. Holonyak telephone interview.

41. Alferov letter.

42. Jane E. Brody, "Bell developing a pocket laser," *New York Times*, Sept. 1, 1970, p. 18. The *Times* article stresses the importance of the new laser for laser technology in general, as well as for communications. It does not mention how laser beams would be transmitted for communications; at that time Bell Labs was still heavily involved with hollow light pipes, and legitimate doubts remained about the feasibility of fiber-optic communications. The contrast between the Bell Labs press conference and Corning's press silence reflects the striking contrast in the two corporate cultures in 1970.

43. A major concern was manufacturing lasers that would meet industrial reliability requirements. "Second thoughts on applications for Bell's c-w room-temperature diode," *Laser Focus* Oct. 1970, pp. 24–26; Panish, interview, July 19, 1995.

44. Selway telephone interview.

45. Isamu Sakuma et al., "Continuous operation of junction lasers at room temperature," *Japan Journal of Applied Physics 10*, pp. 282–283 (1971).

46. Panish telephone interview, June 30, 1995.

47. Panish interview, July 19, 1995.

48. Robert L. Hartman, telephone interview, Dec. 19, 1996.

49. Henry Kressel, "The small, economy-size laser," *Laser Focus 6*, pp. 45–49 (Nov. 1970).

50. Bernard C. DeLoach, Jr., telephone interview, Dec. 18, 1996. Another Bell Labs scientist spotted dark lines in a dead laser at about the same time. B. C. DeLoach, Jr., B. W. Hakki, R. L. Hartman, and L. A. D'Asaro, "Degradation of CW GaAs double-heterojunction lasers at 300 K," *Proceedings of the IEEE 61*, No. 7, pp. 1042–1044 (July 1973).

51. David Newman and Simon Ritchie, "Sources and detectors for optical fibre communications applications: the first 20 years," *IEE Proceedings 133 J*, pp. 213–229 (June 1986).

52. DeLoach telephone interview; follow-up call, Dec. 23, 1996.

53. DeLoach telephone interview and follow-up call.

54. Number from Hartman telephone interview.

55. Henry Kressel and J. K. Butler, *Semiconductor Lasers and Heterojunction LEDs* (Academic Press, New York, 1977; see p. 3).

56. R. L. Hartman, J. C. Dyment, C. J. Hwang, and M. Kuhn, "Continuous operation of GaAs-GaAlAs double-heterostructure lasers with 30°C half-lives exceeding 1000 h," *Applied Physics Letters 23*, No. 4, pp. 181–183 (Aug. 15, 1973).

57. Alan Steventon, interview, Dec. 1, 1994, p. 4.

58. Many historic moments have zipped by me, but this is one that I recall; it became the lead item in the *Laser Focus* product section. "Commercial diodes emit cw at room temperature," *Laser Focus 11*, No. 8, p. 44 (Aug. 1975).

59. W. B. Joyce, R. W. Dixon, and R. L. Hartman, "Statistical characterization of the lifetimes of continuously operate (Al,Ga)As double-heterostructure lasers," *Applied Physics Letters 28*, No. 11, pp. 684–686 (June 1, 1976).

60. Essentially the same paper appeared as R. L. Hartman, N. E. Schumaker, and R. W. Dixon, "Continuously operated (Al,Ga)As double-heterostructure lasers with 70°C lifetimes as long as two years," *Applied Physics Letters 31*, No. 11, pp. 756–759 (Dec. 1, 1977).

61. Bell Labs, "Million hour lifetimes for lasers projected by Bell Labs," press release dated for release June 29, 1977.

62. See, e.g., "Postdeadline reports," *Laser Focus 13*, p. 4 (Aug. 1977).

Chapter 13

1. M. Chown, "Light; the long-distance answer," *New Scientist*, supplement, July 16, 1970, pp. 14–19.

2. D. Gloge, "Optical waveguide transmission," *Proceedings of the IEEE 58*, No. 10, pp. 1513–1522 (Oct. 1970). This paper was submitted in May.

3. J. R. Pierce, "How communications change man's world," in *IEE Centenary Lectures 1871–1971: Electrical Science and Engineering in the Service of Man*, (London, Institution of Electrical Engineers, 1971), pp. 77–91.

4. "Avalanche, Gunn diodes are coming on strongly in the quest for better microwave power sources," *Electronics*, Jan. 5, 1970, p. 128.

5. Chown, "Light."

6. It was founded as the Society of Telegraph Engineers and Electricians, May 17, 1871.

7. There were 65 exhibits. The others included a table-cleaning robot and a tracked hovercraft. J. A. C. King, "The Centenary of the IEE," unpublished manuscript supplied by IEE Archives; Murray Ramsay, interview, Dec. 2, 1994.

8. Murray Ramsay, telephone interview, Dec. 17, 1996.

9. Ramsay interview, Dec. 2, 1994; Martin Chown, interview, Dec. 2, 1994.

10. Martin Chown, telephone interview, Jan. 23, 1997.

11. Chown interview, Jan. 23, 1997.

12. Ramsay telephone interview, Dec. 17, 1996.

13. Murray Ramsay, letter to author, May 27, 1997.

14. Staff report, "Report from Southampton," *Laser Focus*, May 1969, pp. 51–55.

15. Like Queen Elizabeth II, Mountbatten was a descendant of Queen Victoria, but he was far down the line of heirs to the throne.

16. Britain made Reeves a member of the Order of the British Empire in 1945 for his wartime radar work. The Institute of Electrical and Electronics Engineers and the Franklin Institute honored him in the 1960s for pulse-code modulation ("Alec Reeves dies; inventor was 69," *New York Times*. Oct. 14, 1971). In 1969, the British Post Office issued a stamp showing a pulse-code modulation waveform, to celebrate the post office's technological developments. Executives at Standard Telecommunication Labs evidently got some of the stamps and distributed them on specially printed envelopes which identified the subject as pulse-code modulation, and called it a tribute to Reeves. However, official Post Office first-day covers do not identify Reeves or pulse-code modulation (Laszlo Solymar, e-mail to author, April 6, 1998).

17. Reeves wrote little for publication; the invited talks in Scotland and South Africa are among the few exceptions and are invaluable in understanding his ideas. They reveal some strikingly prophetic views of general trends, notably the growth of mobile phones, which he viewed as a better use of radio spectrum than

broadcast television. However, many details are off, such as his prediction of submarine cables made with graded-index fiber and two-kilometer repeater spacing. Alec Reeves, "The future of telecommunications: Bernard Price Memorial Lecture 1969," *Transactions of the South African Institute of Electrical Engineers* (Sept. 1970, pp. 445–465. This is the transcript of a lecture given Sept. 29–Oct. 1, 1969 in South Africa. See also Alec Reeves, "Future prospects in optical communications," John Logie Baird Memorial Lecture, University of Strathclyde, May 30, 1969.

18. "Alec Reeves dies."

19. John Midwinter, interview, Dec. 5, 1994.

20. Electronic circuits long ago replaced electro-mechanical relays in telephone switching systems. You can think of telephone switches as special-purpose computers. Vacuum-tube electronics were used in some early general-purpose computers, but complex general-purpose computers were not possible until the advent of transistors and integrated circuits. The same was true for the special-purpose computers used in telephone switching. Digital telephone transmission also requires fast, inexpensive circuits to convert signals back and forth between digital and analog formats, and those also require transistors and integrated circuits to be practical.

21. William A. Gambling, "Fibres, lasers, and communications," *The Radio and Electronic Engineer 43*, No. 11, pp. 653–654 (Nov. 1973).

22. Martin Chown et al., "Direct modulation of double-heterostructure lasers at rates up to 1 Gbit/s," *Electronics Letters 9*, No. 2, pp. 34–36 (Jan. 25, 1973).

23. Ira Jacobs, "Lightwave system development: looking back and ahead," *Optics & Photonics News*, Feb. 1995, pp. 19–23, 39.

24. Laurence Altman, "Bell's money is riding on millimeter waves for future communications," *Electronics*, Apr. 13, 1970, pp. 96–105; the bit rates are not exact multiples because each level of multiplexing adds overhead bits.

25. Jack A. Baird, "The Picturephone system: forward," *Bell System Technical Journal 50*, No. 2, pp. 219–220 (Feb. 1971), says the first public demonstration of "two-way video telephony" was by Bell Labs in New York on Apr. 9, 1930. While Jenkins and others were trying to broadcast television, Herbert Ives at Bell Labs was thinking of two-way video communication. There was a similar haziness of purpose among early radio developers, many of whom envisioned using radio for two-way conversations.

26. Irwin Dorros, "The Picturephone system: the network," *Bell System Technical Journal 50*, No. 2, pp. 221–233 (Feb. 71).

27. Donald Janson, "Picture-Telephone service is started in Pittsburgh," *New York Times*, July 1, 1970, p. 1.

28. Boyce Rensenberger, "Growth of Picturephones disappoints Bell System," *New York Times*, July 3, 1970, p. 26. This article said that the number of Picturephones had increased from an initial 25 to 33 a year later. The discrepancy with the report a year earlier was unnoticed.

29. Altman, "Bell's money."

30. Gloge, "Optical waveguide transmission."

31. Stewart E. Miller, "Optical communications research progress," *Science 170*, pp. 685–695 (Nov. 13, 1970).

32. I searched both indexes and also those of specialist magazines such as *Electronics* and *Laser Focus. Industrial Research* wrote about research at STL and the British Post Office in January 1971 issue, mentioning the diode laser as a milestone but overlooking Corning's fiber and Bell's interest in communications.

"British push R&D on fiber telecom link" *Industrial Research*, Jan. 1971, pp. 33–34.

33. John N. Kessler, interview Dec. 17, 1996; John N. Kessler, "Fiber optics sharpens focus on laser communications," *Electronics*, July 5, 1971, pp. 46–52.

34. Kessler, "Fiber optics sharpens focus."

35. David Pearson, telephone interview, Feb. 16, 1995.

36. Fiber drawing research was put under Alan G. Chynoweth, director of materials research at Murray Hill. Tingye Li, telephone interview, Jan. 8, 1997; also Paul Lazay, interview, Nov. 12, 1996.

37. Julian Stone, telephone interview, June 28, 1996; his published report plots attenuation on a curve but does not cite the number. See Julian Stone, "Optical transmission in liquid-core quartz fibers," *Applied Physics Letters 20*, pp. 239–240 (Apr. 1, 1972). (The liquid-core fiber had higher loss than the Corning fiber at the 633-nanometer wavelength of a helium-neon laser; the minimum was at longer wavelengths where the Corning fiber suffered from impurity absorption.)

38. Graeme Ogilvie, a metallurgist at the Commonwealth Scientific & Industrial Research Organization (CSIRO), had been working on very transparent organic compounds when he heard of Kao's proposal. Rod Esdaile telephone interview, June 6, 1996. G. J. Ogilvie, R. J. Esdaile, and G. P. Kidd, "Transmission loss of tetrachloroethylene-filled liquid-core-fiber light guide," *Electronics Letters 8*, No. 22, pp. 533–534 (Nov. 2, 1972).

39. D. N. Payne and W. A. Gambling, "The preparation of multimode glass- and liquid-core optical fibres," *Opto-electronics* 5, pp. 297–307 (1973); in a June 9, 1995, telephone interview William Alec Gambling said they eventually reached 4 decibels per kilometer, but this evidently was not published.

40. The practical problems included the many hours needed to fill a long, extremely thin tube with liquid, and the large difference in thermal expansion between glass and liquid. As the temperature changed, the liquid expanded or contracted like mercury in a thermometer, pushing out or pulling away from the ends of the tube [Esdaile, telephone interview; Gambling, interview; W. A. Gambling, "Optical fibres: the Southampton scene," *IEE Proceedings 133 J*, pp. 205–210 (June 1986)].

41. Peter Kaiser, E. A. J. Marcatili, and S. E. Miller, "A new optical fiber," *Bell System Technical Journal 52* pp. 265–269 (Feb. 1973).

42. Ray Jaeger, lecture notes from 1994 talk, personal communication.

43. Peter Kaiser et al., "Spectral losses of unclad vitreous silica and soda-lime silicate fibers," *Journal of the Optical Society of America 63*, pp. 1141–1148 (Sept. 1973).

44. Lazay interview.

45. Henry Marcatili, telephone interview, Jan. 3, 1997.

46. Peter Kaiser, telephone interview, May 17, 1995.

47. Miller, "Optical communications."

48. Charles A. Burrus and B. I. Miller, "Small-area double-heterostructure aluminum-gallium-arsenide electroluminescent diode sources for optical-fiber transmission lines," *Optics Communications 4*, pp. 307–309 (Dec. 1971).

49. Reeves, "The future of telecommunications."

50. Stewart E. Miller, Enrique A. J. Marcatili, and Tingye Li, "Research toward optical fiber transmission systems" (2 parts), *Proceedings of the IEEE*, Dec. 1973, pp. 1703–1751.

51. Reeves, "The future of telecommunications."

52. A. G. Fox and Ivan Kaminow, "Lightwave communication," in S. Millman, ed., *A History of Engineering and Science in the Bell System: Vol. 5. Communication Sciences* (AT&T Bell Laboratories, Indianapolis, 1984, pp. 273–312).

53. Strictly speaking, pulse dispersion is the product of laser spectral width and material dispersion. Narrow-line semiconductor lasers can reduce the problem, but they were not available in the early 1970s. Estimates based on Corning data cited in Jeff Hecht, *Understanding Fiber Optics* (2nd ed.; Sams Publishing, Indianapolis, 1993).

54. Midwinter interview.

55. Ramsay and Chown interviews, Dec. 2, 1994.

56. Charles Lucy, telephone interview, Dec. 3, 1996.

57. Antoni Karbowiak, telephone interview, Feb. 5, 1995.

58. Robert A. Maurer, "Glass fibers for optical communications," *Proceedings of the IEEE 61*, pp. 452–462 (Apr. 1973).

59. Signetics never made money for Corning, and the company sold it in 1975. Chuck Lucy thought it was so much of a fiasco that the culprits from Corning should remain discretely unidentified (Lucy, interview). Timing from Joseph G. Morone, *Winning in High-Tech Markets* (Harvard Business School Press, Boston, 1993); see also Michael McGeary, "Optical fiber for communications," pp. 24–82 in David Roessner, Robert Carr, Irwin Feller, Michael McGeary, and Nils Newman, *The Role of NSF's Support of Engineering in Enabling Technological Innovation, Phase II* (SRI International, Arlington, Va., 1998).

60. Lazay interview.

61. D. B. Keck, R. D. Maurer, and P. C. Schultz, "On the ultimate lower limit of attenuation in glass optical waveguides," *Applied Physics Letters 22*, No. 7, pp. 307–309, (Apr. 1, 1973).

62. Donald Keck, interview, Mar. 7, 1995.

63. Jaeger lecture notes.

64. Olshansky replaced Felix Kapron, who left at the end of 1972. Felix Kapron, interview, June 12, 1995.

65. Robert Olshansky, telephone interview, Feb. 21, 1997.

66. Ibid.; Robert Olshansky and Donald B. Keck, "Pulse broadening in graded-index optical fibers," *Applied Optics 15*, No. 2, pp. 483–491 (Feb. 1976).

67. Tingye Li and Enrique A. J. Marcatili, "Research on optical fiber transmission," *Bell Laboratories Record*, Dec. 1971, pp. 331–337.

68. Lazay interview.

69. Tingye Li, telephone interview, Jan. 8, 1997.

70. Memories differ as to who was involved. The membership may have changed over time, or different people may have represented some groups at different times. In addition to the Crawford Hill and Holmdel managers, members included Barney DeLoach of Murray Hill, Mauro DiDominico from Murray Hill whose detector group paralleled DeLoach's, Mel Cohen from the Western Electric engineering research center in Princeton, Mort Schwartz from cable manufacturing in Norcross, Georgia, and R. E. Mosher of AT&T. Names from Tingye Li telephone interview; Jack Cook, telephone interview, Jan. 2, 1997; Jacobs, "Lightwave system development."

71. Rudolph Kompfner, "Optics at Bell Laboratories—optical communications," *Applied Optics 11*, No. 11, pp. 2412–2425 (Nov. 1972). This paper is an

invaluable check on fallible memories because it summarizes the Bell Labs program when it was written in August, 1972.

72. Although he had already suffered one heart attack, Kompfner took on two professorships, dividing his time between Oxford and Stanford Universities.

73. In Europe, patent applications are published a year after they are filed. In the United States, they are not published until the application is granted, typically a few years after filing.

74. John MacChesney, interview, May 1995.

75. John MacChesney, telephone interview, July 17, 1996.

76. Jaeger lecture notes mention a paper that is probably the MacChesney one listed in n. 77.

77. J. B. MacChesney et al, "Low-loss silica core-borosilicate clad optical fiber waveguide," *Applied Physics Letters 23* pp. 340–341 (Sep. 15, 1973).

78. J. B. MacChesney et al., "Preparation of low loss optical fibers using simultaneous vapor phase deposition and fusion," *Proceedings of the International Congress on Glass* (Vol. 6; American Ceramic Society, Cincinatti 1974, pp. 40–45).

79. Lazay interview.

80. David Hanna, interview, May 22, 1995.

81. D. A. Payne and W. A. Gambling, "New silica-based low-loss optical fiber," *Electronics Letters 10*, pp. 289–290 (July 25, 1974).

82. They made fibers with loss under 5 decibels per kilometer by 1974 and eventually reached 2.5 decibels in the late 1970s, but the process was not adaptable to making graded-index fibers with high transmission capacity. J. R. Tillman, "Research," *Post Office Electrical Engineer's Journal 74*, pp. 282–297 (Oct. 1981); Jim Ainslie and Clive Day, interviews, Dec. 1, 1994.

83. Lucy telephone interview.

84. D. L. Bisbee, "Optical fiber joining technique," *Bell System Technical Journal 50*, pp. 3153–3158 (Dec. 1971).

85. R. B. Dyott, J. R. Stern, and J. H. Stewart, "Fusion junctions for glass-fiber waveguides," *Electronics Letters 8*, No. 11, pp. 290–292 (June 1, 1972).

86. Y. Kohranzadeh, "Hot splices of optical waveguide fibers," *Applied Optics 15*, No. 3, pp. 793–795 (Mar. 1976); D. L. Bisbee, "Splicing silica fibers with an electric arc," *Applied Optics 15*, No. 3, pp. 796–799 (Mar. 1976).

87. Jack S. Cook, "Communication by optical fiber," *Scientific American 229*, No. 5, pp. 28–35 (Nov. 1973).

88. Tingye Li, "Advances in optical fiber transmission research," *Proceedings of the Symposium on Optical and Acoustical Micro-electronics* (Polytechnic Institute of New York, 1974, pp. 97–108).

89. Richard Smith, interview, June 11, 1997; Ira Jacobs, "Lightwave system development."

90. Ira Jacobs, interview, Nov. 19, 1996.

91. Ira Jacobs, "Lightwave system development."

92. Jacobs interview.

93. E. F. O'Neill, ed., *A History of Engineering and Science in the Bell System: Vol. 7. Transmission Technology 1925–1975* (AT&T Bell Labs, New York, 1985; see pp. 665–668).

94. Victor K. McElheny, "Bell testing a high-density phone system" *New York Times*, May 15, 1975, p. 86; in 1970 the target date was 1973.

95. A. G. Fox and W. D. Warters, "Waveguide research," chap. 6 in Millman, ed., *A History of Engineering and Science* (pp. 247–272).

96. William D. Smith, "AT&T forecasts record spending," *New York Times*, Oct. 14, 1971, business section.

97. The Japanese—among the last to abandon millimeter waveguides—supplied the rest of the waveguide needed to finish the array. O'Neill, ed., *A History of Science and Technology* (pp. 623–647 on millimeter waveguide).

98. Roger Heckingbottom, interview, Dec. 1, 1994.

99. D. Merlo, "The millimetric waveguide system: the current situation," *Post Office Electrical Engineer's Journal* 69, pt 1, pp. 34–37 (April 1976).

100. British Telecom Labs interview transcript, Dec. 1, 1994.

101. Richard Dyott, telephone interview, Jan. 11, 1994.

102. O'Neill, ed., *A History of Science and Technology* (p. 647 on millimeter waveguide).

103. Harold Barlow, "Introductory address," *Conference on Trunk Telecommunications by Guided Waves* (Institution of Electrical Engineers, London, 1970, pp. xi–xiv).

Chapter 14

1. Victor K. McElheny, "Threads of glass carry messages in new process," *New York Times*, Nov. 30, 1975, p. 62.

2. F. F. Roberts, "Optical communication today and tomorrow," in *Proceedings Third European Conference on Optical Communications* (VDE-Verlag, Berlin, 1977) pp. 2–7, 1977). This is from the text of an invited opening paper that Roberts prepared before his death.

3. Dorset Police, "Optical fiber communications," press notice, undated, supplied by Nortel.

4. Don Williams, telephone interview, Jan. 15, 1997; also Michael Moncaster, telephone interview, Jan. 15, 1997.

5. Account from Murray Ramsay and Martin Chown interviews, Dec. 2, 1994; Murray Ramsay, telephone interview, Jan. 22, 1997.

6. Martin Chown, telephone interview, Jan. 23, 1997.

7. "Optical fiber link for police computer," *Electronics Weekly*, Sept. 24, 1975, p. 3.

8. D. B. Keck, R. D. Maurer, and P. C. Schultz, "On the ultimate lower limit of attenuation in glass optical waveguides," *Applied Physics Letters* 22, No. 7, pp. 307–309 (Apr. 1, 1973).

9. This intermediate speed was not used in America. The International Telecommunication Union standard speeds used in Europe were 8.4, 34, and 140 million bits per second. The American standard leaped from 45 to 274 million bits per second with no intermediate steps. R. W. Berry and R. C. Hooper, "Practical design requirements for optical fiber transmission systems," *Optical Fiber Communications* (Institution of Electrical Engineers, London, 1975, conference proceedings 132; proceedings of first European Conference on Optical Communications, held Sept. 16–18, 1975, London).

10. K. Kurokawa et al., "A 400 Mb/s experimental transmission system using a graded-index fiber," in *Optical Fiber Communications.*

11. Dick Dyott, telephone interview, Mar. 13, 1997.

12. A series of papers giving formal results appeared two years later in *Bell System Technical Journal 57* (July–Aug. 1978).

13. Rich Cerny, interview, Jan. 9, 1997.

14. "James Godbey dies at 43, a founder of Valtec Corp," *Laser Focus*, Jan. 1979, p. 74.

15. Eric Randall, telephone interview, Feb. 18, 1997.

16. Paul Dobson, telephone interview, Feb. 21, 1997.

17. Ibid.

18. Cerny interview.

19. James Barron, "Irving B. Kahn, 76, a founder of TelePrompTer and cable TV" obituary, *New York Times*, Jan. 25, 1994, p. B8.

20. Account based on industry sources and personal observations.

21. "Insilco plans venture in fiber optics field, will own 51% of firm," *Wall Street Journal*, Dec. 3, 1976, p. 25.

22. Kahn made a killing when he sold his interest back to Insilco in late 1979. The deal was so embarrassing to company officials that Insilco carefully avoided disclosing either the price or the identities of the selling stockholders. ("Insilco buys remainder of firm," *Wall Street Journal*, Sept. 24, 1979, p. 38). However, that was Irving Kahn's last hurrah in fiber optics. He had also started a company to make semiconductor lasers but was never able to take it public—according to industry insiders, because the Securities and Exchange Commission frowned on having a convicted felon as the chairman of a public company. None of his other fiber ventures panned out, and he returned to cable television. His obituary does not mention fiber optics; Barron, "Irving B. Kahn."

23. Ira Jacobs, interview, Nov. 19, 1996.

24. M. I. Schwartz et al., "Bell's Chicago fiber optic test," *Laser Focus 14*, No. 1, pp. 58–64 (Jan. 1978); M. I. Schwartz et al., "The Chicago Lightwave Communications Project," *Bell System Technical Journal 57*, No. 6, pp. 1881–1888 (July–Aug. 1978). Four fibers were broken during cabling, not installation.

25. John Fulenwider, telephone interview, Feb. 7, 1997.

26. "PS-noted in passing: communication applications," *Laser Focus 13*, No. 7, p. 70 (July 1977).

27. John Midwinter, interview, Dec. 5, 1994.

28. Ronald W. Berry, David J. Brace, and Ivon A. Ravenscroft, "Optical fiber system trials at 8 Mbits/s and 140 Mbit/s," *IEEE Transactions on Communications COM-26*, pp. 1020–1027 (July 1978).

29. Midwinter interview.

30. "Europe looks to fiber optics for telephone expansion," *Laser Focus 14*, No. 2, pp. 68–76 (Feb. 1978).

31. Charles Sandbank, ed., *Optical Fiber Communication Systems* (Wiley, Chicester, 1980; see especially chap. 11 by D. H. Hill).

32. R. Bouillie, "Application of optical fibers to existing communication systems," paper presented at the third European Conference on Optical Communications, Sept. 14–16, 1977, Munich. (VDE-Verlag, Berlin, 1977) pp. 231–236.

33. "Europe looks to fiber optics."

34. Midwinter interview.

35. Dyott telephone interview.

36. Strictly speaking, it depends on the inverse fourth power of wavelength, $1/(\text{wavelength})^4$.

37. Keck et al., "On the ultimate lower limit."

38. W. G. French, J. B. MacChesney, P. B. O'Connor, and G. W. Tasker, "Optical waveguides with very low losses," *Bell System Technical Journal 53*, pp. 951–954 (1974).

39. D. N. Payne and W. A. Gambling, "A borosilicate-cladded phosphosilicate-core optical fiber," *Optics Communications 13*, pp. 422–425 (1975).

40. See, e.g., *Optical Fiber Communications*.

41. Strictly speaking, material dispersion is the sum of two factors. One is the change in refractive index of the material as a function of wavelength, thesame effect which spreads colors out into a spectrum after the pass through aprism. The other is the change in the fiber's waveguide properties as a function of wavelength. Both numbers vary with wavelength, and they can have different signs, so their sum can equal zero at a "zero-dispersion" wavelength. This zero point depends both on the glass composition and the fiber structure. Dyott's initial estimate of 1.23 micrometers was a bit off the now-accepted value of 1.31 micrometers for single-mode fiber [Dyott, telephone interview; R. B. Dyott, "Group delay in glass fiber waveguides," in *Proceedings, Trunk Telecommunications by Guided Waves* (Institution of Electrical Engineers, London, 1970, pp. 176–181)].

42. Felix Kapron, interview, June 12, 1995; F. P. Kapron and D. B. Keck, "Pulse transmission through a dielectric optical waveguide," *Applied Optics 10*, No. 7, pp. 1519–1532 (July 1971).

43. D. N. Payne and W. A. Gambling, "Zero material dispersion in optical fibers," *Electronics Letters 11*, pp. 176–178 (1975).

44. M. Horiguchi, "This week's citation classic," in Arnold Thackray, *Contemporary Classics in Engineering and Applied Science* (ISI Press, Philadelphia, 1986, p. 165).

45. M. Horiguchi and H. Osanai, "Spectral losses of low-OH-content optical fibers," *Electronics Letters 12*, pp. 310–312 (June 10, 1976).

46. Forgotten for nearly 20 years, gallium nitride has recently become important for blue diode lasers.

47. In Hsieh's quaternary compounds, the total number of indium and gallium atoms had to equal the number of phosphorus and arsenic atoms. When he started with indium phosphide, that meant that gallium replaced some indium and arsenic replaced some phosphorus. J. Jim Hsieh, interview, Jan. 30, 1996.

48. Ibid.

49. J. J. Hsieh, J. A. Rossi, and J. P. Donnelly, "Room temperature cw operation of GaInAsP/InP double-heterostructure lasers emitting at 1.1 μm," *Applied Physics Letters 28*, No. 12, pp. 709–711 (June 15, 1976).

50. Hsieh, interview; J. J. Hsieh and C. C. Shen, "Room-temperature cw operation of buried-stripe double-heterostructure GaInAsP/InP diode lasers," *Applied Physics Letters 30*, No. 8, p. 429–431 (Apr. 15, 1977).

51. H. Osanai et al., "Effects of dopants on transmission loss of low-OH content optical fibers," *Electronics Letters 12*, pp. 549–550 (Oct. 14, 1976).

52. Tatsuya Kimura and Kazuhiro Daikoku, "A proposal on optical fiber, transmission systems in a low-loss 1.0–1.4 μm wavelength region," *Optical and Quantum Electronics 9*, pp. 33–42 (1977).

53. Hsieh interview.

54. M. I. Schwartz, W. A Reenstra, and J. H. Mullins, "Bell's Chicago fiber optic test," *Laser Focus 14* pp. 58–64 (Jan. 1978).

55. Ira Jacobs, "Fiber optics technology and application in the Bell System," *Proceedings Fiber Optics and Communications (FOC '78)* (Information Gatekeepers Inc., Brookline, Mass., 1978, pp. 12–13).

56. John Midwinter, "Optical fiber communications systems development in the UK," *IEEE Communications Magazine*, Jan. 1982, pp. 6–11. The contracts were signed in Mar. 1979.

57. Richard Epworth, interview, Dec. 2, 1994; Richard Epworth, "25 years of optical communications," unpublished manuscript.

58. Richard Epworth, "The phenomenon of modal noise in analog and digital optical fiber systems," in Charles K. Kao ed., *Optical Fiber Technology II* (IEEE Press, New York, 1981, pp. 260–269). Originally presented at Fourth European Conference on Optical Communications, Genoa, Italy, September 1978.

59. Richard Epworth, paper ThD1 (revised summary, in postdeadline section), *Technical Digest, Topical Meeting on Optical Fiber Communications, Mar 6–8, 1979, Washington* (Optical Society of America, Washington); Clemens Baack et al., "Modal noise and optical feedback in high-speed optical systems at 0.85 μm," *Electronics Letters 16* 592–593 (1980).

60. Double-crucible fibers never quite caught up with other processes, and the technology worked poorly for both single-mode fibers and long-wavelength fibers. Clive Day, British Telecom Labs interview, Dec. 1, 1994.

61. Robert Olshansky, telephone interview, Feb. 21, 1997.

62. "Japan shows its strength in optical communications," *Laser Focus* December 1977, pp. 48–52

63. T. Miya, Y. Terunuma, T. Hosaka, and T. Miyashita, "Ultimate low-loss single-mode fiber at 1.55 μm," *Electronics Letters 15* 106–108, 15 February 1979.

64. The reference is to George Orwell's novel *1984*. Bill Plummer, telephone interview, Mar. 21, 1997; Will Hicks, interview, Feb. 4, 1994.

65. Clive Foxall and Sidney O'Hara made the decision to change materials; O'Hara also pushed single-mode fiber. Alan Steventon, interview, Dec. 1, 1994.

66. "An optical link is installed in Las Vegas for independent telephone meeting," *Laser Focus* January 1978, pp. 64–66.

67. Dobson telephone interview.

68. Figures from Kessler Marketing Intelligence circa 1978, presented at the Newport Conference on Fiberoptic Markets. While I no longer have the original source, I vividly remember the story.

69. *Wall Street Journal* February 24, 1978, p. 14; March 31, 1978, p. 29.

70. "Wire maker to acquire 13% of Valtec, specialist in optical communications," *Laser Focus* April 1977, p. 36; Rich Cerny, interview, January 9, 1997; "Two new entries in fiber optic market: Siemens-Corning and Canada Wire," *Laser Focus* February 1978, pp. 76–78.

71. "Valtec seeks entry to cable-TV market with acquisition of metal-cable maker," *Laser Focus* December 1977, pp. 43–44.

72. "James Godbey dies at 43, a founder of Valtec Corp.;" *Laser Focus* January 1979, p. 74.

73. Eric Randall telephone interview, February 18, 1997.

74. Postdeadline reports," *Laser Focus World*, February 1980, page 4; also *Wall Street Journal* January 8, 1980, p. 18.

75. "M/A Com's purchase of Valtec expands resources for development," *Laser Focus* August 1980, pp. 52–54; *Wall Street Journal* June 18, 1980, p. 14; September 24, 1980, p. 39.

76. Interviews with Ken Nill, Jim Hsieh, Jan. 30, 1996.

77. "A permanent fiber trunk in Fort Wayne is expected to save GTE $1.5 million," *Laser Focus 15*, p. 48 (Jan. 1979).

78. Allen Kasiewicz, interview, Jan. 9, 1997.

79. Jacobs interview.

80. "Nippon Telephone transmits at 1.27 μm through 53 km of fiber without repeater," *Laser Focus 14*, pp. 52–54 (Nov. 1978).

81. Jeff Hecht, "Preview of OSA's fiber meeting," *Laser Focus 15*, pp. 34–40 (Mar. 1979).

82. M. G. Blankenship et al., "High phosphorous containing P_2O_5-GeO_2-SiO_2 optical waveguide," postdeadline paper PD-3 in *Technical Digest Optical Fiber Communication* (Optical Society of America, Washington, D.C., 1979).

83. Detlef Gloge and Tingye Li, "Multimode fiber technology for digital transmission," *Proceedings of the IEEE 68*, No. 10, pp. 1269–1275 (Oct. 1980).

84. Marshall C. Hudson and Lewis C. Kenyon, "Installation and measurement of an 18.7-and 23.5-km unrepeatered fiber-optic interoffice trunk," paper TuDD1 in *Technical Digest: OFC '82, Topical Meeting on Optical Fiber Communication* (Optical Society of America, Washington, D.C., 1982).

85. The company had the cable-television franchise for the whole province. "3400-km fiber trunk network planned to provide cable TV on Canadian plains," *Laser Focus*, Sept. 1979, pp. 64–66; Graham C. Bradley, "Wobbly DC to megablinking light in 3.8 billion seconds," Saskatchewan Telecommunications June 1995, unpublished; Graham Bradley, telephone interview, Feb. 6, 1997.

86. John Midwinter, "The year of optical fibers," *National Electronics Review 1980/81* (National Electronics Council, London, 1981, pp. 8–13).

87. Midwinter interview.

88. Ibid.

89. D. R. Smith, "Advances in optical fiber communications," *Physics Bulletin 33*, pp. 401–403 (1982).

90. Paul Lazay, interview, Nov. 12, 1996.

91. Midwinter interview.

92. Ira Jacobs, "Lightwave system development: looking back and ahead," *Optics & Photonics News*, Feb. 1995, pp. 19–23, 39; "Bell System plans $79M fiber trunk to serve Northeast Corridor by 1984," *Laser Focus*, Mar. 1980, pp. 60–62.

93. Will Hicks, interview, Feb. 4, 1994.

94. C. David Chaffee, *The Rewiring of America: The Fiber Optics Revolution* (Academic Press, Boston, 1988, p. 35).

95. Jacobs, "Lightwave system development."

96. The lower loss at 1.3 micrometers would offset the lower power from the LED, so all three channels would have repeaters every seven kilometers, at the same points as the old coaxial cable. Jacobs interview.

97. Jacobs, "Lightwave system development"; Kasiewicz interview.

98. Jacobs, "Lightwave system development."

99. See papers in *Technical Digest: OFC '82, Topical Meeting,*" including postdeadline papers; the AT&T quote is from D. P. Jablonowski, "Present and future fiber manufacturing outlook," paper TuGG3.

100. The American telephone industry and its suppliers were accustomed to following the lead set by AT&T, which was firmly in the graded-index camp. So did industry analysts and reporters.

101. After graduating from the Harvard Business School, McGowan spent

three years selling the Todd-AO wide-screen movie system to theaters, an unusual start at a time when most Harvard graduates went directly to large companies. Larry Kahaner, *On the Line: The Men of MCI—Who Took on AT&T, Risked Everything, and Won* (Warner Books, New York, 1986, pp. 47–48).

102. Ira Magaziner and Mark Patinkin, *The Silent War: Inside the Global Business Battles Shaping America's Future* (Vintage Books, New York, 1989; see chap. 9 on Corning).

103. Chaffee, *The Rewiring of America.* Chaffee covered the fiber industry during its growth heyday in the 1980s, as editor of *Fiber Optics News,* and his account tells the business side of the story.

104. The actual distances were 1400 kilometers for AT&T vs. 6800 kilometers for MCI. Jeff Hecht, "Singlemode fibers go to market," *Lasers & Applications 2,* pp. 53–57 (May 1983).

105. Bradley, "Wobbly DC."

106. The cross-licenses also gave Corning access to Bell Labs technology. While Corning held the dominant patents, it benefited from not having to battle AT&T patents.

107. Phil Black, telephone interview, Jan. 28, 1997.

108. Magaziner and Patinkin, *The Silent War* (see chap. 9 on Corning).

109. The decree was dated July 27, 1981; "Corning wins patent suit against ITT," *Laser Focus,* Sept. 1981, pp. 104–106.

110. Black telephone interview; Randall telephone interview.

111. *New York Times,* July 24, 1982, p. 36; citation from *New York Times Index 1982.*

112. "M/A COM sells stake in Valtec to partner," *Wall Street Journal,* Mar. 1, 1983, p. 8.

113. Dobson telephone interview.

114. Lazay interview.

115. For a more detailed account of the business side of fiber development, especially in the 1980s, see Chaffee, *The Rewiring of America.*

116. Ken Nill, interview, Jan. 30, 1996.

117. Hsieh interview; Nill interview.

118. Cerny interview quoted $6 million. Others have mentioned the same figure.

119. Hicks interview.

120. Bill Plummer, telephone interview, Mar. 21, 1997.

121. Graded-index fiber did not fade away entirely because it can carry a billion bits per second over limited distances, a kilometer or less. It remains in use for short systems within buildings or between buildings on a university or industrial campus.

Chapter 15

1. Alec H. Reeves, "The future of telecommunications: Bernard Price Memorial Lecture 1969," *Transactions of the South African Institute of Electrical Engineers,* Sept. 1970, p. 445–465.

2. The SE-ME-WE 3 cable, to be finished the end of 1998, spans 38,000 kilometers (23,600 miles) and cost $1.3 billion. Adele Hars, "Suppliers selected for world's longest submarine cable," *Lightwave,* May 1997, pp. 7, 9.

3. Arthur C. Clarke, *How the World Was One: Beyond the Global Village* (Bantam, New York, 1992).

4. "Sailing ship to satellite: The transatlantic connection," exhibit brochure, Institution of Electrical Engineers, London (1994). The long wave band allowed one channel; short waves allowed 15. C. W. Hansell was instrumental in developing this generation of technology.

5. Richard J. O'Rorke, Jr., *1990 World's Submarine Telephone Cable Systems* (US Government Printing Office, Washington, D.C., 1991).

6. E. F. O'Neill, ed., *A History of Engineering and Science in the Bell System: Vol. 7. Transmission Technology 1925–1975* (AT&T Bell Laboratories, New York, 1985, pp. 337–371).

7. Early germanium transistors were short-lived and much more sensitive to temperature than modern silicon transistors, so there was some reason for the delay. R. D. Ehrbar, "Undersea cables for telephony," in Peter K. Runge and Patrick R. Trischitta, eds., *Undersea Lightwave Communications* (IEEE Press, New York, 1986 pp. 3–22).

8. O'Rorke, *1990 World's Submarine*."

9. I use only statute (land) miles here; repeater spacing is often given in nautical miles, which are slightly longer. The spacing for TAT-6 was 2.5 nautical miles. Ehrbar, "Undersea cables."

10. Peter Runge, "A high-capacity optical-fiber undersea cable system" paper presented at Conference on Laser and Electro-Optic Systems, Feb. 1980, San Diego.

11. A few newer systems, notably Motorola's Iridium system and the Teledisic system, use many satellites in lower orbit to provide continuous coverage of points on the ground, so they suffer negligible delay.

12. Alec H. Reeves, "The future of telecommunications."

13. Peter Runge, interview, July 19, 1995.

14. Runge, "A high-capacity."

15. O'Neill, ed., *A History of Engineering* (pp. 337–371).

16. Phil Black, telephone interview, Jan. 28, 1997.

17. "Britain and an ITT subsidiary begin tests of undersea fiber optic cable," *Laser Focus*, May 1980, pp. 66–68, quotes paper presented by P. Worthington, "Application of optical fiber system in undersea service," at the *International Conference on Submarine Telecommunication Systems* conducted Feb. 1980 at the Institution of Electrical Engineers, London.

18. A third pair of single-mode fibers were included in the cable as spares, in case of fiber failure. That caution was reflected in the design of the entire TAT-8 system. "Bell plans a single mode fiber system for transatlantic installation by 1990," *Laser Focus*, May 1980, pp. 68–70.

19. Paul Lazay, interview, Nov. 12, 1996; other leaders of single-mode fiber development were Dave Pearson and Bill French.

20. Peter K. Runge, "Deep-sea trial of an undersea lightwave system," paper MD2 presented at the Topical Meeting on Optical Fiber Communications, Feb. 28–Mar. 2, 1983, New Orleans.

21. Runge interview.

22. Naoya Uchida and Naoshi Uesuge, "Infrared optical loss increase in silica fibers due to hydrogen," *Journal of Lightwave Technology LT-4*, pp. 1132–1137 (Aug. 1986).

23. Don Keck, interview, Mar. 7, 1995.

24. John Midwinter, interview, Dec. 5, 1994.

25. Murray Ramsay, interview, Dec. 2, 1994.

26. Uchida and Uesuge, "Infrared optical loss."

27. Kiyofumi Mochizui et al., "Influence of hydrogen on optical fiber loss in submarine cables," in Runge and Trischitta, eds., *Undersea Lightwave Communications* (pp. 177–188).

28. Keck interview.

29. Murray Ramsay and Martin Chown, interviews, Dec. 2, 1994.

30. Keck interview.

31. Mochizui et al., "Influence of hydrogen."

32. The number quoted varies; an initial (May 16, 1983) AT&T press release says 29 companies are involved; 1990 directory of submarine cables (O'Rorke, *1990 World's Submarine*) lists 23.

33. Martin Chown, interview, Dec. 2, 1994.

34. AT&T Press release, Nov. 16, 1983.

35. "World submarine telecommunication systems," map (STC Submarine Cables Ltd., London, Sept. 1984).

36. O'Rorke, *1990 World's Submarine.*

37. Press release, STC Submarine Cables, dated Oct. 30, 1986.

38. George A. Heath and Martin Chown, "The UK-Belgium no. 5 optical fiber submarine system," in Runge and Trischetta, eds., *Undersea Lightwave Communications* (pp. 129–141).

39. Press release, STC Submarine Cables.

40. Hiroshi Fukinuki et al., "The FS-400M submarine system," Runge and Trischetta, eds., *Undersea Lightwave Communications* (pp. 69–82).

41. Gordon Reinholds was the engineer, Runge, interview.

42. Runge, interview.

43. Lazay interview.

44. Runge, "A high-capacity."

45. Fridolin Bosch, G. M. Palmer, Charles D. Sallada, and C. Burke Swan, "Compact 1.3μm laser transmitter for the SL undersea lightwave system," in Runge and Trischetta, eds., *Undersea Lightwave Communications* (pp. 445–458).

46. O'Rorke, *1990 World's Submarine.*"

47. See, e.g., Michael Kinsley, "Is AT&T hamstringing Comsat?" *New York Times*, June 13, 1976, p. 14, section III.

48. R. A. Williamson, project director, *International Cooperation and Competition in Civilian Space Activities* (Office of Technology Assessment, Washington, D.C., July 1985; OTA-ISA-239).

49. The technology for dispersion-shifted fibers was refined gradually by fine-tuning the structure of the fiber. V. A. Bhagavatula. M. S. Spotz, W. F. Love, and Donald B. Keck, "Segmented-core single-mode fibers with low loss and low dispersion," *Electronics Letters 19*, No. 9, pp. 317–318 (Apr. 25, 1981); B. J. Ainslie et al., "Monomode fibre with ultra low loss and minimum dispersion at 1.55 microns," *Electronics Letters 18*, pp. 842–844 (1982).

50. Runge interview.

51. Leonard Heymann, "Intelsat mounts defensive campaign against cables," *Communications Week*, Sept. 22, 1986, p. 51.

52. "TAT-8 schedule slips two months," *Lasers & Optronics*, August 1988, p. 29.

53. Kathleen Killett, "Fiber optic cable approaches transatlantic service cutover," *Communications Week*, Dec. 5, 1988, p. 32.

54. David Zielenziger, "Trans-Atlantic link goes on call," *Electronic Engineering Times*, Dec. 26, 1988, p. 16.

55. Notes in author's files Apr. 6, 1989.

56. Charles J. Koester and Elias Snitzer, "Amplification in a fiber laser," *Applied Optics 3*, No. 10, pp. 1182–1186 (Oct. 1964).

57. Charles P. Sandbank, ed., *Optical Fiber Communication Systems* (Wiley, Chichester, 1980, p. 19).

58. John W. Hicks, Jr., US Patent 4,616,898, "Optical communications using Raman repeaters and components therefor," filed Sept. 28, 1983, issued Oct. 14, 1986.

59. R. J. Mears et al., "Low-noise erbium-doped fiber amplifier operating at 1.54 μm," *Electronics Letters 23*, p. 1026 (1987); also interview with David Hanna, May 22, 1995. Emmanuel Desuvire independently developed erbium-doped fiber amplifiers at the same time at AT&T Bell Labs. The doped fibers are excited by diode lasers emitting at 980 or 1480 nanometers.

60. W. Christopher Barnett, "The TPC-5 cable network," *IEEE Communications Magazine 34*, pp. 36–40 (Feb. 1996).

61. Runge interview.

62. Ibid.

Chapter 16

1. Isaac Asimov, science column, *Magazine of Fantasy and Science Fiction*, Aug. 1962.

2. David E. Fisher and Marshall Jon Fisher, *Tube: The Invention of Television* (Counterpoint, Washington, D.C., 1996, p. 339).

3. William H. Dutton, Jay G. Blumler, and Kenneth L. Kraemer, eds., *Wired Cities: Shaping the Future of Communications* (G. K. Hall and Co., Boston, 1987), summarizes the history of the program.

4. William H. Dutton, telephone interview, Feb. 13, 1997.

5. Lee B. Becker, "A decade of research on interactive cable," Dutton et al., (pp. 102–123).

6. John Fulenwider, telephone interview, Feb. 7, 1997.

7. Standard cable television sends all video signals to every subscriber but scrambles premium channels so that they require decoding by a special set-top box.

8. John Fulenwider, "Study of an all-optical communications system for trunking, switching, and distribution of wideband signals," paper presented at *21st International Wire and Cable Symposium* Atlantic City, Dec. 1972 (pp. 35–46).

9. Fulenwider interview.

10. Carol Davidge, "America's talk-back television experiment, QUBE," in Dutton et al. *Wired Cities.*

11. *A Summary Version of the Comprehensive Report on Hi-OVIS Project July '78–Mar '86* (New Media Development Association, Tokyo, 1988).

12. The fibers had a 150-micrometer core clad with plastic, a type called plastic-clad silica. At the time the system was designed in 1976, they were inexpensive and attractive for short communication systems, but they suffered much larger pulse spreading than graded-index fibers. (This sort of pulse spreading was why graded-index fibers were invented.) *A Summary Version.*

13. *Hi-OVIS Project: Interim Report, Hardware/Software Experiments: July '78–*

March 79 (Visual Information System Development Association, MITI, Tokyo, 1979); see also other sources, such as Masahiro Kawahata, "Hi-OVIS," Dutton et al., *Wired Cities* (pp. 179–200).

14. "The fibered city as social laboratory: a progress report on Japan's Hi-OVIS," *Laser Focus*, Jan. 1980, pp. 58–60.

15. Jeff Hecht, talk delivered at the 1980 Newport Conference on Fiber-Optic Markets, Kessler Marketing Intelligence, Newport, R. I.; Jeff Hecht, "Marketing broad band services," *Laser Focus*, Dec. 1980, p. 6.

16. Masahiro Kawahata, e-mail to author, Apr. 10, 1997.

17. "The fibered city."

18. Masahiro Kawakata, e-mail to author, Apr. 10, 1997

19. Quote from Tetsuhiko Ikegama, senior vice president, NTT, talk delivered at the Massachusetts Institute of Technology, Apr. 14, 1994. Like most NTT accounts of fiber-optic development in Japan, Ikegama's talk did not mention Hi-OVIS by name, although he did praise MITI's early work. I have seen other NTT reports in which the omission of Hi-OVIS was much more striking.

20. Letter from Akiko Kato, NTT PR Dept., Dec. 5, 1997. Stringing fiber to distribution nodes near homes by the earlier date is much less expensive than stringing fiber directly to all homes and businesses. NTT evidently plans to run fiber links to individual homes as requested or needed.

21. Canada earlier laid fibers to 40 homes in the exclusive Yorkville district of Toronto, but the local cable-television system blocked tests of video services. The Manitoba Telephone System had rights to offer cable television in Elie and Ste. Eustache, so video tests could proceed there P. H. Chouinard, M. Larose, F. M. Banks, and J. R. Berry, "Description and performance review of the Yorkville integrated services fiber optic trial," paper 28.3 in *International Conference on Communications, 1979, Boston, Mass* (IEEE Conference Digest, Piscataway, N.J., 1979).

22. Figures are in US dollars; the budget was $9.5 million Canadian. Brian B. McCallum, "Fiber optic subscriber loops: a look at the Elie system," *Laser Focus*, Nov. 1980, pp. 64–67.

23. Jeff Hecht, "Fiberopolis," *Omni*, Dec. 1982, pp. 120–128.

24. Rod Kachulak, telephone interview, Mar. 3, 1997.

25. Joel Stratte-McClure, "French telecommunications," advertising supplement to *Scientific American*, Sept. 1980.

26. Michel Triboullet, telephone interview, Feb. 13, 1997.

27. "Biarritz optical fiber system," paper circa 1980, from France, not specifically attributed.

28. "Postdeadline reports," *Laser Focus*, Dec. 1980, p. 4.

29. "Postdeadline reports."

30. William H. Dutton, "Driving into the future of communications? Check the rear-view mirror," paper presented at "POTS to PANS, social issues in the multimedia evolution from plain old telephony services to Picture and Network Services," Mar. 28–30, 1994, Hintlesham, Suffolk, UK (paper available via World Wide Web).

31. Thierry Vedel, "Local policies for wiring France," in Dutton et al., *Wired Cities* (pp. 255–278).

32. E.g., at the September 1979 European Conference on Optical Communications in Amsterdam, there were 36 Japanese papers, 17 each from Britain and the United States, 12 from Germany, and a mere 6 from France. See Charles

Sandbank, "Opening address," *Sixth European Conference on Optical Communication* (Institution of Electrical Engineers, London, 1980 Conference Publication 190, pp. 1–3).

33. Subscribers received black-and-white cameras because color was too expensive, although the screens were color. (The official reason was "because lighting in private homes and apartments tends to be inadequate for good color filming.") Francois Gerin, "Video communications: images, sounds and data in freedom!" manuscript dated Oct. 1985, circulated by Direction Generale des Telecommunications, Ministry of Post, Telephone and Telephone.

34. Triboullet telephone interview.

35. Marc P. Duchesne, e-mail to author Feb. 12, 1997.

36. Roland Goarin, "Component reliability results from the Biarritz field trial and from "plan cable" volume deployment," paper 17.1, in *Technical Digest, IEEE Globecom 1991*, (IEEE, Piscataway, N.J., 1991)

37. Triboullet telephone interview.

38. Kachulak telephone interview.

39. Paul Shumate, telephone interview, Mar. 29, 1997.

40. GTE press release, Apr. 13, 1988.

41. *The Cerritos Project: 1992 Annual Report* (GTE Telephone Operations, 1992).

42. Robert Olshansky, telephone interview, Feb. 21, 1997.

43. Paul Shumate, "What's happening with fiber to the home," *Optics & Photonics News*, Feb. 1996, pp. 16–21, 75.

44. Shumate, telephone interview.

45. "NTT opticalizes access network," information supplied by Akiko Kato, NTT Public Relations Department, Mar. 29, 1996.

46. Will Hicks, "Where fiber optics should go," paper presented at Fiberoptics Futures conference, Waltham, Mass., sponsored by New England Fiberoptics Council, May 26, 1993.

Chapter 17

1. Alec Reeves, "Future prospects in optical communication," John Logie Baird Memorial Lecture, University of Strathclyde, May 30, 1969.

2. John Midwinter, interview, Dec. 5, 1994.

3. Donald Keck, interview, Mar. 7, 1995.

4. Charles H. Burrus, Herweg Kogelnik, and Tingye Li, "Obituaries: Stewart E. Miller," *Physics Today*, Nov. 1990, pp. 102–103.

5. Jack Cook, telephone interview, July 9, 1997.

6. For a history of the industrial growth, see C. David Chaffee, *The Rewiring of America: The Fiber Optics Revolution* (Academic Press, Boston, 1988).

7. Jeff Hecht, *Understanding Fiber Optics* (Sams Publishing, Indianapolis, 1987).

8. Ira Jacobs, interview, Nov. 19, 1996.

9. Sharks got all the press, but the only cable they damaged was the one in the Canary Islands. Gophers are serious hazards to buried cables. Like other rodents, their front teeth grow continually, so they instinctively gnaw anything they can get their teeth around. Fiber cables are particularly vulnerable because they are small, so cables buried in areas where gophers live are armored with heavy metal wires. To test their protection, telephone and electric power industries developed a standard gopher test, which involved running cable through a simulated

gopher burrow and watching what live gophers did to the cable (Donna Cunningham, AT&T PR Dept., fax to author Apr. 27, 1995).

10. John Fulenwider, telephone interview, Nov. 7, 1988, for article in *South* magazine.

11. Peter Wranik, telephone interview, Oct. 31, 1988, for article in *South* magazine.

12. Stewart E. Miller and Ivan P. Kaminow, eds., *Optical Fiber Telecommunications II* (Academic Press, Boston, 1988).

13. Linn F. Mollenauer, interview, July 19, 1995; Linn F. Mollenauer and K. Smith, "Demonstration of soliton transmission over more than 4000 km in fiber with loss periodically compensated by Raman gain," *Optics Letters 13*, No. 8, pp. 675–677 (Aug. 1988).

14. Raman amplification transfers energy from a strong laser beam at one wavelength to a weaker beam at a second.

15. L. F. Mollenauer, R. H. Stolen, and M. N. Islam, "Experimental demonstration of soliton propagation in long fibers: loss compensated by Raman gain," *Optics Letters 10*, No. 5, pp. 229–231 (May 1985).

16. M. Nakazawa et al., "Experimental demonstration of soliton data transmission over unlimited distances with soliton control in time and frequency domains," postdeadline paper PD7 in *Postdeadline Papers, Conference on Optical Fiber Communications, February 21–26, 1993* (Optical Society of America, Washington, D.C., 1993).

17. L. F. Mollenauer et al., "Demonstration, using sliding-frequency guiding filters, of error-free soliton transmission over more than 20,000 km at 10 Gbit/s, single-channel, and over more than 13,000 km at 20 Gbit/s in a two-channel WDM," postdeadline paper PD8, *Postdeadline Papers.*

18. Mollenauer interview.

19. Andrew Chraplyvy, interview, July 19, 1995.

20. A. R. Chraplyvy et al., "One-third terabit/s transmission through 150 km of dispersion-shifted fiber," *IEEE Photonics Technology Letters 7*, p. 98 (1995).

21. Chraplyvy interview.

22. A. H. Gnauck et al., "One terabit/s transmission experiment," Postdeadline paper 20 in *Conference on Optical Fiber Communications OFC '96 Technical Digest* (Optical Society of America, Washington, D.C., 1996).

23. H. Onaka et al., "1.1 Tb/s WDM transmission over a 150 km, 1.3 μm zero-dispersion single-mode fiber," postdeadline paper 19 in *Conference on Optical Fiber Communications.*"

24. T. Morioka et al., "100 Gbit/s × 10 channel OTDM/WDM transmission using a single supercontinuum WDM source," postdeadline paper 21 in *Conference on Optical Fiber Communications.*"

25. Information posted on Project Oxygen web site, http://www.oxygen.org in November 1998; also Robert Poe, Project Oxygen, private communication.

26. Ben Harrison, "Quad-WDM technology hikes MCI's network capacity," *Lightwave*, July 1996, pp. 6–8.

27. Paul Palumbo, "MCI hits 40 Gbit in OC-192 trial," *Lightwave*, Mar. 1997, pp. 1, 17.

28. Lucent Technologies press release. No company has yet installed any system that uses all the available wavelength channels, so actual operating speeds are less.

29. Tom Geyer, conversation at FiberFest 97, Framingham, Mass., Mar. 31, 1997.

30. Peter Runge telephone interview, April 11, 1997

31. Paul Mortensen, "NTT projects lower-cost fiber to the home in 1997," *Lightwave*, Dec. 1996, pp. 1, 26.

32. Paul Shumate, telephone interview, Mar. 28, 1997.

33. Jason Stark, telephone interview, Feb. 24, 1997.

Index